VOICE OVER IP CRASH COURSE

Voice Over IP Crash Course

Steven Shepard

McGraw-Hill
New York Chicago San Francisco Lisbon
London Madrid Mexico City Milan
New Delhi San Juan Seoul Singapore
Sydney Toronto

McGraw-Hill

A Division of The **McGraw-Hill** *Companies*

3 4 5 6 7 8 9 0 DOC/DOC 0 9 8 7 6

ISBN 0-07-226241-9

The sponsoring editor for this book was Jane Brownlow and the production supervisor was David Zielonka. It was set in Century Schoolbook by MacAllister Publishing Services, LLC.

Printed and bound by R. R. Donnelley & Sons.

This book is printed on recycled, acid-free paper containing a minimum of 50 percent recycled de-inked fiber.

**This book is for my children,
Cristina and Steven.
Is there any greater gift?**

CONTENTS

ACKNOWLEDGMENTS

I often tell people that writing isn't something I *do,* but rather it's something that I *am.* I write these books because I enjoy the craft—it's as simple as that. I'm also fascinated by telecommunications and IT technologies, not so much by how they work but rather because of the immense social, economic, and competitive changes they bring about if deployed properly. Consequently, I don't write for engineers; I write for managers, strategists, senior decision-makers. I *don't* target the bitweenies because plenty of highly technical and very good titles are already written for them. I am sometimes criticized by reviewers because my books are not as technical as others—in fact, one recent review observed, "the author is clearly not an engineer because this book reads more like a business book than it does a technical book." I took that comment as a compliment—because it's true. I'm *not* an engineer, I don't *target* engineers, and I'm far more interested in the business implications of technology deployment than I am in the names of the fields that make up the underlying protocols. Now, I am honored to count large numbers of engineers as my friends and clients, and I am technically trained and do strive to be as technically accurate as I possibly can by relying on an army of talented resources who serve as fact checkers, content auditors, and reality injectors. These friends and colleagues help me achieve one goal with all my books: to produce a collection of work that is not only timely and technically accurate, but also enjoyably readable. Most important, when my readers finish one of my books, I want them to be able to go away and do something better because of what they read.

That being said, I have to once again give my thanks to all those people who helped me convert an idea—and an odd one at that—into another book.

First and foremost, my thanks go out to the students and personnel of Champlain College in Burlington, Vermont. Without them this book would not exist. You opened your wiring closets, your dorm rooms, your raised floor, your classrooms, and your offices. By doing so you opened my eyes. Most especially I thank Aaron Videtto, Paul Dusini, Dave Whitmore, and Gary Kessler, all of whom took time out of their busy schedules to help me with my endless questions.

My thanks also go to Phil Asmundson, Kim Barker, Paul Bedell, Jane Brownlow, Joe Candido, Phil Cashia, Steve Chapman, Dick Dadamo, Bruce Degn, Jonathan Dunne, Donna Epps, Jack Gerrish,

Jack Garrett, Steve Green, Dave Hill, Lisa Hoffmann, Alfred Hsi, Ron Hubert, Andy Jentis, Tony Kern, Naresh Lakhanpal, Tracey Lewis, Viki MacMillan, Dee Marcus, Roy Marcus, Alvaro Marques, Gary Martin, John Martin, Bob Maurer, Dennis McCooey, Paul McDonagh-Smith, Alan Nurick, Katy O'Connor, Chris O'Gorman, Richard Parlato, Girish Pathak, Dick Pecor, Kenn Sato, Dave Stubbs, Elvia Szymanski, Philip Takken, Calvin Tong, Jack Tongue, Lisa Watson-Krause, and Craig Wigginton. You all helped in countless ways; thank you.

I also thank my wife, Sabine, and my kids, Cristina and Steve. Thanks for—well, everything.

FOREWORD

"We've got a project we want to talk with you about." Thus began a conversation with my McGraw-Hill editors, Jane Brownlow and Steve Chapman. We were sitting in a restaurant in Las Vegas during the USTA Telecom 2004 show, and the project was this book. At first I was a bit reluctant because several very good books were already on the market about IP telephony, and it didn't make sense to me to write another one. So we agreed after much discussion that I would go away and look closely at the available titles before making a decision, which I did. The result was an idea that began to germinate, and before long I had accepted the opportunity and was furiously doing interviews and market research and creating the manuscript.

My initial observation confirmed what I already knew—indeed there were good books on the market for those readers wishing to dive deep into the technical aspects of VoIP, and any attempt on my part to write another one would be a poor use of my time and McGraw-Hill's resources. I looked at the existing titles, I read through many of them, and as I did a thought came to me. All of these books did a superb job of explaining the inner workings of TCP/IP protocols, the differences between the PSTN and the evolving IP-based infrastructure, the pros and cons of old and new, and the equipment that could be found in a typical VoIP network. They spent far less time, however, examining VoIP evolution as a case study. After all, this shift in technology infrastructure represents a major change in the way that networks operate and the way that services are delivered to enterprise and residence customers. Wouldn't it make sense to look at a real implementation of VoIP, from initial concept all the way through to installation and turn-up? The more I thought about this model, the better I liked it, and so did Jane and Steve. So I proposed a book design that would offer the same valuable technology descriptions found in other titles, but I added a twist. I would present the technology within the framework of a real VoIP conversion effort, showing not only the advantages that the change brought, but also the problems and challenges that made the conversion complex and difficult. Of course, now I had to go out and find a company that had substantially converted to VoIP so that I could study it, a company willing to share the trials and tribulations of such a major undertaking.

Luck was with me. For the last eight years I have had the honor of serving as a member of the Board of Trustees of Champlain College in

Burlington, Vermont, near where I live. About one year ago the IT staff at Champlain completed a campus-wide VoIP changeover. This was an all-or-nothing decision: If they were going to do this, they were going to do it correctly and all the way, because they had done their homework and knew that an IP-based campus infrastructure had the potential to bring significant advantages to the school in many diverse ways. As a trustee, I was aware of the conversion because I had to vote for the capital and expense dollars required to finance the new system. However, the pressures of my job kept me from digging more than peripherally into the conversion process as it was occurring. When I finally did, while writing this book, I was overwhelmed by the enormity of the project and the commitment it took to make it successful on the part of a lot of people. On the other hand, I was astonished by the level of functionality it brought to the people who live and work on campus and by the diversity of new applications it made possible.

This book, then, is about the *business* of undertaking a VoIP conversion. It offers detailed overviews of the underlying technologies and protocols, the players that are involved in the VoIP space, the application sets that reside upon the network, and the pros and cons of IP convergence. It also discusses the thinking that goes into the go-no-go decision, the vendor selection process, the study of available application sets, the financial implications of VoIP, the human resources issues associated with convergence, and a host of other details that are at least as important as a fundamental understanding of the technology itself. In essence, I wrote this book so that companies that are considering making the change to VoIP can do so with both eyes open and all of their questions asked and answered before proceeding too far down the convergence road. There is no question that VoIP represents the future of telephony. However, it is important to understand that the question is not *whether* to go with VoIP, but rather *when*. Timing, as they say, is everything.

This book is divided into six chapters. Chapter 1 dives into the reasons that lie behind the decision to convert to a VoIP infrastructure—or not. Chapter 2 examines the inner workings of the PSTN and other legacy technologies. Chapter 3 then provides a migration path to the emerging IP network, describing the underlying technologies and their similarities and differences relative to the PSTN. Chapter 4 talks about logistics: the timeframe that should be selected for the conversion, the interdependencies that exist, and the order in which the processes should occur. Chapter 5 describes the operational considerations of the conversion, and Chapter 6 discusses the human resources considerations of a VoIP network overhaul.

I think you'll find this approach interesting and useful because it presents VoIP within a real business application context.

One last thing: Although this book is really about the *business* of putting in a VoIP network and application set, there's a lot of technology here. However, because I know that some readers are looking for one or the other, I have done everything I can to structure the book so that those not interested in the underlying technological details can skip them without affecting the readability or effectiveness of the book.

As always I invite your comments and suggestions; thanks for reading. All the best.

Steven Shepard
Williston, Bangkok, Harlow, Las Vegas

[illegible]

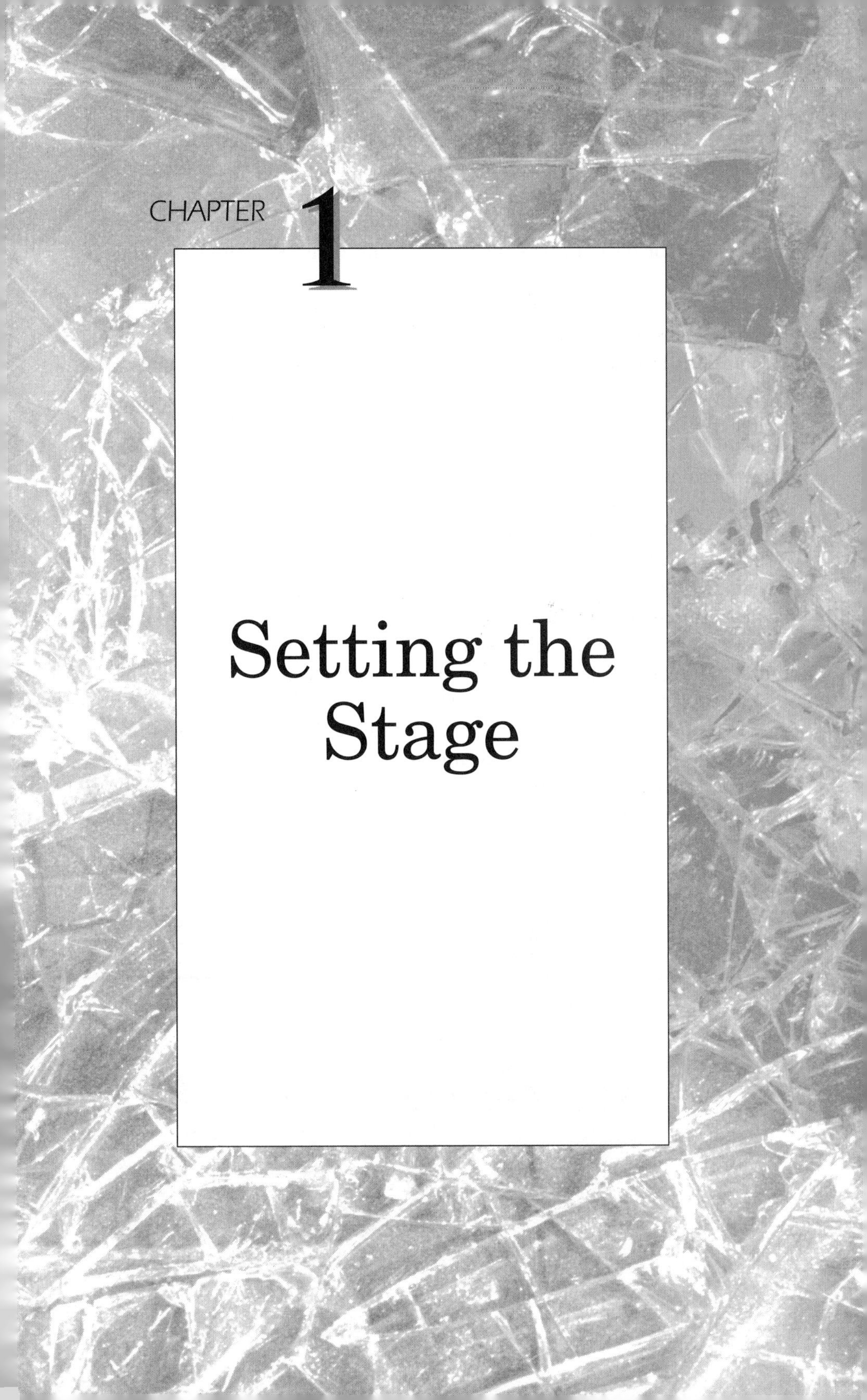

CHAPTER 1

Setting the Stage

Vermont is a state of 600,000 people nestled snugly against the northeastern shore of Lake Champlain (see Figure 1-1). It is something of a study in contrasts: On the one hand, it is home to one of IBM's largest semiconductor fabs as well as to such high-tech entities as Infineon (semiconductor design), IDX (health care software), Polhemus (virtual-reality devices), Resolution (video and media fulfillment), and Husky Industries (injection molding). On the other hand, Vermont is reputedly home to more cows than people, has massive, well-run dairy farms ("Some days I make $50,000, and some days I don't. But the days I do far outnumber the days I don't," one farmer told me), and a quirky sense of humor that demands thought before laughter ("Have you lived here all your life?" I asked the old man, rocking in his chair on the front porch of the general store. "Not yet," he replied nonchalantly, without taking his eyes off the road in front of him.).

This is a hard place to live for much of the year. Winter is harsh ("Eleven months of winter and one month of really bad skiing," say the sweatshirts sold here), with temperatures plummeting into the 20-below-zero range for weeks at a time. Some years Lake Champlain freezes shore to shore with ice so thick that motor homes cluster in the center of the lake's bays, forming ice fishing villages for months at a time. Pickup trucks drive across the lake to the Adirondacks in New York. Drivers heading north on Highway 89 (the state's only real highway of any consequence) toward Canada, 30 minutes from Burlington, sometimes notice the sign 3 miles before the border crossing that marks the 45th parallel, the midpoint between the equator and the north pole.

Figure 1-1 Burlington Harbor

Figure 1-2
Burlington

The skiing and snowboarding in the Green Mountains are marvelous. And although they don't compare to the majesty of the Rocky Mountains, these mountains have a charm and luster of their own that attracts visitors from all over the world. When the Von Trapp family, made famous in Rodgers and Hammerstein's *Sound of Music,* fled Austria, they settled here, in the Green Mountains, because the landscape looked so much like the place they left behind. And when the trees burn like fire in the fall, there is no place on Earth more beautiful.

At 39,000 people, Burlington (see Figure 1-2) is the state's largest city. It is a charming, old-world place with pedestrian walkways, world-class shops and restaurants, downtown parks, and a beautiful Lake Champlain waterfront (www.hazecam.net/burlington.html). It's also a safe city. Many people leave their keys in their cars while shopping, and I have neighbors who don't *own* house keys.

Burlington is also home to one university, the University of Vermont, and three acclaimed colleges, Champlain College, St. Michael's College, and Burlington College. As a result the town enjoys the eclectic flavor and endless cultural variety of a typical college town. It reminds me a great deal of my own days at UC Berkeley: lots of dogs, lots of dreadlocks, lots of creative, talented, smart people. It's invigorating.

Visitors who drive east from the waterfront on Main Street toward the Green Mountains will soon begin to climb as they enter the Hill Section of Burlington, shown in Figure 1-3. On the left is the University of Vermont, on the right, Champlain College. The neighborhood behind Champlain is a collection of old Victorians, immaculately kept, beautifully landscaped. Once the homes of lumber barons, bankers, and

Figure 1-3
The Hill Section of Burlington

diplomats, they are now either private homes, college offices, or elegant dormitories for the students at Champlain College. And though the buildings are from a bygone era, their technology infrastructure is anything but. Their connectivity architecture is as modern as it can possibly be, but elegantly so.

Our Case Study: Champlain College

Champlain College (see Figure 1-4) was founded in 1878 as the Burlington Collegiate Institute. In 1884 it was acquired and its name was changed to Burlington Business College. For the next few years, it changed locations several times until 1958, when it was moved to its present location in the Hill Section of Burlington. Originally founded as a secretarial skills institute, it soon began to grow beyond that original narrow definition of purpose. From a certificate base it became a 2-year school offering associates' degrees, to a 4-year school offering baccalaureate degrees, to a full graduate school in 2002, offering a master's degree in Managing Innovation in Information Technology. In 1993 the school inaugurated its acclaimed Champlain College Online, and in 1995 it opened the first of what would become numerous offshore satellite campuses (Israel, India, the United Arab Emirates). Ranked in the top tier of "the best comprehensive colleges in the north" by *U.S. News' America's Best Colleges in 2005*, Champlain continues to grow and change.

Figure 1-4
Champlain College

Today the college comprises 40 buildings on a rolling, wooded 23-acre campus. The buildings, some of which are shown in Figure 1-5, are Victorian and modern, high tech, and high touch. A state-of-the-technical-art optical network interconnects all campus buildings, and a diversely routed and securely dispersed server farm serves the campus community, faculty and students alike. Wireless Fidelity (WiFi) hubs wash the campus in wireless; this place is *connected.* The campus data network is as good as it can currently be and has been for several years.

The voice network, however, has historically been a different story. As the campus has expanded and added buildings—a combination of restored Victorians and purpose-built structures—the school's IT

Figure 1-5
The Miller Information Commons, Champlain College

organization connected them as required to make them part of the greater connected campus. For the most part this meant modifying the cable plant to provide multiple voice lines into buildings that were historically private homes, a relatively straightforward process. Over time Ethernet connections were added as well, giving students full data connectivity in their dorm rooms.

Campus Functionality

Of course, running a college campus requires more than catering to the communications requirements of the student population, which in the case of Champlain College is approximately 1,700 people. The college is a competitive business, after all, and requires internal communications capabilities to support the myriad functions involved in running the place. These include such applications as enrollment support, marketing, student services, various help desks, security, data warehousing, Web hosting, Internet access, health care, and internal communications. Furthermore, as the school's services have evolved, its communications needs have evolved in lockstep. When the school was primarily a 2-year school offering associate degrees, the need for a complex data infrastructure was significantly lower than today, as the school offers online degree programs, 4-year baccalaureate degrees and masters' degrees in addition to 2-year programs. The school draws students from a broad geography that is global in scope. At one time the student body was exclusively from Vermont; today, because of the addition of online programs and the expanded curriculum, students hail from all over the world. In fact, I had an interesting experience a few years ago that reflects the global nature of the school's population. While working on a project in Tokyo, I met a gentleman who worked for my client company. At lunchtime one day, a friend of his dropped by to say goodbye because she had been admitted to college in the United States and was leaving the next day for—you guessed it—Burlington, Vermont, where she would pursue a degree in hotel and restaurant management at Champlain College. The world doesn't get much smaller than that!

This increasing complexity began to put undeniable pressure on the school's existing voice and data infrastructure around 2001 and was exacerbated in the years that followed by increasing enrollment, the addition of new majors (including such technology-dependent fields as computer forensics, network security, and e-gaming design), the construction of several new buildings, the expansion of the online degree programs, the addition of the master's program, and a greater need for both voice and data-based communications with the outside world. By 2003 it became clear to the school's director of IT services that the existing infrastructure was simply not up to the task. He and his team took a long, hard look at the factors they knew they had to consider before they made a decision about the technological direction they should choose.

Planning for Change

Paul Dusini, shown in Figure 1-6, is everything you would want in a chief technology officer (CTO). He is young but has hands-on experience with technology infrastructures. He is possessed of an experimental mindset but is professionally skeptical of innovations that lack a functional track record. He's a voracious reader of trade journals and an unstoppable devourer of data, but he is an active member of several industry groups where he has the opportunity to get his hands dirty, play with the newest toys from Geekdom and compare notes with other CTOs facing the same challenges that he sees at Champlain. When the idea of considering a voice over IP (VoIP) solution was presented to him, he was initially skeptical.

"The principal strategic driver, that is, the thing that drove us to even consider VoIP as an alternative, was this," Dusini told me in an interview

Figure 1-6
Paul Dusini

in his gabled office on campus. "We had an old PBX [or private branch exchange] that was nearing the end of its useful life. We had watched that situation for a couple of years and had in fact decided to wait a while before making a decision about replacement technologies, particularly so that we could watch voice over IP [Internet Protocol] to see how the technology and the applications springing up around it developed. In the end we waited 2 years, and the question that drove us was this: Do we replace the old PBX system, or do we think about convergence and consider all of the applications (e-mail, IM [instant messenger], Web browsers, etc.) and the value that converging them on a single platform could bring? Then we asked ourselves this: How could we bring in VoIP and make it part of a single platform that would run all of the applications I just mentioned, but run them in a way that they would enable added capabilities?

"We knew that sooner or later we had to make a PBX decision. We were faced with the reality that we had a legacy system that provided us with traditional phone services, but that was *all* it did. But we also recognized that we had a responsibility to the college to accept the fact that the future was going to be information- and knowledge-driven and therefore had to include unified messaging and call centers with more capabilities than they had at the time, both of which were becoming important revenue centers and service centers on campus."

The fact that support for emerging applications was a critical consideration in Dusini's planning for the future was a key indicator of the need for a converged solution. "We also had the advantage—at least I think it was an advantage!—that at the time we began to think seriously about the conversion, we were in the middle of an ERP [enterprise resource planning] implementation. This meant that we had a bunch of technologists onsite looking at Java and open-source capabilities and trying to determine what it would take to simultaneously create a student Web portal, an employee Web portal, and so on. What was interesting about our discussions with the technologists was that we realized that voice had kind of slipped into the background, yet the entire discussion revolved around voice and IP as the enabling forces. Because our eventual plan told us that all of the students and employees would ultimately have an IP phone on their desks or in their dorm rooms with display screens on them, we asked our technology partners to look at how we could take advantage and push information out to the dorms and offices on the campus using XML and other tools."

For the period during which they monitored the PBX situation, Dusini and his team visited and revisited the forces driving the need to change their technology game plan, and while a few minor priority shifts occurred, nothing substantive did.

"After all that time, and after hours and hours of analysis, we came to the same conclusion. Ultimately, applications were the real reason we looked at VoIP and a converged network. The fact is that when we began to look at upgrading our campus infrastructure, traditional PBXs were rapidly becoming a thing of the past because of IP and convergence. Furthermore, our timing was perfect: We had lots of choices. There were new, full-feature PBXs appearing on the market, and hybrid PBXs had begun to get attention. So we had choices in front of us that were easy to make because by the time we had to make them we had already done the work required to make those decisions properly and in an informed way."

As an aside, the hybrid PBX that Dusini refers to comprises IP phones and cards that provide access to the traditional PBX switch. The IP cards have onboard digital signal processors (DSPs), which convert IP packets to circuit traffic and transmit proprietary signaling information and feature data as IP packets for delivery to an IP telephone. It has become a major bridging mechanism for VoIP conversions.

Making Choices

The choices that faced Dusini and his staff were not as easy to make as he modestly describes. Compelling reasons existed to simply invest in a new PBX and stick with the status quo. After all, for the most part the old PBX was still doing precisely what it was designed to do. And while choosing to keep the incumbent technology platform might have been the easiest road to take, Champlain, to its credit, has never operated that way. Case in point: A few years ago, after careful, emotion-free analysis, the school made the very difficult decision to end its organized sports programs in favor of fitness programs that would benefit the entire campus community rather than just the small community of athletes. And while the decision was not celebrated by all, everyone recognized how difficult it was to make and that it was the *right* decision, no matter how emotionally hard it was.

On the other hand, the economics of evolution leaned toward the correct side of the balance sheet. "The switch was so old that the vendor didn't even manufacture new parts for it anymore," says Aaron Videtto (see Figure 1-7), Champlain College's telecommunications administrator and the overall project manager for the school's VoIP conversion effort.[1] "All

---

[1]Aaron Videtto recently accepted a position with Citrix where he is a customer network engineer. I wish him the best.

Figure 1-7 Aaron Videtto

of our technical support was coming from reserviced equipment. In fact, when a component failed on our PBX, our support technician was stealing parts from scrapped PBXs that he had stored in his barn! It was ugly. And to add insult to injury, our PBX decided to start randomly generating static electricity toward the end of its life, apparently frying boards for its own amusement. Bottom line? It was ready to be turned off."

The particular point that pushed Dusini and Videtto to make a decision was a major on-campus building project that started a few years ago. Expansion of the college's curriculum and existing programs resulted in a need for three new buildings: the Main Street Suites and Conference Center (see Figure 1-8), a dorm and corporate conference facility; the Center for Global Business and Technology (see Figure 1-9), a state-of-the-art building designed to house classrooms, computer labs, and meeting facilities; and a student life complex (see Figure 1-10) offering dining services, meeting facilities, a student union hall, and a fitness center.

Figure 1-8 Main Street Suites and Conference Center, Champlain College

Figure 1-9 Center for Global Business and Technology, Champlain College

Figure 1-10 Student Life Complex, Champlain College

"The three new buildings really changed our thinking," says Videtto. "We kicked off an RFI [Request for Information] and sent it to a collection of vendors, and from their responses we concluded that it would cost us about $300K to upgrade our PBX—and that's just the hardware, not the wiring for the three new buildings or any other physical plant upgrades that might have to take place. More importantly, responses to the RFI and our own research *also* told us that we could get a completely new system for the entire campus for about $500K. So we put together a formal RFP [Request for Proposal] for a campuswide VoIP network and sent it off to a smaller list of vendors. At that point we were off and running."

Things Began to Move

At that point things began to move relatively quickly. "Once we made our decision, things began to fall into place," says Dusini, smiling. "But I have to tell you, we really wrestled with this project. In fact, for a period of time there, we actually considered getting out of the phone business in the dorms altogether! After all, these are Millennial kids and they all come to school fully wired. They don't want our phones—they've told us so—but they *do* want to be connected. So for a while there, we considered something as simple as dropping a handful of analog lines in the dorm rooms, then telling the students to go down to the bookstore and spend $9.00 on a phone. Needless to say, we didn't get far down that path!"

Millennials are an interesting generation of people. Born between 1982 and 2004, the oldest of them entered the workplace last year; the youngest were born last year. They are confident, highly team oriented, and remarkably, refreshingly conventional. Unlike the Gen-Xers that came before them, it's okay to be smart, and *also* unlike the Gen-Xers, the Millennials actually *like* their parents. They tend to be strong achievers if they are properly motivated, optimistic about all things, sociable, highly moral, and street smart. Finally, and interestingly, they are *absolutely oblivious* to authority. They are so morals-driven, so self-policing, that they just don't understand the need for outside authority in their lives. This is a puzzlement to law enforcement. This is as much because of law enforcement's obligation to enforce laws that the Millennials don't understand the need for, as their need to understand the behavior and drivers of Millennials coming into law enforcement as employees.

It's interesting to note that the Millennials are the first generation in the 54-year history of television that are watching *substantially* less of it than any previous generation, choosing to get their entertainment content elsewhere. This is a serious, vexing concern for television advertisers. Millennials are also the first generation born into a world that has *always* had the Internet/Web and has *always* had cellular telephony. As a result their technology loyalties are quite different from those of preceding generations: More and more of them, for example, are choosing to use their mobile phone as their primary mode of communication, eschewing the perceived safe harbor of 911 service and carrier-class voice quality in favor of the freedom of mobility and on-demand connectivity. Needless to say, this poses a substantial challenge for incumbent telephone companies that have invested untold billions of dollars into their in-place, wireline networks and now watch as their costs remain the same and revenues decline. It also presents a challenge for places like Champlain, who must take into account these behavior modalities as the college designs the network and its services. The same holds true in the enterprise space as well.

This group is also extraordinarily social. They like interactive activities and work well in small groups. Their learning preferences include teamwork and experiential activities, especially if there is a technology element—particularly one that involves gaming technology. Their strengths include collaboration and multitasking. Witness the typical teenager sitting in front of a computer with four or five IM sessions going, MP3 music playing, e-mail responses in various stages of completion, while talking on the cell phone—and doing homework—well. They are goal oriented and can remain strikingly positive in the face of difficulty or adversity.

Interestingly, this group is typically racially and ethnically diverse, but more important they are typically racially and ethnically *oblivious.* There is hope for this world yet.

All of these characteristics should lead the reader to understand that this is a distinctly different group of people than those found in the typical corporation today. They are a rich source of capability and energy, and as long as they are motivated properly they will prove to be a formidable component of the workforce. When Dusini and Videtto looked closely at what they were trying to do, they came to the realization that this evolutionary exercise they were considering could be one of the best things that ever happened to the school.

"In the final analysis we asked ourselves this," says Dusini, steepling his fingers and looking intently into the distance. "If we could move voice and data over the same network, and we could create an application to push database content out to screens on the phones, and we could control those phones and make them do things—make them light up, ring, display special text messages, [and allow] students to look things up and download data, then it struck us that *that* would add clear value to the way that we communicate with students. After all, it's increasingly hard for us to communicate with Millennial students—they walk around with multiple cell phones, multiple e-mail accounts, and multiple IM accounts. Complicating that relationship is the fact that they're kind of elusive: They keep changing those accounts, and we don't know which ones to use to reach them most effectively. In fact, they're driving some of the professors nuts because they don't want to come to office hours for a face-to-face meeting. Instead they want to engage in an IM or SMS [short message service] session with the professor on an ad hoc, do-it-when-they-need-help basis. Culturally, this is 180 degrees out of whack with everything that education has been about for 100 years. But it's reality, and we have to respond to it. This is, after all, a business, and the students are our customers.

"So we began to ask ourselves how we could blend the traditional culture of academia with the new Millennial culture. For example, what if we—and 'we' could be a professor, an advisor, or even a financial aid counselor—could engage in a phone-based conversation with a student, but could control his or her browser while talking with him or her, pushing appropriate content out in real time to augment and enrich the conversation?" (See Figure 1-11.)

"We spent a lot of time talking about these kinds of new advantages that campus convergence might bring, but we also saw the potential of some of the classic benefits. For example, we know that in the long run a VoIP infrastructure is cheaper. Also, we're a very dynamic campus, and with VoIP we know that moves, adds, and changes are trivial, because a student or employee can unplug their IP phone, move it to a new office, plug it back in, and all of the right features follow—and no one knows it happened. And if done correctly, and of course we had no choice in this, the system will support E911 capability."

> "What we're running here is basically a network made of Legos™. It's incredibly flexible."
>
> — *Paul Dusini*

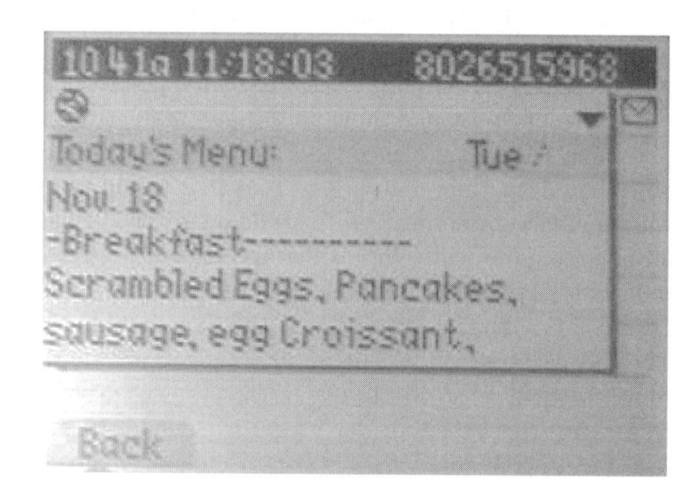

Figure 1-11 Using the push application to engage in more meaningful interactions with clients

"So what were we looking at? We saw the potential for a network that would make it easier to move people around the campus in a seamless fashion, a network that would save money in the long term, and we saw a model emerging that would cause the technology to adapt to us instead of the other way around."

Counting on Convergence

Champlain College is not unique in its evolution or in its desire to use technology as a business facilitator. In fact, the key advantage of the IP evolution is convergence, a fundamental change in the way technology capabilities and business goals come together.

Depending on the strategic and operational goals of the business, a scalable, converged communications architecture can lead to enhanced innovation in the business and a measurable acceleration in business success. Convergence itself comes in five "flavors" and is typically implemented in several operational phases.

The Five Forms of Convergence

Convergence actually transcends the technological coming-together that is inherent in an IP network. *Enterprise convergence* refers to the process of bringing together business applications such as productivity tools (customer relation management [CRM], enterprise resource planning [ERP], and supply chain management) and unified messaging as a way to improve the business of doing business. By bridging the logical gaps that exist between these applications and creating common shared databases, businesses can dramatically affect their relationships with customers in a powerful, positive way. Lou Platt, the former chair of Hewlett-Packard, once observed in a speech that, "If HP knew what HP knows, we'd be three times more profitable." His quote is a corollary to the old "knowledge is power" quote and is equally true. In today's rapid-fire competitive marketplace, the company that gets to market first with the right product wins the game. "If we build it, they will come," doesn't work any longer. A better quote might be, "*If* we understand what they want, and *if* we build it first, and correctly, and *if* we sell it for the price that they want to pay, they *might* show up."

IT convergence simply observes that there is a great deal of value to be gained by aligning business processes as a way to lower the total cost of ownership of IT resources. IT is a support function, after all, and the degree to which it actually provides support is a measure of how well it will be funded and supported by its constituency. And as we will see in a later place in the book, IT and telecom are converging—and dramatically changing the rules of the technology game.

Network technology convergence recognizes that in-place networks comprise many different protocol structures operating across a variety of modalities, and that these networks will not be dismantled simply because something new (i.e., IP) comes along. The typical corporation has invested heavily in its enterprise network, a network that most likely includes various types of wired and wireless Ethernet, some form of bridging and routing infrastructure, optical plant in the metro transport environment, Asynchronous Transfer Mode (ATM), frame relay or private line in the core, not to mention the vast investment it undoubtedly has in its voice network. The expectation that convergence will present a compelling enough reason to engage in a "forklift upgrade" is a fantasy.

Protocol convergence speaks to the critical role of IP as the bridging technology between disparate protocols that are inevitably coresident on modern networks. Some technology historians refer to IP as the "four-foot-eleven-and-a-half-inch protocol." Any questions?

The distance between the wheels of Roman chariots (see Figure 1-12) was precisely four feet, eleven and a half inches. We know this because the iron-rimmed wheels of those chariots ground deep tracks into the stones of Roman Empire roads that are still in use today. Stage coach wheels? Four feet, eleven and a half inches. Standard-gauge railroad-tracks? You guessed it. As long as a rail car manufacturer builds its wheel trucks to that standard, then its cars will roll safely on any system's tracks. Of course, those pesky narrow-gauge *protocol violations* exist that either require specially built train cars or wheel trucks capable of operating on both gauges (protocol conversion). When I was a teenager growing up in Spain, my friends and I used to take weekend trips to various European destinations. (It was cheap and unbelievably safe in those days. After all, Generalissimo Francisco Franco was in power, and if you did something bad, you just . . . went away.) If we went to Paris, we would leave Madrid on the train in the early evening, arriving at the Pyrenees border crossing around midnight—at which point we would have to wake up, gather our things, cross the border on foot, and

Figure 1-12 Roman chariot with standard wheelbase

change trains. It seems that Francisco Franco was smart: Trains were a major transportation option in the early twentieth century, and he ordered Spain's system to be based on a narrower-than-the-standard gauge—to prevent foreign armies from using the railroads as an invasion mechanism. Clever, eh?

In the data world, IP is the four-foot-eleven-and-a-half-inch protocol because it is universally accepted and is largely payload agnostic. It interfaces seamlessly with virtually every other protocol and can transport virtually any payload. Think of IP as a data bucket that can transport anything that fits in it. This is the basis for how protocols work; for a better understanding of the interplay among protocols, please refer to Appendix C and read the overview of Open Systems Interconnection (OSI), the best-known protocol model, that you will find there.

Finally, we have *value convergence,* which is the ultimate goal of convergence in the enterprise. Enterprise, IT, network technology, and protocol convergence set the stage for value convergence, which is the ultimate goal of the convergence phenomenon because it is at this point that the magic of business improvement happens. This is where we begin to see the promise of unified messaging and productivity, where business-team cross-collaboration is enhanced, where mobility is added to the connectivity mix wherever it makes sense, where stronger and more intimate customer relationships arise, and where productivity enhancements start to show up in the workplace. By building a converged network infrastructure and making a functional commitment to what it can do, the workplace benefits and prospers.

At Champlain, those benefits showed up in a variety of interesting ways. "What we were really looking for more than anything else was a quick and easy way to broadcast messages out to the entire student body," says Paul Dusini. "With our old PBX, we were limited to using the voicemail system to broadcast messages to students, staff, and faculty. And in fact, we used it *a lot*. What we found, though, was that the students weren't listening to the messages. And in fact, we learned that the only thing we were really doing was making them mad. Each mailbox was limited by the PBX to a maximum of five messages. And since we (the school) were filling up their mailboxes with broadcast messages, the students got frustrated—and rightly so—because there was no room available for their family or friends to leave messages, some of which were far more important than what we were broadcasting.

"We also wanted the ability to communicate quickly with students on an ad hoc basis, in addition to communicating with our staff. We're in the Northeast after all, which means we are often subject to rapid and extreme changes in weather (see Figure 1-13). If an ice storm is imminent or if classes get cancelled for the day, we want to be able to let people know immediately. And of course, as we were talking through all of the what-if scenarios, IT jumped on the same bandwagon, recognizing that this would be a great tool for sending a message to make people

Figure 1-13 Spring in Vermont

aware of a virus or other disruptive event that could affect the campus network."

In fact, this scenario occurred shortly after network conversion: A burst water pipe in the building that housed the main server room threatened to flood the basement, a condition which would have resulted in the loss of the entire campus network because the servers were all down there. IT was able to send out a campuswide message, telling everyone that they had precisely 5 minutes to save the files they were working on and log off before the system was shut down to be moved above the water line. It worked.

Implementing Convergence

Convergence typically occurs in three phases: network convergence, IT infrastructure convergence, and the deployment of a set of applications specifically designed to take advantage of the converged infrastructure.

Convergence is a series of technological and managerial interdependencies that result in a "perfect marriage" between the underlying technology infrastructure and the "overlying" process of running the business. In the case of Champlain College, the issue to be dealt with was improving the school's ability to communicate with its student body, staff, and faculty. In the first phase, Dusini, Videtto, and their staff had to determine the nature of the problem they needed to fix. Second, they had to address the physical changes to the network that would have to take place if their conversion effort was to be successful. Third, they had to take a hard, emotionless look at the software and hardware tools the school required to do its job as well as it could be done. And finally, once they had addressed each of those issues, they turned their attention to the application set.

This four-layer structure (see Figure 1-14) is typical of a well-run IT organization. "We had vendors coming out of our ears, telling us about their one-of-a-kind solution," laughs Dusini, shaking his head. "The problem was that they couldn't tell us what our business issue was. It would irritate them when I asked them to describe our problem and how their unique solution would uniquely fix it. Typically, they couldn't describe what we were facing in more than the vaguest terms. And what bothered us most about that was that our problems are far from unique—in fact, with a few minor exceptions, our challenges are pretty typical."

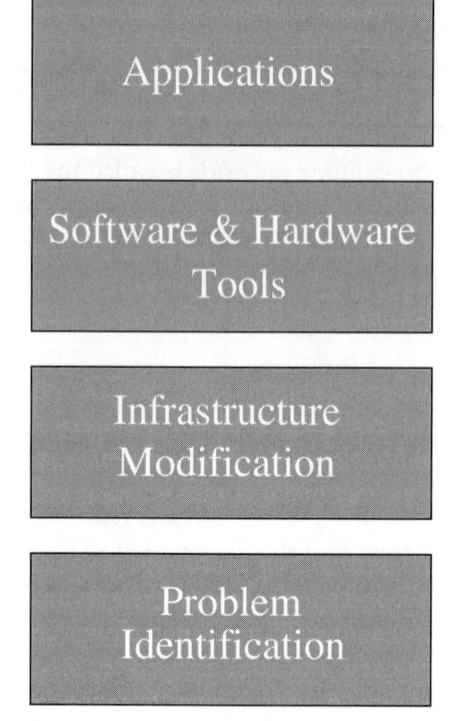

Figure 1-14 The four-layer process for considering a major infrastructure change

Application Enhancement

The school identified four critical applications that required enhancement: more effective use of voicemail, which we have already briefly discussed; campuswide message distribution; improved directory access; and enhanced information sharing. When Champlain began the process of searching out the various vendors that could provide the application set that would take best advantage of the revitalized network and dovetail more or less seamlessly with the school's existing applications, the list narrowed quickly to a handful of companies. One of those was Net6, now a division of Citrix Corporation. The firm's Web site lists the company's prime initiative as "Enabling employees and partners to find, organize, and collaborate on applications and information—yielding higher productivity, better customer service and improved decision-making. Equally important, Citrix solutions increase overall access security and immediately deliver measurable returns and benefits." Those are certainly in alignment with Champlain's goals, one of the reasons that the school selected Net6 as a partner.

"Net6 offers a suite of software applications that are very powerful," explains Aaron Videtto. "The one that really caught our eye is a product called the Net6 Application Gateway. It basically allows us to distribute information on a campuswide basis and display that information on the screens of the desktop phones. And to make that process even easier, we use another product called the Net6 Design Studio to quickly convert application content and information for display on the telephone screens. Design Studio keeps us from having to hire application programmers, because we simply point the application at the content we want converted, and it pulls it in and converts it for us. It really makes a difference. And what our faculty, staff, and students see is information that looks like it was custom-developed for them. Students, for example, now see daily cafeteria menus, information about events happening on campus or downtown, weather reports, schedule information, and other data that is actually useful to them. And because it's done online, we don't tie up their voicemail with stuff that is important to us but perceived to be less important by them. It's a much more customer-focused way of doing business. We can also create specific distribution lists so we can send out different messages to different groups of people, thus making our communications efforts that much more effective and efficient."

Another area where Champlain's convergence effort has played out positively is in campus help desk operations. Visual Voice Mail, another Net6 application, allows phone users to see on the screen a list of their messages as well as a set of controls that look like what you would find on a typical VCR. "The phones," explains Videtto, "have programmable soft keys [see Figure 1-15], which means that we can make them do whatever makes most sense for the application at hand—there's no need to memorize a set of application-specific commands like you typically have to do with traditional voicemail systems. And because we get a lot of new people moving into the dorms or working on campus every year, we use the screensaver capability of the Push Server to send telephone and voicemail instructions to display on the screen of the telephone, so that questions can be addressed without the need for a call to the help desk. This had the added advantage of reducing the amount of training required to operate the phones and retrieve voicemail messages, which in turn reduced the number of help desk calls."

Another application that is embedded in the Net6 Voice Office Application Suite is the Push Server, which gives administrators the ability to

Figure 1-15 Programmable soft keys, shown at bottom of screen

transmit the audio, text, and graphical alerts that we have just described to IP phone displays. Aaron Videtto took advantage of the screensaver capabilities of the Net6 Push Server in a rather unique way. "Instead of sending out the traditional broadcast voicemail messages described earlier," he explains, "which are (1) not listened to and (2) a major irritant to students, we set up the messages as push content that rotates on the screens of the IP phones when they're sitting idle. I have the ability to use a feature called the Idle URL setting of the Cisco CallManager IP PBX to set the duration of idle time before the telephone goes into screensaver mode and seeks out the IP address for the Push Server. As soon as it seeks the address, it begins to display whatever messages are hosted there. I simply have the support folks type the messages that we want users to see displayed, insert whatever graphics are required, and use the technology to push the messages out to the telephone displays. Once the phone is seen to be idle for whatever time period we specify, the messages begin to be displayed on the phone in the same way a screensaver appears on a PC. These messages simply rotate until the phone goes off-hook or until we change the messages.

"The best part is that the Push Server allows us to create different message sets for different user groups. We have total control."

Dusini and Videtto were also looking for enhancement in the area of directory assistance. The original system they deployed was cumbersome and complex; as a result campus operators and help desk personnel were fielding far more calls of that particular type than they were staffed to handle. They researched the available application set for enhanced directory management and settled on the Voice Office Quick Complete Directory. With this application, the Lightweight Directory Access Protocol (LDAP)[2] directory is loaded directly into the Application Gateway, which eliminates the need to repeatedly synchronize the call-manager directory every time a change occurs. It also eliminates the need for the call manager to do real-time LDAP lookup over the network each time the directory is used, thus reducing the total traffic flowing across the production environment.

Because of the way the Voice Office Quick Complete Directory "prunes the directory tree" when a user types a name into the system, fewer key strokes are required to find the requested name. The application requires a single keystroke per letter.

Important—But Not Here Yet

The advantages that VoIP offers to companies that consider it for their infrastructure needs can be compelling. However, they don't all arrive at once, and in some cases don't arrive at all. "One area that we are looking forward to as things evolve—but for which we don't yet have an immediate need—is savings on long distance," Dusini told me in a follow-up meeting. "A lot of large businesses go to VoIP because of the perception that they will save a great deal on long-distance dollars, and for some of them that becomes a very pleasant reality. This is not yet a big driver for us. We may eventually find a reason to take advantage of LD savings, but initially our interest is in making the organization easier to change, supporting work-at-home employees and students, and taking care of our out-of-country people.

"There are also some cultural considerations around expectations that the network is rock-solid that we have to take into account. VoIP is very good, but it does have its own unique set of peculiarities that are service

---

[2]LDAP is used to access information directories. It is based on the standards contained within the OSI X.500 standard but is significantly simpler. And unlike X.500, LDAP supports Transmission Control Protocol (TCP)/IP. LDAP is not yet widely implemented but will ultimately make it possible for an application running on virtually any computer to obtain and act on directory information. And because LDAP is an open standard, applications have no need to be concerned about the type of server hosting the application.

affecting. Luckily, the ubiquity of cell phones helps a lot because they don't offer perfect service and an expectation has been established that universal access is a fair tradeoff for the occasional quality glitch."

Dusini and his team of IT and communications professionals saw VoIP as a gateway to opportunities that the school could take advantage of as part of its evolution and growth strategy. They also had the right perspective: VoIP would clearly make a difference for them, but it would not result in untold diminution of cost. They looked beyond the cost to the less tangible (but no less real) advantages that a converged network infrastructure would bring to them in terms of customer intimacy and service delivery.

Of course, they also looked closely at the infrastructure. This would not be cheap; they would have to spend significant dollars on hardware, software, and enhancements to human capital, but they knew that in the final analysis the benefits would far exceed the costs.

Driving Forces

The decisions made by Champlain College are in alignment with similar decisions being made all over the world by enterprises looking to upgrade their IT and telecom infrastructures. Numerous factors cause businesses to look at VoIP and convergence, and to truly understand the decisions being made it is important to have a clear awareness of where we are so that we can better understand where we're going. Before we dig too deeply into the gross anatomy of the telephone network in the next chapter, let's pause to consider the forces shaping the industry as we know it today. Something of a perfect storm is under way, and while plenty of sunshine is on the horizon we're still rising and falling between wave troughs at the moment. As you read the following vignettes, think of them as follows: Individually, they may appear to be inconsequential pieces of fabric in a quilt and in some cases off base, but when combined they create something complex and dazzling. Consider them in aggregate.

- Many factors have combined to create the telecom industry as we know it today. One of the earliest is the arrival of widespread mobile telephony and the "training" of the marketplace to accept lower quality of service (QoS) in exchange for the freedom of mobility. But *most* important is the training of the market to accept lower quality service—*period*.

- Equally critical is the growth in broadband access availability and the widespread use of and reliance on the Internet and World Wide Web. In lockstep comes the promise of convergence. And as the technology advances and demonstrates increasing levels of capability, demands for better QoS lead to better control of jitter, latency, and packet loss.
- The ongoing convergence of technologies leads to the convergence of IT and telecom in the enterprise space. The promise of application integration, bundling, and unified messaging leads to experimentation with packet voice and ultimately to the introduction of VoIP.
- Bottleneck reduction becomes the order of the day. Broadband access, high-speed routing at the edge of the network, the rise of Multiprotocol Label Switching (MPLS), and growing support by vendors such as Nortel, Cisco, Lucent, and Juniper for carrier-grade routers move the bottleneck inexorably out of the network and into the application.
- The burst telecom bubble, beginning in March 2001, creates intense marketplace heartburn. The love affair with technology ends abruptly; a romance with services and business enhancement begins. Customer intimacy, solution selling, customer relationship management (CRM), ERP, data mining, and knowledge management become the most oft-heard phrases in the business lexicon. Technologists jump out of the driver's seat (or, more likely, are pushed); the customer jumps in with the following big four requirements on their short list. The big four are remarkably simple in their demands, impossibly difficult in their execution: (1) *If I buy from you, will your product or service measurably enhance my revenues?* (2) *Will they reduce my overall costs?* (3) *Will they somehow enhance or at least stabilize my competitive position in the market?* (4) *And will they help to mitigate the downside risk I face, operating in an increasingly and aggressively competitive market?* Scratching their information-superhighway road rash, service providers and manufacturers look to reinvent themselves as solution providers.
- Calls for customer service reach an all-time high. Call centers and contact centers grow like mushrooms on a summer lawn. Enterprise speed and agility, particularly as they relate to customer service, become prime differentiators. The ability to maintain customer contact and continuity is viewed as a critical contributor to customer satisfaction—and to revenue assurance.

- Recovery from the burst bubble leads to inevitable industry consolidation, buy-down, and intensifying competition. Customers react predictably, pitting one vendor against another in a pricing and service delivery frenzy.
- Regulatory reform favors the incumbent local exchange carriers (ILECs) and facilities-based competition, leading to an upsurge in cable, wireless, and power companies offering voice over their newly deployed IP infrastructures. They do it because they can and because voice was, is, and always will be the killer application. Suddenly IP represents the fundamental underpinning of the triple play (voice, video, and data) and a growing realization that the ILEC is no longer the only game in town for carrier-grade service.
- VoIP and related technologies continue to evolve and improve as networks become broader, faster, and more capable. In response to the broadside attack on their service bastion by cable, wireless, and others, the telcos counter, announcing service packages that include entertainment and broadcast content delivered over DSL—in effect their own triple (and in some cases, quadruple) play.
- The game has become one of offering boutique services in a commodity market. Basic technology has indeed become a commodity, as have Internet access, storage, wireline voice, long distance, content of many types, switching, routing, and wireless.
- As if commoditization isn't enough, the industry reels from attacks by the modern-day equivalent of Visigoths and Vandals. Blows against the beleaguered empire come from Skype and Vonage, Yahoo! and Google, Virgin Wireless, Microsoft, and an increasingly demanding and technologically adept customer base. Skype and Vonage successfully undermine the ILECs' positions of circuit-switched power, demonstrating just how good—*and free*—VoIP (over Internet) can be. Yahoo! and Google make IP-based mail, chat, and storage applications available to the masses—at no charge. Google's GMail offers a gigabyte of storage to every user. Virgin Wireless proves (1) the power of brand and (2) the increasing irrelevance associated with owning the network—and the power inherent in owning the customer and working the brand loyalty game instead.

Microsoft codifies everything, extending its desktop and set-top-hungry tentacles into every possible customer touch point. Microsoft is desktop, set-top, palm-top, mobile, gaming, content, application, operating system, and central office. They don't want the infrastructure; they just want the customer. So they revamp Windows XP, call it Longhorn, and upgrade their embedded messenger service, codenamed

Istanbul, to include a carrier-grade VoIP client. And the customers? They . . . just . . . want . . . more.

These forces—Skype and Vonage, Yahoo! and Google, Virgin Wireless, Microsoft, and the customers—behave like biological viruses. They infect and multiply, and the world around them irrevocably changes.

- Devices converge, get smaller, and become more capable. Desktop phones now integrate IP-based Web browsers. Mobile devices combine full-function personal digital assistants (PDAs) and phones and connect to the network at high speed. Vendors roll out Wi-Fi telephones, further facilitating convergence over the enterprise network and offering untold bypass opportunities.
- The consumer and enterprise markets bump and rub against one another as corporations send waves of employees home to work. These small-office/home-office (SOHO) workers, together with the burgeoning small-to-medium business (SMB) market, create demand for centralized service quality delivered over a fully distributed network to increasingly remote workers. Follow-me and find-me services, sometimes called "presence applications," become highly desirable as mechanisms for logical convergence. Employees and their employers become enamored of the increasing irrelevance of physical location.
- The convergence of IT and telecom nears completion. In the enterprise, where VoIP has become widely deployed, voice has morphed into just another data application. It is, after all, nothing more than a sporadic contribution of packets to the overall network data stream, and with the broad deployment of edge bandwidth and digital compression, an increasingly smaller percentage of that data is devoted to voice. As this realization dawns, CTOs throughout the enterprise domain make a bold move: They issue mandates that cause their IT organizations to absorb the functions and responsibilities of their historically dedicated voice service organizations. This move does two things: It formalizes enterprise recognition that VoIP is *clearly* an enterprise application and further recognizes that *voice* is *not.* Consider this: The packets emerging from a traditional data application such as e-mail are remarkably forgiving of delay and jitter because the asynchronous application that creates and consumes them is equally forgiving. If the message is delayed in its arrival by an additional 10 minutes, for the most part no one cares. However, e-mail is extraordinarily intolerant when it comes to packet loss. The loss of a single packet can result in catastrophic data corruption.

Voice, on the other hand, has precisely the opposite behavior. The human ear is such a poorly engineered listening device that 40 percent packet loss can often go undetected. But introduce 30 milliseconds of delay into the packet stream and customers begin to turn surly.

So the coming together of voice and data is a powerful and compelling thing as far as the customer is concerned. For the service provider, however, the result is a complex architectural and logistical problem, the management of which is not optional.

- Many companies have jumped on the VoIP bandwagon, but not always for the right reasons. Rationale ranges from "gotta go VoIP because it's cool" to "huge reduction in expense" to "unified messaging." Implementers, however, must never forget one thing: *Voice* is the *application*. *VoIP* is a *technology option*. The sounds that emanate from a standard telephone and the sounds that emanate from an IP-based device are identical—and they'd better be, because if they're not the fish won't bite.

See the trend? This isn't going to stop. The reasons for the evolution to VoIP from traditional circuit switching represent a powerful and compelling argument for change.

IP Telephony: Making the Case

So what are the motivating factors that make the case for an enterprise move to an IP infrastructure?

Number One: Cost Savings

IT personnel looking to make an infrastructure change involving IP typically examine it for the following reasons. First, a widely held belief exists that significant cost savings can be realized through the conversion from a traditional circuit-switched infrastructure to an IP-packet infrastructure. And although the outlay of real capital and expense dollars may not be as greatly reduced as many would like to think, the savings do come, and in a variety of ways. In a network that comprises multiple locations, packet switching reduces the cost of voice transmission. Furthermore, when voice migrates to IP, it becomes one more data application and now has the ability ride the same network as the enterprise's data traffic. The result is far more efficient network and traffic management. After all, IT personnel now manage one network, not two. They also now combine voice-based network changes with those of data activity, which results in a measurable cost reduction.

One area where VoIP results in significant savings is in the area of staffing and the need to physically locate staff members in specific locations. Because IP telephony is location independent and does not require collocation with the local switch or PBX, the resources of the network and support personnel can be located wherever it makes sense to put them, regardless of the location of the network itself.

Number Two: Productivity Gains

VoIP systems today typically include a range of productivity tools such as voicemail, e-mail, fax forwarding, and embedded directory services. The value that is brought by embedding these features in the system shows up as gains in efficiency and productivity: By managing all traffic types through a single graphical user interface (GUI), employees can avoid the time lost (and resulting customer frustration) that stems from having to switch between applications and disparate databases. And by taking advantage of so-called presence applications, a caller can tell whether the called party is available and can direct the system to track them down through a variety of contact modalities such as follow-me service to a mobile phone, VoIP running on a PDA, and so on. This capability, implemented through the Session Initiation Protocol (SIP), which will be discussed later in the book, eliminates the need for the customer to "track down" the called party. Instead, the system does so using a rule set created by the called party and based on whether he or she is in a position to be contacted, and if so, the best way to do so. If a caller were to place a call to a person in a business and the person is not at his or her desk, the called party may invoke a rule set that uses the following routing information:

1. Call received by main number at desk.
2. If no answer after three rings, redirect call to mobile number.
3. If no answer after five rings, redirect call to VoIP account on PDA.
4. If no answer, send to voicemail with directions to send a SMS message to a specific identity (account).

The result is that the customer needs a single number to make contact, and the system takes care of intrasystem routing, taking whatever steps it needs to establish contact between the caller and the called party.

Number Three: Customer Intimacy

I occasionally get strange looks for my propensity to quote Tom Peters on a fairly regular basis. In the early 1980s Peters wrote *In Search of Excellence,* in my mind a book that should be required reading for anyone who works for a company with customers. Peters captures the essence of this issue with the following quote, which introduces one of the chapters of the book:

> Probably the most important management fundamental that is being ignored today is staying close to the customer to satisfy his needs and anticipate his wants. In too many companies, the customer has become a bloody nuisance, whose unpredictable behavior damages carefully made strategic plans, whose activities mess up computer operations, and who stubbornly insists that purchased products should work.

Customer relations are what business is all about, and management of those relationships is what makes one company better than its closest competitor. Customer relationship management, or CRM, is one of the most fundamental and important activities that companies can engage in. And IP-based communications systems have the ability to enhance this function dramatically. Central to good customer relationships is the ease with which customers have the ability to do business with you. Automatic call routing, mentioned earlier, is one way that companies improve their customer affinity practices; another is through the critical analysis of customer behavior: buying patterns, calling patterns, how they use purchased products, and so on. Because VoIP systems have the ability to generate detailed reports that can be analyzed to yield this kind of information, managers can take action based on the reported data to reduce hold times, provide online information, and, therefore, reduce the total number of dropped calls. Interactive voice response units (IVRs) give the called party the ability to manage incoming calls by routing them to the shortest queue, the operator most informed about his or her particular geography, a specific language requirement, a product expert, or the same person he or she spoke with on a prior call. Similarly, callers who choose to use e-mail or a Web chat as their preferred mode of communications can do so and still enjoy the proper degree of appropriate routing.

Number Four: Applications

There is a widely believed myth that the single greatest advantage of engaging in an all-IP conversion effort is the tremendous cost savings that will result. Perhaps . . . but more likely not. There will unquestionably be some degree of cost savings, but the real advantage, the culminating event that makes a VoIP conversion so powerful, is the convergence of applications.

Already converged application sets exist on the market that facilitate a range of capabilities, some of which were described earlier by Aaron Videtto in his explanation of the Net6 capabilities. One of the best known of these is "click-though" or "click-to-call," a technique that allows a user to seamlessly move from an IM or SMS into a voice session with a single click. Already the major software houses—PeopleSoft, Siebel, Oracle, Microsoft—are developing applications that will integrate seamlessly into the voice environment, allowing for interactive voice-enabled database queries. A customer might be able, for example, to "ask the system" for today's weather in a particular area, and the system will respond appropriately. In essence the user carries on a conversation with the system—there is no human in the loop.

Caveat Emptor: Quality of Service (QoS)

One of the inherent downsides of VoIP is that unlike the public switched telephone network (PSTN), QoS in VoIP networks is not assured. With VoIP, QoS becomes really nothing more than a configurable (and billable) feature. After all, the mobile phone has made it possible to offer an inferior grade of phone service that customers are eager to pay for! Paul Dusini agrees: "In a lot of ways we got lucky, because the widespread use of cell phones and their sometimes questionable service quality meant that we could get away with a lot of things during our initial startup phase. Eventually, we took care of the glitches that were service affecting, but because we had that soft window, we were able to take advantage of the perception that less than perfect service is, at least temporarily, okay. We certainly didn't build our service-delivery model around that, but it bought us some much needed breathing room."

Voice Quality

The enterprise must therefore identify and make plans to deal with those issues that will have an impact on QoS. Failure to do so can result in chronic service-quality problems. In addition, it is critical to put into place an ongoing management program for identification and resolution of service-quality issues.

For example, audio quality can be measured from both the user's perspective as well as in more qualitative and quantitative technical terms. For a very long time, voice-quality measures have been based on ITU P.800, the worldwide standard that establishes service-quality benchmarks regarding how listeners perceive and understand speech quality in public networks. It uses an opinion-based scoring technique that measures listeners' comprehension under highly controlled conditions. This system is referred to as a *mean opinion score* (MOS) and is widely used for the evaluation of telecommunications infrastructure components.

Quality measurements in the technical domain include *call setup time*, which measures the time it takes for an end-to-end call to be placed; *post dialing delay,* which measures the interval between the last dialed digit and the first ringing tone; *call completion ratio,* which is a measure of the number of calls actually completed versus those that are unsuccessful; and *dial tone delay,* which measures the interval between going off-hook and hearing a dial tone.

What is interesting, of course, is that these measures were originally created for use in the legacy telephone network. As the migration to VoIP continues, network digitization grows, and the mix of services becomes more complex, these measures must be augmented and modified. IP voice, for example, relies on digital compression to reduce bandwidth requirements in the network. Qualitative measures must therefore also include *jitter*, the (typically varying) interval between packets as they arrive at the router; *packet loss*, the inevitable result of operating in a connectionless packet environment; and *delay,* a measure of the end-to-end transit time for packets in the network. Delay is fine, but it must be consistent from packet to packet to avoid quality loss.

Because of the fundamental way in which IP works, IP networks are not overly concerned with the bandwidth assigned to each conversation. As a result so-called wideband phones, recently introduced in the IP voice market, actually improve the quality of audio carried at frequencies higher than 3.3 KHz (the lower limit of the so-called voice band) and as high as 7 KHz. In fact, audio quality is directly related to the sampling rate, that is, the frequency at which audio signals are sampled (see Figure 1-16) as part of the digitization and packet creation process. DVD-quality audio, for example, is sampled at 192 KHz. On the other hand, CD-quality audio is sampled at 44.1 KHz, and classic telephony is sampled at 8 KHz. Higher sampling rates enable the network to capture harmonics that provide tone and richness, facilitating better perceived quality for applications such as music on hold. For people with hearing difficulties, wideband phones may improve productivity and enhance the customer's experience using the telephone. Higher sampling rates do, however, require more bandwidth; implementers must consider this as they make plans for their own system design.

The standards for wideband audio include G.711 adaptive differential pulse code modulation (ADPCM) and G.729 compression. ADPCM is described later in the book; G.729 relies on a relatively complex signal-encoding scheme called *code-excited linear-prediction coding* (CELP) to encode speech signals at 8 Kbps. The coder operates on speech frames of 10 ms that correspond to 80 samples at a sampling rate of 8,000 samples per second. The encoder includes voice activity detection and comfort noise generation (VAD/CNG), both important QoS additions for digital voice systems, and the coder/decoder (CODEC) is also capable of processing silence frames. G.729 offers near toll-quality performance and is

Figure 1-16 Audio sampling

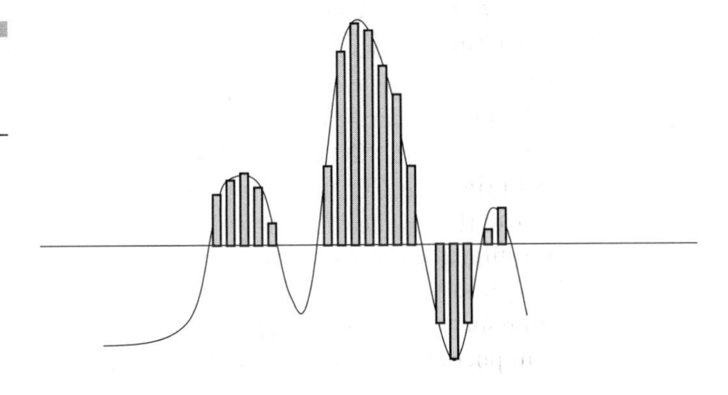

the default CODEC for voice-over-frame relay, as described by the Frame Relay Forum. It also works very well in VoIP applications.

To improve application performance and to provide remote users with greater end-to-end control of the session, compression control should most effectively reside in the IP phone rather than at a central shared router.

CODEC Standards

The word CODEC is a contraction of the two words "coder" and "decoder," the two software functions required for voice digitization and in some cases compression. Coding is the process of analog-to-digital conversion; decoding is the opposite, clearly a set of functions that sits between analog and digital systems such as the analog access to the PSTN and an IP-based network. Many CODECs are available, and CODEC selection is typically dependent on choices related to voice quality and processing speed. Figure 1-17 lists the most commonly used CODECs and their characteristics. As an example, in a situation where call quality is of the utmost importance such as in a customer support center, a G.711 CODEC would be an ideal selection because of its extremely high quality characteristics. This is the voice digitization standard used throughout the PSTN to create eight-bit pulse code modulation (PCM) samples at the rate of 8,000 samples per second. The G.711 CODEC creates a 64 Kbps bitstream and has two forms. A-Law G.711 PCM converts 13-bit PCM samples into eight-bit compressed PCM samples. μ-Law (Mu-Law) PCM converts 14-bit PCM samples into eight-bit compressed PCM samples. A-Law is used throughout most countries in the world; μ-Law is used in the United States, Canada, and a handful of other countries.

On the other hand, in situations where voice across the WAN will be used as a way to reduce the cost of long distance, a G.729a CODEC may be perfectly adequate, because voicemail may be the most common form of traffic encoded using the CODEC. Today, vendors perform most of the leg work associated with CODEC selection, choosing the one that best meets the needs of the application they will serve with their installation.

Special cases must be considered during the selection process, however, one of which is known as *transcoding.* If two or more parties are communicating using different CODECs, such as a cellular user communicating with a VoIP device, the network must translate (or transcode) between the two disparate coding schemes. For example, if the mobile user is talking on a Global System for Mobile communications (GSM) phone, the PSTN will have to translate between G.729a and the full rate or half-rate supported by GSM networks. Each conversion introduces delay and the possibility of introducing noise into the system, which can result in a perceptible degradation in QoS.

CODEC Considerations

The ITU specifies a maximum acceptable round-trip delay of 300 ms for voice applications, 150 ms in each direction. Any delay that exceeds a quarter of a second will most likely be noticed by the user and, more important, deemed to be unacceptably low quality.

In modern VoIP systems where devices (routers) operate at wire speed and have deep memory and processing resources available to them, delays caused by switches and routers are almost unheard of. If delays occur in the network, they more than likely occur at the interface between the local area network (LAN) and wide are network (WAN), where bandwidth can be low (overutilized because of the choke-point phenomenon) and congestion can be high.

The first order of business, then, is to ensure that adequate bandwidth will be available at all critical junction points. In IP networks, because of decreasing packet size and enhanced levels of compression, the packet header (not the data behind the packet header) becomes the bulk of the transmitted traffic.

Figure 1-17 The most commonly used CODECs

CODEC	Data Rate	Voice Quality
G.711	64 Kbps	High
G.723.1	6.4, 5.3 Kbps	Low
G.726	40, 32, 24, 16 Kbps	Medium
G.728	16 Kbps	Medium
G.729	8 Kbps	Medium

Prioritization can also be a powerful tool for the preservation of voice service quality. Several techniques make this possible; one is to prioritize based on source and/or destination address, such as the address of the call server. Another is to prioritize packets based on User Datagram Protocol (UDP) port numbers, which will be described later. Either way, placing voice (and ultimately video) at higher priority levels than their less demanding data counterparts will result in more well perceived QoS in the applications that suffer most from network latency.

The Assessment Process

So what is the process a company should go through before making the decision to go with a converged solution? The first step is to consider the evolving needs of the enterprise. In almost every case, the desire to evolve to a VoIP network has to do with one or more of the following three things: a desire to lower total capital expenditure (CAPEX) and operating expenditures (OPEX) spending, a desire to raise enterprise productivity levels, and a desire to stabilize or improve competitive positioning.

Controlling CAPEX and OPEX Spending

Although not always the most important issue, spending control is always the first and most visible desired outcome when the discussion of VoIP arises. The industry has been awash in the promise of IP and convergence for so long now that there is an expectation that one of its greatest contributions to the enterprise will be cost reduction. And although it certainly can help, the additional advantages a converged infrastructure brings about cannot be ignored. That being said, there are unquestionably ways in which VoIP can directly affect spending. First, because it represents the best that convergence has to offer, it can significantly reduce CAPEX and OPEX due to the overall reduction in network and protocol complexity that it brings about. Second, it can reduce the cost of long-distance charges through Internet telephony and toll bypass made possible by the implementation of an IP PBX environment. Third, by implementing desktop VoIP in the enterprise, the cost of moves, adds, and changes is substantially reduced because of the ability of the network to auto-identify an IP telephone when it is unplugged in one location and plugged into another.

VoIP can also have an impact on calling-card charges from SOHO and other remote workers, particularly those who work internationally. And because of the hand-in-glove complementary and functional relationship that exists between VoIP and broadband access services, it can also reduce the cost of mobile telephony through the substitution of VoIP over broadband.

IT personnel should be able to create a list of potential cost reduction areas that are under their control. For example, by taking advantage of more and more commonly deployed broadband access options, servers that use toll-free (to the user, not the enterprise!) 800-number access can be eliminated in favor of secure access by remote workers using cable modems, digital subscriber line (DSL), or broadband wireless through a secure virtual private network (VPN). They should also investigate the viability of either pure VoIP or IP delivered through an IP PBX as a way to eliminate or at least substantially reduce long-distance charges.

Because VoIP can (and should) result in a convergence of data applications onto a single converged infrastructure, a number of items should be examined by IT personnel as they move toward a VoIP decision. One is application consolidation: moving voice, video, and data onto a single, more easily manageable network. Second is the possibility of centralizing e-mail and voicemail servers as a way to reduce spend. Third is consideration for the use of server-based applications such as audio and videoconferencing and the spending reduction that can result from bringing those applications in house rather than buying them from a service vendor.

Please understand: Not all of these will be appropriate considerations for every enterprise, but *all of them* should be considered. One key to success here is to question everything. If the slightest possibility exists that an idea could result in reduced CAPEX and OPEX spending, consider it.

Enterprise Productivity

Productivity is one of those weasel words that is so overused that it has lost a great deal of its meaning. Productivity—a measure of the ability to *produce* something—can be measured in a plethora of ways. But to meaningfully affect enterprise productivity, a new technology must have impact across a broad spectrum of functional entities and at all levels of the corporation—not just in a call center or among the bit-weenies. Furthermore it must work seamlessly within a wide range of preexisting technology foundations, including wireless LANs, and it must operate over multiple access devices, including both wireline and wireless telephones and PDAs with embedded applications such as Web access, SMS/IM, and e-mail. All users—mobile and remote workers, executive personnel, sales teams, engineers, and HR—must be able to take advantage of the productivity gains that a converged VoIP network brings. These include (but are certainly not limited to) support for virtual offices with integrated PC and phone, both with access to directory services; access to and more efficient use of enterprise information resulting from the deployment of unified messaging applications and their databases; faster virtual workgroup decision making brought about by universal connectivity, multimedia access, and access device agnosticism; the deployment of wireless LAN access throughout the enterprise to facilitate anytime, anywhere connectivity for roaming and mobile employees; efficiency gains that result from the deployment of secure VPN connectivity over broadband, thus allowing remote users to work effectively at a remote location, either permanently or on an ad hoc basis; advantages stemming from the deployment of a converged messaging system that allows users to access voicemail, SMS/IM, and e-mail using any device they want, including a traditional telephone (text-to-voice message conversion); and finally the ability to deliver "computer information" even when there is no computer to deliver it to. For example, Champlain College's IP telephones have multimedia screens with access to the Web, allowing information from an application gateway to reach an individual phone, all phones, or groups of phones segmented by VLAN addressing.

Improved Competitive Positioning

Given the increasingly competitive nature of the business world today, it's no wonder that companies are scrambling to protect and defend whatever competitive advantage they have in their chosen marketplace. Competitive advantage is often a fleeting, ephemeral thing, so any technological tool that can extend the advantage is well received. VoIP is clearly one of those.

Because of the distributed nature of the typical VoIP environment, the services it delivers can be equally distributed, allowing the enterprise to be flexible and fast and to reconfigure itself as required to match the movement of the market. Virtuality is, ironically, a very real thing when its capabilities are deployed within the corporation. For example, a virtual call center, where there is no bricks-and-mortar call center but rather a collection of agents working out of their homes, provides tremendous reconfiguration capability and adaptability to changing call volumes, language requirements, time of day shifts, continuity requirements, resource availability, and general customer demand. By matching the call center configuration to the market, customer intimacy and satisfaction are maintained.

Consider the case of JetBlue Airways, one of the best examples of a corporation making effective use of technology to change the way the firm does business. The airline was created 4 years ago with plans for a cost structure that could support inordinately low fares, and it worked. While its competitors lost billions of dollars, JetBlue posted profits out of the gate. Technology proved to be the company's greatest competitive advantage: JetBlue created the industry's first "paperless cockpit," giving laptops to pilots and first officers and saving 4,800 personhours per year in paper processing. The company also installed VoIP connections in the homes of 600 reservation agents who interact with customers, taking reservations and providing services that would traditionally be offered only from a costly call center.

A converged IP-based network can improve competitive vigor in numerous ways. For example, many companies deploy multimedia servers that host multiple customer contact applications such as instant messaging, fax-back service, e-mail, and voicemail, allowing customers to have immediate access to subject-matter experts while at the same time giving the enterprise the ability to deliver information to customers rapidly and flexibly. Web services provide customers with a variety of communication options *on the customer's terms, not the company's.* The availability of a multimedia communications portal reduces the number of abandoned transactions and hang-ups, improves overall customer satisfaction, and leads to a greater sense of caring. Furthermore, by supporting both wireline and wireless access, customer contact can be maintained under all circumstances, regardless of the location of the called agents or of the access device they choose (or are forced) to use.

The decision to go with a converged network must be based on solid business requirements. In some cases they will be the direct result of a compelling and unmet need that exists between the customer and the enterprise; in others, they will be the result of an economic event, such as the obsolescence of a significant hardware component (Champlain's PBX) or the expiration of a major lease. Still other companies may do the math and conclude that the move to a converged infrastructure will not only reduce spending but also result in a significant reduction in technology-related risk, a future-proofing process, if you will.

Chapter Summary

The choice of VoIP as an end-state solution is a complex one that must be made in stages, taking into account a long list of factors. These include application evolution, the needs of the business, the needs of the customer, economic considerations, staffing concerns, and a host of others. The consideration process must be based on a clear and emotionless understanding of the inner workings of a typical telephone network, how it differs from a VoIP network, and the pros and cons of each relative to application delivery and customer contact. In the chapter that follows, in keeping with our discussion of things biological, we turn our attention to the anatomy and physiology of a far more complex creature: the purpose-built telephone network.

In the introduction I promised you that this is a book about VoIP but that it will not linger too long on the inner workings of the underlying technologies. That being said, I also told you that this is a book about the *business* side of VoIP and the powerful impact it can have on a business when it is properly deployed. The power that comes from a wisely thought-out VoIP migration is all about knowledge: *knowing* why you are choosing to evolve; *knowing* what to expect; *knowing* when to make the change; *knowing* how it will all happen; *knowing* where the technology, service, application, human resources, and management changes will take place; and *knowing* how to respond to them all. There's a wonderful line, reputedly from *Alice in Wonderland*, that says, "If you don't know where you're going, any road will take you there." And of course the only way to know where you're going is to know where you've been. So the next chapter covers the waterfront, past to present, and may even stick its toe into the warm and inviting waters of the future.

We begin with the PSTN: what it is, how it works, why it does what it does. We then morph into a similar discussion about the evolving IP-based network so that the two can be compared as equals.

We often make the observation in our industry that "the question is not whether the move from circuit to packet will occur, but rather when." True enough. But "when" is an amorphous timeframe and I can promise you that the circuit-switched PSTN isn't going away anytime soon. In fact, for all its perceived faults and all the whispered chatter about its age and failing capabilities, it remains the most complex and functional machine ever built. It does precisely what it was designed to do, and will for some time to come. Parallel knowledge bases, showing the relative roles of the PSTN and the VoIP network, are valuable constructs.

Here we go.

CHAPTER 2

A Remarkable Machine: The Public Switched Telephone Network (PSTN)

Before we dig into the Internet Protocol (IP) as a delivery modality, let's spend some time on the legacy circuit-switched telephone network. It is a remarkable system that delivers high-quality services to everyone it touches, and yet here we are, preparing to monkey with it by introducing an alternative system that has the audacity to offer the option of a lower grade service! By understanding the PSTN we gain a better understanding and awareness of exactly what the customer will expect from a voice over IP (VoIP) system when we deploy it.

When a customer makes a telephone call, a complex sequence of events takes place that ultimately leads to the creation of a temporary end-to-end connection. To begin, let's follow a typical phone call through the network.

When a caller picks up the telephone, the act of lifting the handset[1] closes a circuit, which allows current to flow from the switch in the local central office that serves the telephone. The switch electronically attaches an oscillator to the circuit called a *dial tone generator*, which creates the customary sound that we all listen for when we place a call.[2] The dial tone serves to notify the caller that the switch is ready to receive the dialed digits.

The caller now dials the desired telephone number by pressing the appropriate buttons on the phone. Each button generates a *pair of tones* (listen carefully and you can hear them) that are slightly dissonant. This is done to prevent the possibility of a human voice randomly generating a frequency that could cause a misdial. The tone pairs are carefully selected so that they cannot be naturally generated by the human voice. This technique is called *dual tone multifrequency* (DTMF). It is not, however, the only way.

You may have noticed that there is a switch on many phones with two positions labeled "TONE" and "PULSE." When the switch is set to the tone position, the phone generates DTMF. When it is set on pulse, it generates the series of clicks that old dial telephones made when they were

---

[1]It doesn't matter whether the phone is corded or cordless. If cordless, pushing the TALK button creates a radio link between the handset and the base station, which in turn closes the circuit.

[2]It's interesting to note that all-digital networks, such as those that use the integrated services digital network (ISDN) as the access technology, don't generate a dial tone because packets don't make any noise! Instead, the ISDN signaling system (Q.931 for those of you collecting standards) transmits a "ring the phone" packet to the distant telephone. The phone itself generates the dial tone, and the ringing bell! The same is true in VoIP environments. Packets don't ring!

used to place a call. When the dial was rotated—let's say to the number three—it caused a switch contact to open and close rapidly three times, sending a series of three electrical pulses (or more correctly, interruptions) to the switch. DTMF has been around since the 1970s, but switches are still capable of being triggered by pulse dialing.

Back to our example. The caller finishes dialing the number, and a digits collector in the switch receives the digits. The switch then performs a rudimentary analysis of the number to determine whether it is served out of the same switch. It does this by looking at the area code (*numbering plan area*, or NPA) and prefix (NXX) of the dialed number.

A telephone number comprises three sections, shown below.

802-555-7837

The first three digits are the NPA, which identifies the called region.

The second three digits are the NXX, which identifies the exchange, or office, that the number is served from, and by extension the switch to which the number is connected. Each NXX can serve 10,000 numbers (555-0000 through 555-9999). A modern central office switch can typically handle as many as 15 to 20 exchanges of 10,000 lines each.

In our example, the called party is served out of a different switch than the caller is, so the call must be routed to a different central office. Before that routing can take place, however, the caller's local switch routes a query to the signaling network, known as *Signaling System 7* (SS7). SS7 provides the network with intelligence. It is responsible for setting up, maintaining, and tearing down a call, while at the same time providing access to enhanced services such as *custom local area signaling services* (CLASS), 800-number portability, *local number portability* (LNP), *line information database* (LIDB) lookups for credit card verification, and other enhanced features. In a sense it makes the local switch's job easier by centralizing many of the functions that formerly had to be performed locally.

The original concept behind SS7 was to separate the actual calls on the public telephone network from the process of setting up and tearing down those calls, as a way to make the network more efficient. This had the effect of moving the intelligence out of the PSTN and into a separate network where it could be somewhat centralized and therefore made available to a much broader population. The SS7 network, shown in Figure 2-1, consists of packet switches (*signal transfer points*, or STPs)

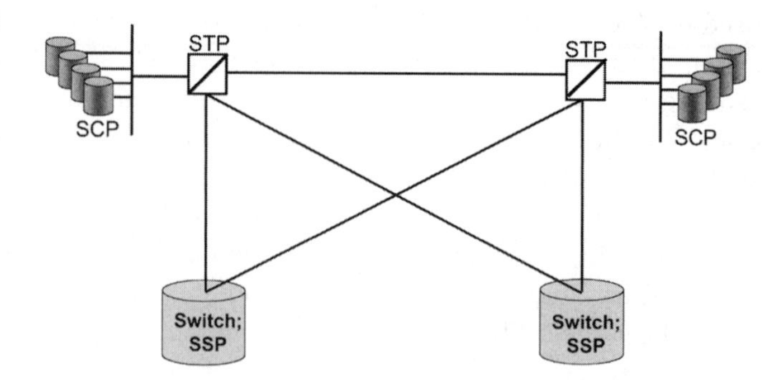

Figure 2-1 The SS7 network

and intelligent database engines (*service control points*, or SCPs), interconnected to each other and to the actual telephone company switches (*service switching points*, or SSPs) via digital links, typically operating at 56 to 64 Kbps.

When a customer in an SS7 environment places a call, the following process takes place. The local switching infrastructure issues a software interrupt via the SSP so that the called and calling party information can be handed off to the SS7 network, specifically an STP. The STP in turn routes the information to an associated SCP, which performs a database lookup to determine whether any special call-handling instructions apply. For example, if the calling party has chosen to block the delivery of caller ID information, the SCP query will return that fact.

Once the SCP has performed its task, the call information is returned to the STP packet switch, which consults routing tables and then selects a route for the call. Upon receipt of the call, the destination switch will cooperate with SS7 to determine whether the called party's phone is available, and if it is, it will ring the phone. If the customer's number is not available due to a busy condition or some other event, a packet will be returned to the source indicating the fact and SS7 will instruct the originating SSP to put a busy tone or reorder in the caller's ear. A busy tone indicates that the phone is already engaged in a call; reorder, sometimes called "fast busy," simply indicates that the call failed to complete through the network. At this point, the calling party has several options, one of which is to invoke one of the many CLASS services such as *automatic ringback*. With automatic ringback, the network will monitor the called number for a period of time, waiting for the line to become available. As soon as it is, the call will be cut through, and the calling party will be notified of the incoming call via some kind of distinctive ringing.

Thus, when a call is placed to a distant switch, the calling information is passed to SS7, which uses the caller's number, the called number, and SCP database information to choose a route for the call. It then determines whether there are any special call-handling requirements to be invoked, such as CLASS services, and instructs the various switches along the way to process the call as appropriate.

These features comprise a set of services known as the *Advanced Intelligent Network* (AIN), a term coined by Telcordia (formerly Bellcore, sold to Science Applications International Corporation, recently sold again to Providence Equity and Warburg Pincus). The SSPs (switches) are responsible for basic calling, while the SCPs manage the enhanced services that ride atop the calls. The SS7 network, then, is responsible for the signaling required to establish and tear down calls and to invoke supplementary or enhanced services. It is critically important, even more so today as the network begins the complex process of migrating from a circuit-switched model to a VoIP packet-based model. Later we will discuss how SS7 signaling networks interact with IP-based signaling protocols such as the *Session Initiation Protocol* (SIP) and H.323. One major way in which they differ is topology: SS7 is a centralized system, whereas SIP and H.323 are far more distributed.

Network Topology

Before we enter the central office, we will explain the topology of the typical network (please refer to Figure 2-2). The customer's telephone, and most likely, his or her PC and modem, are connected via *house wiring* (sometimes called inside wire) to a device on the outside of the house or office building called a *protector block* (see Figure 2-3). It is nothing more than a high-voltage protection device (a fuse, if you will) designed to open the circuit in the event of a lightning strike or the presence of some other high-voltage source. It protects both the subscriber and the switching equipment and provides the demarcation point between the telephone company's network assets and those of the customer inside the house.

The protector connects to the network via *twisted pair wire.* Twisted pair is literally that—a pair of wires that have been twisted around each other to reduce the amount of crosstalk and interference that occur between wire pairs packaged within the same cable. The number of

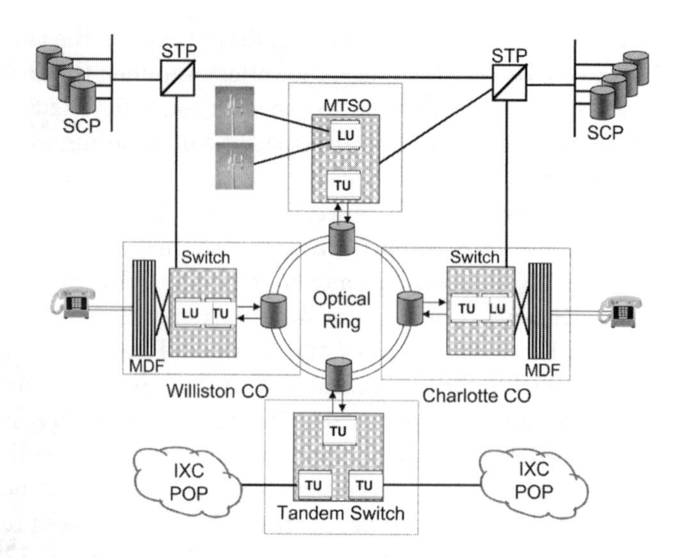

Figure 2-2 Typical network

Figure 2-3 Protector block, shown at left

twists per foot is carefully chosen. An example of twisted pair is shown in Figure 2-4.

The twisted pair(s) that provides service to the customer arrives at the house on what is called the *drop wire*. It is either aerial, as shown in Figure 2-5, or buried underground. We note that there may be multiple pairs in each drop wire because even today many households have a second line for a home office, computer, fax machine, or teenager.

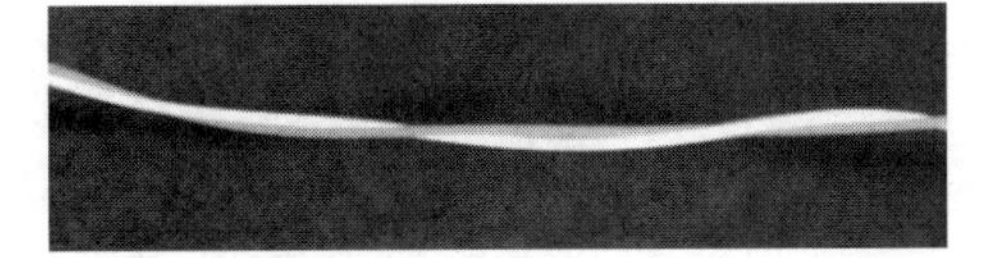

Figure 2-4 Twisted pair wire

Figure 2-5 Aerial drop wire

Once it reaches the edge of the subscriber's property, the drop wire typically terminates on a terminal box, such as that shown in Figure 2-6. There, all of the pairs from the neighborhood are cross-connected to the main cable that runs up the center of the street. This architecture is used primarily to simplify network management. When a new neighborhood is built today, network planning engineers estimate the number of pairs of wire that will be required by the homes in that neighborhood. They then build the houses, install the network and cross-connect boxes along the street, and cross-connect each house's drop wire to a single wire pair. Every pair in the cable has an appearance at every terminal box along the street. This allows a cable pair to be reassigned to another house elsewhere on the street, should the customer move. It also allows cable pairs to be easily replaced should a pair go bad. This design dramatically simplifies the installation and maintenance process, particularly given the high demand for additional cable pairs today.

Figure 2-6 Terminal box, sometimes known as B-box

This design also results in a challenge for network designers. These multiple appearances of the same cable pair in each junction box are called *bridged taps.* They create problems for digital services because electrical signal echoes can occur at the point where the wire terminates at a pair of *unterminated* terminal lugs. Suppose, for example, that cable pair number 201 is assigned to the house at 377 Long View Drive. It is no longer necessary, therefore, for that cable pair to have an appearance at other terminal boxes because it is now assigned. Once the pair has been cross-connected to the customer's local-loop drop wire, the technician *should* remove the appearances at other locations along the street by terminating the open wire appearances at each box. This eliminates the possibility of a signal echo occurring and creating errors on the line, particularly if the line is digital. ISDN is particularly susceptible to this phenomenon.

When outside plant engineers first started designing their networks, they set them up so that each customer was given a cable pair from his or her house all the way to the central office. The problem with this model was cost. It was very expensive to provision a cable pair for each customer. With the arrival of time division multiplexing (TDM), however, the "one dog, one bone" solution was no longer the only option. Instead, engineers were able to design a system under which customers could share access to the network as shown in Figure 2-7. This technique, known as a *subscriber loop carrier,* uses a collection of T1 carriers to combine the traffic from a cluster of subscribers and thus reduce the amount of wire required to interconnect them to the central office. The best known carrier system is called the SLC-96 (pronounced *SLICK)*, originally manufactured by Western Electric/Lucent, which transports traffic from 96 subscribers over four 4-wire T-carriers (plus a spare). Thus, 96 subscribers can be served by 20 conductors between their neighborhood and the central office instead of the 192 wires that would otherwise be required.

The only problem with this model is that customers are by and large restricted to the 64 Kbps of bandwidth that loop-carrier systems assign to each subscriber. That means that subscribers wishing to buy *more* than 64 Kbps—such as those that want DSL—are out of luck. And because it is estimated that as many as 70 percent of all subscribers in the United States are served from loop carriers, this poses a problem that service providers are scrambling to overcome. New versions of loop-carrier standards and technologies, such as GR-303 and optical remotes that use fiber instead of copper for the trunk line between the remote terminal and the central office terminal, go a long way toward solving this problem by making bandwidth allocation far more flexible. There is still quite a ways to go, however.

Typically, as long as a customer is within 12,000 feet from a central office, he or she will be given a dedicated connection to the network, as shown in Figure 2-8. If he or she is farther than 12,000 feet from the central office, however, he or she will normally be connected to a subscriber loop-carrier system of one type or another.

Figure 2-7 Subscriber loop carrier system. Customers share access to the network via a collection of multiplexed facilities to reduce outside plant cost.

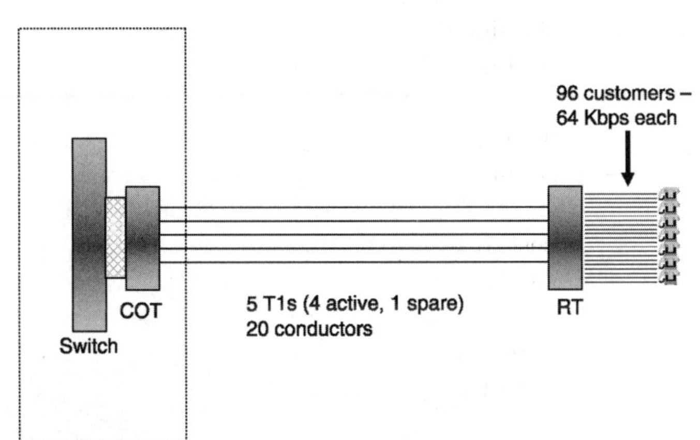

Either way, the customer's local loop, whether as a standalone pair of wires or as a timeslot on a carrier system, makes its way over the physical infrastructure on its way to the network. It may be aerial, as shown in Figure 2-9, or underground, as shown in Figure 2-10. If it has to travel a significant distance, it may encounter a load coil along the way, which is a device that "tunes" the local loop to the range of frequencies required for voice transport and extends the distance over which the signals can travel. A *load pot* comprises multiple load coils and performs loading for

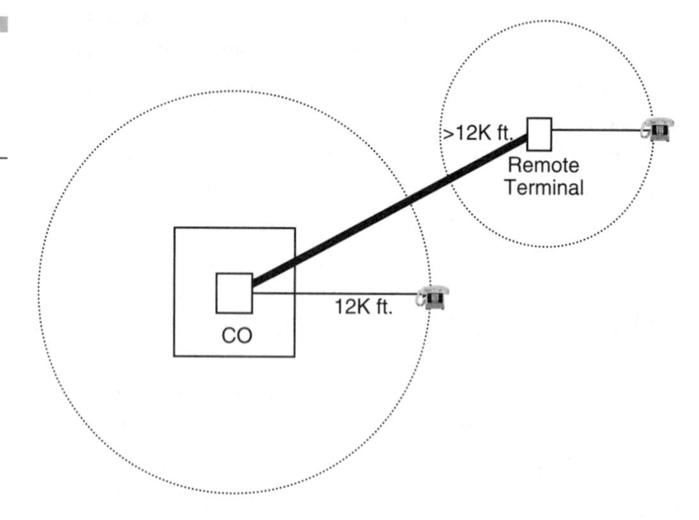

Figure 2-8 Carrier serving area (CSA) architecture

Figure 2-9 Aerial plant. Telephony is the lowest set of facilities on the pole.

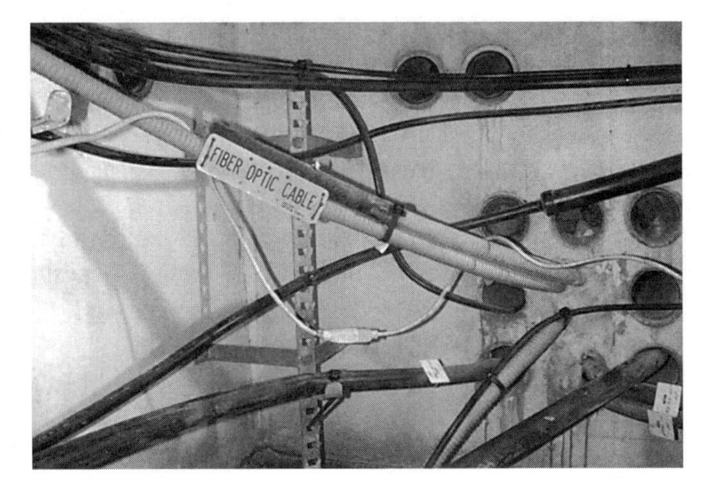

Figure 2-10 Underground plant

all the cable pairs in a cable that require it. It may also encounter a repeater if it is a digital loop carrier; the repeater receives the incoming signal, now weakened and noisy from the distance it has traveled, and reconstructs it, amplifying it before sending it on its way.

As a cable approaches the central office, its pairs are often combined in a splice case (see Figure 2-11) with those of other approaching cables to create a larger cable. For example, four 250-pair cables may be combined to create a 1,000-pair cable, which in turn may be combined with others to create a 5,000-pair cable that enters the central office. Once inside the office, the cables are broken out for distribution. This is done in the cable vault; an example is shown in Figure 2-12. The large cable on the left side of the splice case is broken down into the collection of smaller cables exiting on the right.

Figure 2-11 Splice cases (two on left, one on right)

Figure 2-12 In the cable vault, the large cables shown on the left are broken down into the smaller cables on the right for distribution throughout the office.

The Central Office

By now our cable has traveled over aerial and underground facilities and has arrived at the central office. It enters the building via a buried conduit that feeds into the *cable vault*, the lowest sub-basement in the office. Because the cable vault is exposed to the underground conduit system, a danger always exists that noxious and potentially explosive gases (methane, mostly) could flow into the cable vault and be set afire by a spark. These rooms are therefore closely monitored by the electronic equivalent of the canary in a coal mine and will cause alarms to sound if gas at dangerous levels is detected. A cable vault is shown in Figure 2-13.

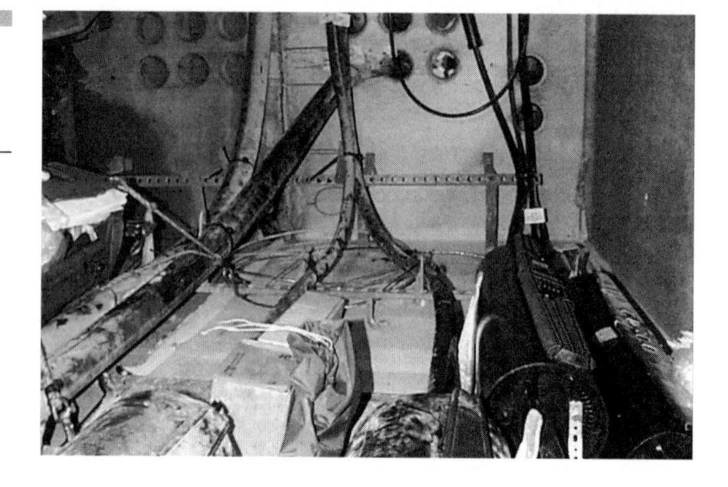

Figure 2-13 Another view of the cable vault

The cables that enter the cable vault are large and encompass hundreds if not thousands of wire pairs, as shown in Figure 2-14. Their security is obviously critical, because the loss of a single cable could mean loss of service for hundreds or thousands of customers. To help maintain the integrity of the outside plant, large cables are pressurized with very dry air, fed from a pressurization pump and manifold system in the cable vault. The pressure serves two purposes: It keeps moisture from leaking into a minor breach in the cable and serves as an alarm system in the event of a major cable failure. Cable plant technicians can analyze the data being fed to them from pressure transducers along the cable route and very accurately determine the location of a break. As soon as the large cables have been broken down into smaller pair bundles, they leave the cable vault on their way up to the main distribution frame, or MDF.

Before we leave the basement of the central office, we should discuss power, a major consideration in an office that provides telecommunications services to hundreds of subscribers.

When you plug in a laptop computer, the power supply does not actually power the laptop. The power supply's job is to charge the battery (assuming the battery is installed in the laptop); the battery powers the computer. That way, if commercial power goes away, the computer is unaffected because it was running on the battery to begin with.

Central offices work the same way. They are powered by massive wet cell battery arrays, shown in Figure 2-15, that are stored in the basement. The batteries are quite large—about 2 feet tall—and are connected together by massive copper bus bars. Meanwhile, commercial power is

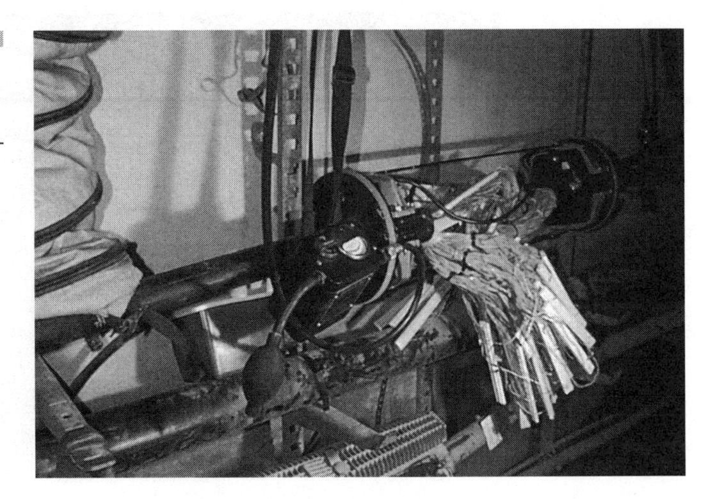

Figure 2-14 Wire pairs in the cable vault

Figure 2-15 Central office battery array

fed to the building from several sources on the power grid to avoid dependency on a single source of power. The AC power is fed into an inverter and filter that converts it to 48 volts DC, which is then trickled to the batteries. The voltage maintains a constant charge on them. In the event that a failure of commercial power should occur, the building's *uninterruptible power supply* (UPS) equipment, shown in Figure 2-16, kicks in and begins a complex series of events designed to protect service in the office. First, it recognizes that the building is now isolated and is on battery power exclusively. The batteries have enough current to power a typical central office for several hours before things start to fail. Second,

Figure 2-16 Turbine in rooftop enclosure

technicians begin overseeing the process of restoration and of shutting down nonessential systems to conserve battery power. Third, the UPS system initiates startup of the building's turbine, a massive jet engine that can be spun up in about a half-hour. The turbine is either in a soundproof room in the basement or in an enclosure on the roof of the building (see Figure 2-17). Once the turbine has spun up to speed, the building's power load can slowly be shed onto it. It will power the building until it runs out of kerosene, but most buildings have several days' supply of fuel in their underground tanks.

The 48 volts created by the power array in the central office is fed to traditional telephones via the wire pairs that connect them to the network. When a call needs to be placed to a customer's line, ringing current (significantly higher than 48 volts) is placed on the line until the phone is either answered or the caller hangs up. When the call is answered, the ringing current is removed.

Of particular importance is the fact that the phones on the PSTN are powered by a central office battery. In the event of a power failure, the phones still work because the uninterruptible central office power continues to operate.[3]

This is one of the key considerations when implementing a VoIP network. "Packets" do not have the ability to supply power to a telephone;

Figure 2-17 Uninterruptible power supply (UPS)

[3]However, keep one thing in mind. Most households today rely heavily on cordless phones (not cellular), which have a base station and a cordless handset. In the event of a failure of commercial power, these phones *will not work* because the base station radio relies on commercial power. I recommend that if you don't have at least one wired phone in your home, *buy one*.

the operating voltage of a packet network is very low. It is critical, therefore, that implementers think about how they will power the phones in the network. After all, customers have come to expect carrier-grade service, which includes power. In some cases individual uninterruptible power supplies are installed in remote buildings to avoid service interruption. In other cases, such as at Champlain College, power is supplied via a dedicated pair of wires on the Ethernet interface that feeds the phones. This technique will be discussed in a later chapter.

The cables leave the cable vault via fireproof foam-insulated ducts and travel upstairs to whatever floor houses the MDF. The MDF, shown in Figure 2-18, is a large iron frame that provides physical support for and technician access to the thousands of wire pairs that interconnect customer telephones to the switch. The cables from the cable vault ascend and are hardwired to the *vertical side* of the MDF, shown in Figure 2-19. The ends of the cables are called the *cable heads.* The vertical side of the MDF has additional overvoltage protection, shown in the diagram. The arrays of plug-ins are called "carbons;" they are simply carbon fuses that open in the event of an overvoltage situation and protect the equipment and people in the office.

The vertical side of the MDF is wired to the *horizontal side.* The horizontal side, shown in Figure 2-20, is ultimately wired to the central office switch. Notice the mass of wire lying on each shelf of the frame: These are the cable pairs from the customers served by the office. Imagine the complexity of troubleshooting a problem in all that wire.

Figure 2-18 Main distribution frame (MDF)

Figure 2-19 Vertical side of MDF, showing protectors

Figure 2-20 Horizontal side of MDF

The horizontal side also provides craft/technician access for repair purposes. Each of the white panels shown in the photo flips open to reveal a set of terminals for each wire pair, to which a test set can be connected.

From the MDF, the cable pairs are connected to the local switch in the office. Figure 2-21 shows an Ericsson local switch. Remember that the job of the local switch is to establish temporary connections between two or

Figure 2-21 An Ericsson local switch

more customers that wish to talk or between two computers that wish to spit bits at each other. The *line units* (LUs) on the switch provide a connection point for every cable pair served by that particular switch in the office. Conceptually, then, it is a relatively simple task for the switch to connect one subscriber in the switch with another subscriber in the same switch. After all, the switch maintains tables—a directory, if you will—so it knows the subscribers to which it is directly connected.

Far more complicated is the task of connecting a subscriber in one switch to a subscriber in another. This is where network intelligence becomes critically important. As we mentioned earlier, when the switch receives the dialed digits, it performs a rudimentary analysis on them to determine whether the called party is locally hosted. If the number is in the same switch, the call is established. If it resides in another switch, the task is a bit more complex. First, the local switch must pass the call on to the local tandem switch, which provides access to the *points of presence* (POPs) of the various long-distance carriers that serve the area. The tandem switch typically does not connect directly to subscribers; it connects to other switches only. It also performs a database query through SS7 to determine the subscriber's long-distance carrier of choice so that it can route the call to the appropriate carrier. The tandem switch then hands the call off to the long-distance carrier, which transports it over the long-distance network to the carrier's switch in the remote (receiving) office. The long-distance switch passes the call through the remote tandem, which in turn hands the call off to the local switch that the called party is attached to.

SS7's influence once again becomes obvious here. One of the problems that occurred in earlier telephone system designs was the following. When a subscriber placed a call, the local switch handed the call off to the tandem, which in turn handed the call off to the long-distance provider. The long-distance provider seized a trunk, over which it transported the dialed digits to the receiving central office. The signaling information, therefore, traveled over the path designed to produce revenue for the telephone company, a process known as *in-band signaling*. As long as the called party was home, and wasn't on the phone, the call would go through as planned and revenue would flow. If the party wasn't home, however, or if he or she was on the phone, then no revenue was generated. Furthermore, the network resources that another caller might have used to place a call were not available to them, because they were tied up transporting call setup data—which produces no revenue.

SS7 changes all that. With SS7, the signaling data travels across a dedicated packet network from the calling party to the called party. SS7 verifies the availability of the called party's line, reserves it, and then—*and only then*—seizes a talk path. Once the talk path has been created it rings the called party's phone and places a ringing tone in the caller's ear. As soon as the called party answers, SS7 silently drops out until one end or the other hangs up. At that point it returns the path to the pool of available network resources.

Interoffice Trunking

We have not discussed the manner in which offices are connected to one another. As Figure 2-22 illustrates, an optical fiber ring with add-drop multiplexers interconnects the central offices so that interoffice traffic can be safely and efficiently transported. The switches have *trunk units* (TUs) that connect the back side, called the trunk side, of the switch to the WAN, in this case the optical ring.

Trunks that interconnect offices have historically been four-wire copper facilities (a pair in each direction). Today they are largely optical, but are still referred to as "four-wire" because of their two-way nature.

An interesting point about trunks. In the 1960s and early 1970s, most interoffice trunks were analog rather than digital. To signal, they used single-frequency tones. Because these trunks did not "talk" directly to customers, there was no reason to worry about a human voice inadvertently emitting a sound that could be misconstrued as dial tone. There was therefore no reason to use DTMF dialing; instead, trunk signaling was performed using single-frequency tones. Specifically, if a switch wished to seize a long-distance trunk, it issued a single-frequency 2,600 Hz tone, which would signal the seizure to take place. Once the trunk seizure had occurred, the dialed digits could be outpulsed and the call would proceed as planned.

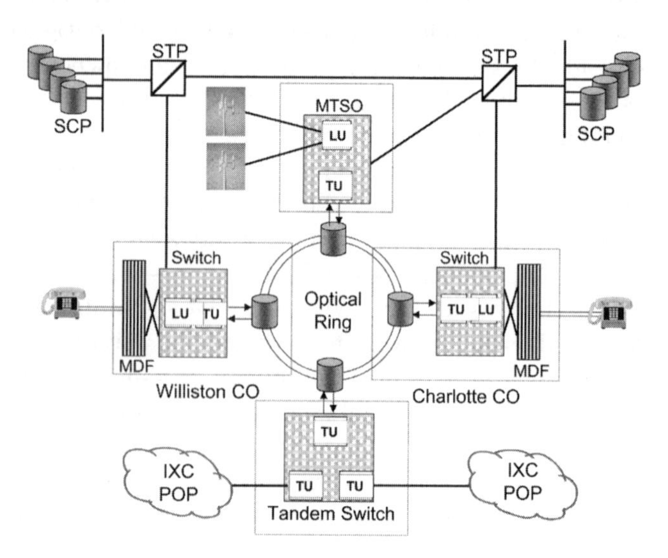

Figure 2-22 Typical network showing fiber ring interconnecting offices

In 1972, John Draper, better known by his hacker name "Captain Crunch," determined that the toy plastic bosun's whistle that came packed in boxes of Cap'n Crunch cereal emitted—you guessed it—a 2,600 Hz tone. Anyone who knew this could steal long-distance service from AT&T by blowing the whistle at the appropriate time during call setup. After word of this capability became common knowledge, Cap'n Crunch cereal became the breakfast food of choice for hackers all over the country.

Returning to our example again, let's retrace our steps. Our caller dials the number of the called party. The call travels over the local loop, across aerial and/or underground facilities, and enters the central office via the cable vault. From the cable vault the call travels up to the MDF and then on to the switch. The number is received by the switch in the calling party's serving central office, which performs an SS7 database lookup to determine the proper disposition of the call and any special service information about the caller that it should invoke. The local switch routes the call over intraoffice facilities to a tandem switch, which in turn connects the call to the POP of whichever long-distance provider the caller is subscribed. The long-distance carrier invokes the capabilities of SS7 to determine whether the called phone line is available and, if it is, hands the call off to the remote local service provider. When the called party answers, the call is cut through and it progresses normally.

There are, of course, other ways that calls can be routed. The local loop could be wireless, and if it is, the call from the cell phone is received by a cell tower (see Figure 2-23) and is transported to a special dedicated cellular switch in a central office called a *mobile telephone switching office* (MTSO). The MTSO processes the call and hands it off to the wireline network via interoffice facilities. From that point on, the call is handled like any other. It is either terminated locally on the local switch or handed to a tandem switch for long-distance processing. The fact of the matter is that the only part of a cellular network that is truly wireless is the local loop—everything else is wired.

Before we wrap this chapter, we should take a few minutes to discuss carrier systems, voice digitization, and multiplexed transport, all of which take place (for the most part) in the central office. In spite of all the hype out there about the Internet and IP magic, *plain old telephone service* (POTS) remains the cash cow in the industry. In this final section we'll explain T- and E-carrier, synchronous optical network (SONET), synchronous digital hierarchy (SDH), and voice digitization techniques.

Figure 2-23 A cell tower, disguised as a palm tree!

Conserving Bandwidth: Voice Transport

The original voice network, including access, transmission facilities, and switching components, was exclusively analog until 1962 when T-carrier emerged as an intraoffice trunking scheme. The technology was originally introduced as a short-haul, four-wire facility to serve metropolitan areas and was a technology that customers would *never* have a reason to know about—after all, what customer could ever need a meg and a half of bandwidth? Over the years, however, it evolved to include coaxial cable facilities, digital microwave systems, fiber, and satellite, and of course became a premier access technology that customers knew a great deal about.

Consider the following scenario: A corporation builds a new headquarters facility in a new area just outside the city limits. The building will provide office space for approximately 2,000 employees, a considerable number. Those people will need telephones, computer data facilities, fax lines, videoconferencing circuits, and a variety of other forms of connectivity.

The telephone company has two options it can exercise to provide access to the building. It can make an assessment of the number of pairs of wire that the building will require and install them; or, it can do the same assessment but provision the necessary bandwidth through carrier systems that transport multiple circuits over a shared facility. Obviously, this second option is the most cost effective and is, in fact, the option that is most commonly used for these kinds of installations. This model should sound familiar: Earlier we discussed loop carrier systems and the fact that they reduce the cost of provisioning network access to far-flung neighborhoods. This is the same concept; instead of a residential neighborhood, we're provisioning a corporate neighborhood.

The most common form of multiplexed access and transport is T-carrier, or E-carrier outside the United States. Let's take a few minutes to describe them.

Framing and Formatting in T1

The standard T-carrier multiplexer, shown in Figure 2-24, accepts inputs from 24 sources, converts the inputs to pulse code modulation (PCM) bytes, then time division multiplexes the samples over a shared four-wire facility, as shown in Figure 2-25. Each of the 24 input channels yields an eight-bit sample, in round-robin fashion, once every 125 microseconds (8,000 times per second). This yields an overall bit rate of 64 Kbps for each channel (8 bits per sample × 8,000 samples per second). The multiplexer gathers one eight-bit sample from each of the 24 channels and

Figure 2-24 T-carrier multiplexer channel banks, showing DS-0 cards

Figure 2-25 A time division multiplexer in action

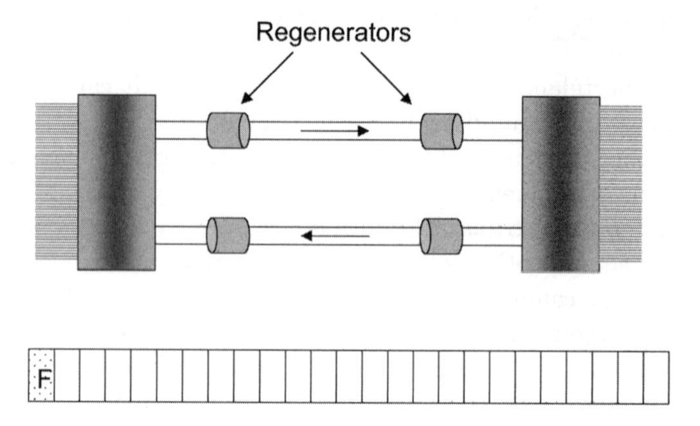

aggregates them into a 192-bit frame. To the frame it adds a frame bit, which expands the frame to a 193-bit entity. The frame bit is used for a variety of purposes that will be discussed in a moment.

The 193-bit frames of data are transmitted across the four-wire facility at the standard rate of 8,000 frames per second, for an overall T1 bit rate of 1.544 Mbps. Keep in mind that 8 Kbps of the bandwidth consists of frame bits (one frame bit per frame, 8,000 frames per second); only 1.536 Mbps belongs to the user.

Beginnings: D1 Framing

The earliest T-carrier equipment was referred to as D1 and was considerably more rudimentary in function than modern systems are (see Figure 2-26). In D1, every eight-bit sample carried seven bits of user information (bits 1 through 7) and one bit for signaling (bit 8). The signaling bits were used for exactly that: indications of the status of the line (on-hook, off-hook, busy, high and dry, etc.), while the seven user bits carried encoded voice information. Because only seven of the eight bits were available to the user, the result was considered to be less than toll quality (128 possible values, rather than 256). The frame bits, which in modern systems indicate the beginning of the next 192-bit frame of data, toggled back and forth between zero and one.

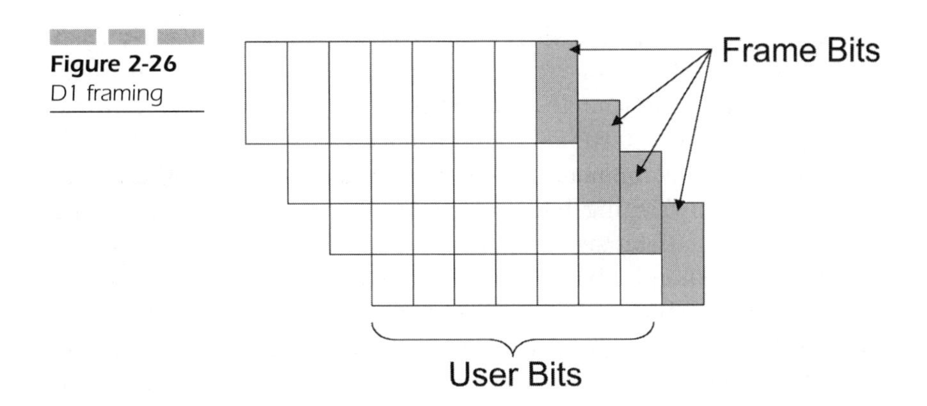

Figure 2-26
D1 framing

Evolution: D4

As time went on and the stability of network components improved, an improvement on D1 was sought and found. Several options were developed, but the winner emerged in the form of the D4 or superframe format. Rather than treat a single 193-bit frame as the transmission entity, superframe "gangs together" twelve 193-bit frames into a 2,316-bit entity (see Figure 2-27) that obviously includes 12 frame bits. Please note that the bit rate has not changed; we have simply changed our view of what constitutes a frame.

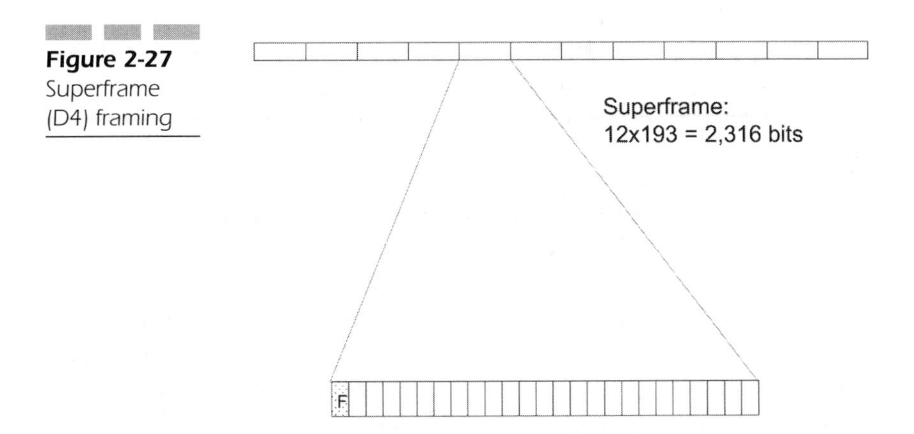

Figure 2-27
Superframe (D4) framing

Because we now have a single (albeit large) frame, we clearly don't need 12 frame bits to frame it; consequently, some of them can be redeployed for other functions. In superframe, the six odd-numbered frame bits are referred to as terminal framing bits and are used to synchronize the channel bank equipment. The even framing bits, on the other hand, are called signal framing bits and are used indicate to the receiving device *where* robbed-bit signaling occurs.

In D1, the system reserved one bit from every sample for its own signaling purposes, which succeeded in reducing the user's overall throughput. In D4, that is no longer necessary; instead, we signal less frequently and only occasionally rob a bit from the user. In fact, because the system operates at a high transmission speed, network designers determined that signaling can occur relatively infrequently and still convey adequate information to the network. Consequently, bits are robbed from the sixth and eighth iteration of each channel's samples, and then only the least significant bit from each sample. The resulting change in voice quality is negligible.

Back to the signal framing bits: Within a transmitted superframe, the second and fourth signal framing bits would be the same, but the sixth would toggle to the opposite value, indicating to the receiving equipment that the samples in that subframe of the superframe should be checked for signaling state changes. The eighth and tenth signal framing bits would stay the same as the sixth but would toggle back to the opposite value once again in the twelfth, indicating once again that the samples in that subframe should be checked for signaling state changes.

Today: Extended Superframe (ESF)

Although superframe continues to be widely utilized, an improvement came about in the 1980s in the form of *extended superframe* (ESF), shown in Figure 2-28. ESF groups 24 frames instead of 12 into an entity, and like superframe it reuses some of the frame bits for other purposes. Bits 4, 8, 12, 16, 20, and 24 are used for framing, and they form a constantly repeating pattern (001011 . . .). Bits 2, 6, 10, 14, 18, and 22 are used as a six-bit *cyclic redundancy check* (CRC) to check for bit errors on the facility. Finally, the remaining bits—all of the odd frame bits in the frame—are used as a 4 Kbps facility data link for end-to-end diagnostics and network management tasks.

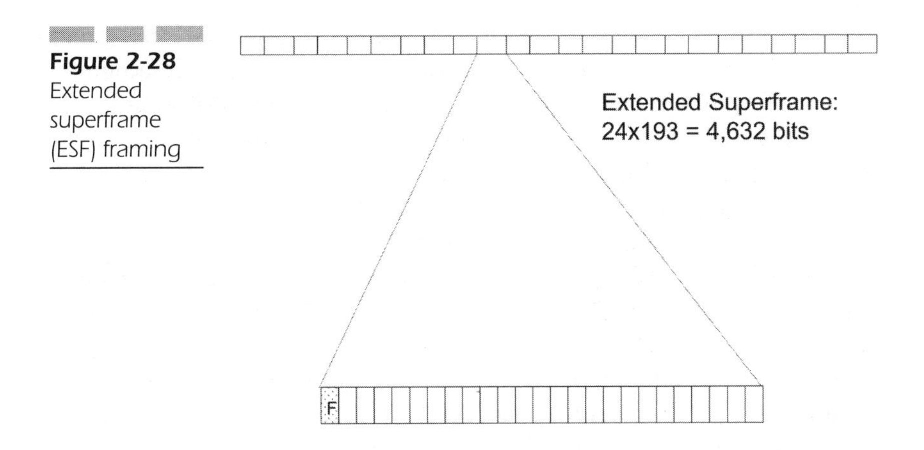

Figure 2-28 Extended superframe (ESF) framing

ESF provides one major benefit over its predecessors: the ability to do nonintrusive testing of the facility. In earlier systems, if the user reported trouble on the span, the span would have to be taken out of service for testing. With ESF, that is no longer necessary because of the added functionality provided by the CRC and the facility data link.

The Rest of the World: E1

E1, used for the most part outside of the United States and Canada, differs from T1 on several key points. First, it boasts a 2.048 Mbps facility rather than the 1.544 Mbps facility found in T1. Second, it utilizes a 32-channel frame rather than a 24-channel frame. Channel 1 contains framing information and a four-bit cyclic redundancy check (CRC-4); channel 16 contains all signaling information for the frame; and channels 1 through 15 and 17 through 31 transport user traffic. The frame structure is shown in Figure 2-29.

A number of similarities exist between T1 and E1 as well: Channels are all 64 Kbps, and frames are transmitted 8,000 times per second. And

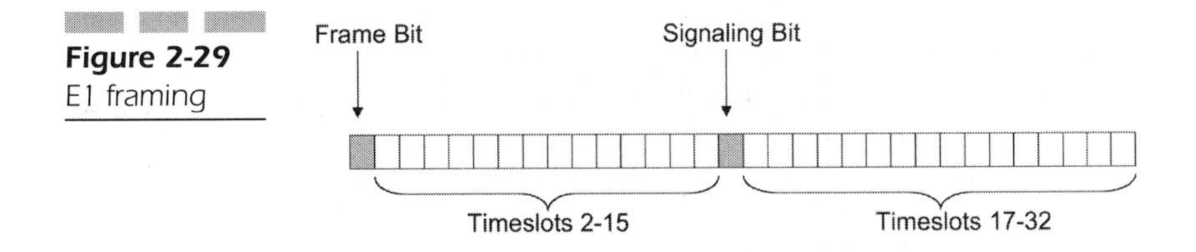

Figure 2-29 E1 framing

whereas T1 gangs together 24 frames to create an ESF, E1 gangs together 16 frames to create what is known as a European Telecommunications Standards Institute (ETSI) multiframe. The multiframe is subdivided into two submultiframes; the CRC-4 in each one is used to check the integrity of the submultiframe that preceded it.

A final word about T1 and E1: Because T1 is a departure from the international E1 standard, it is incumbent upon the T1 provider to perform all interconnection conversions between T1 and E1 systems. For example, if a call arrives in the United States from a European country, the receiving American carrier must convert the incoming E1 signal to T1. If a call originates from Canada and is terminated in Australia, the Canadian originating carrier must convert the call to E1 before transmitting it to Australia.

Up the Food Chain: From T1 to DS3 . . . and Beyond

When T1 and E1 first emerged on the telecommunications scene, they represented a dramatic step forward in terms of the bandwidth that service providers now had access to. In fact, they were *so* bandwidth rich that neither the public nor the industry had any concept that a customer would ever need access to them. What customer, after all, could ever have a use for a million and a half bits per second of bandwidth?

Of course, that question was rendered moot in short order as increasing requirements for bandwidth drove demand that went well beyond the limited capabilities of low-speed transmission systems. As T1 became mainstream, its usage went up, and soon requirements emerged for digital transmission systems with capacity greater than 1.544 Mbps. The result was the creation of what came to be known as the North American Digital Hierarchy, shown in Figure 2-30. The table also shows the European and Japanese hierarchy levels.

From DS1 to DS3

We have already seen the process employed to create the DS1 signal from 24 incoming DS0 channels and an added frame bit. Now we turn our attention to higher bit-rate services. As we wander our way through this explanation, pay particular attention to the complexity involved in

Figure 2-30 North American Digital Hierarchy

Hierarchy Level	Europe	United States	Japan
DS0	64 Kbps	64 Kbps	64 Kbps
DS1		1.544 Mbps	1.544 Mbps
E1	2.048 Mbps		
DS1c		3.152 Mbps	3.152 Mbps
DS2		6.312 Mbps	6.312 Mbps
E2	8.448 Mbps		32.064 Mbps
DS3	34.368 Mbps	44.736 Mbps	
DS3c		91.053 Mbps	
E3	139.264 Mbps		
DS4		274.176 Mbps	
			397.2 Mbps

creating higher-rate payloads. This is one of the great advantages of SONET and SDH.

The next level in the North American Digital Hierarchy is called DS2. And although it is rarely seen outside the safety of the multiplexer in which it resides, it plays an important role in the creation of higher bit-rate services. It is created when a multiplexer *bit interleaves* four DS1 signals, inserting as it does so a control bit, known as a C-bit, every 48 bits in the payload stream. Bit interleaving is an important construct here, because it contributes to the complexity of the overall payload. In a bit-interleaved system, multiple bit streams are combined on a bit-by-bit basis as shown in Figure 2-31. When payload components are bit interleaved to create a higher-rate multiplexed signal, the system first selects bit 1 from channel 1, bit 1 from channel 2, bit 1 from channel 3, and so on. Once it has selected and transmitted all of the first bits, it goes on to the second bits from each channel, then the third, until it has created the super-rate frame. Along the way it intersperses C-bits, which are used to perform certain control and management functions within the frame.

Figure 2-31 Bit-interleaved system

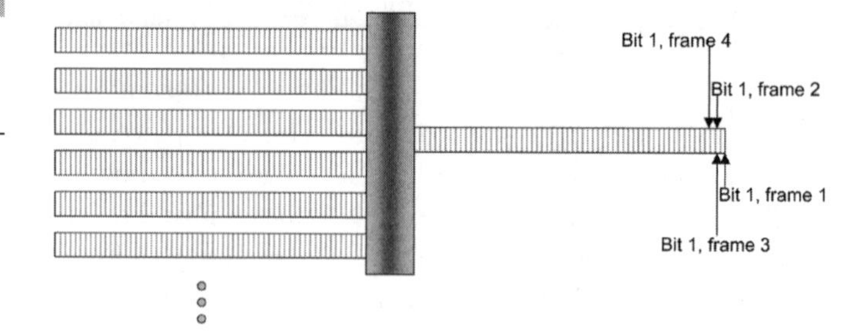

Once the 6.312 Mbps DS2 signal has been created, the system shifts into high gear to create the next level in the transmission hierarchy. Seven DS2 signals are then bit interleaved along with C-bits after every 84 payload bits to create a composite 44.736 Mbps DS3 signal. The first part of this process, the creation of the DS2 payload is called *M12 multiplexing;* the second step, which combines DS2s to form a DS3, is called *M23 multiplexing.* The overall process is called *M13* and is illustrated in Figure 2-32.

The problem with this process is the bit-interleaved nature of the multiplexing scheme. Because the DS1 signal components arrive from different sources, they may be (and usually are) slightly off from one another in terms of the overall phase of the signal—in effect, their "speeds" differ slightly. This is unacceptable to a multiplexer, which must

Figure 2-32 M13 multiplexing process

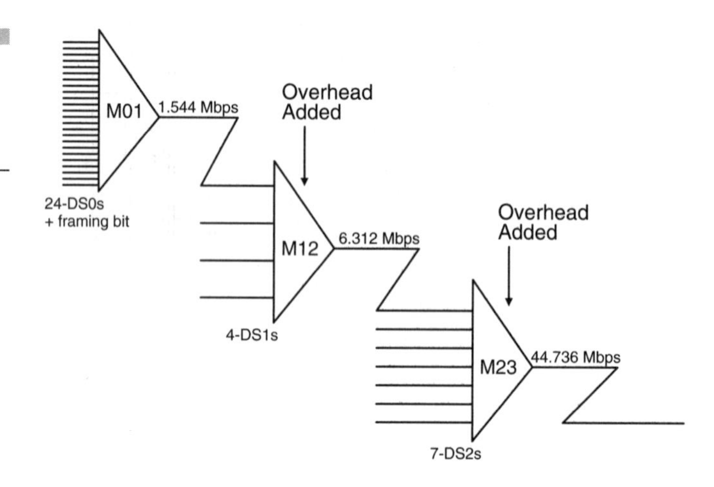

rate-align them if it is to properly multiplex them, beginning with the head of each signal. In order to do this, the multiplexer inserts additional bits, known as "stuff bits," into the signal pattern at strategic places serving to rate-align the components. The structure of a bit-stuffed DS2 frame is shown in Figure 2-33; a DS3 frame is shown in Figure 2-34.

The complexity of this process should now be fairly obvious to the reader. If we follow the left-to-right path shown in Figure 2-35, we see the rich complexity that suffuses the M13 signal-building process. Twenty-four 64 Kbps DS0s are aggregated at the ingress side of the T1[4] multiplexer, grouped into a T1 frame, and combined with a single-frame bit to

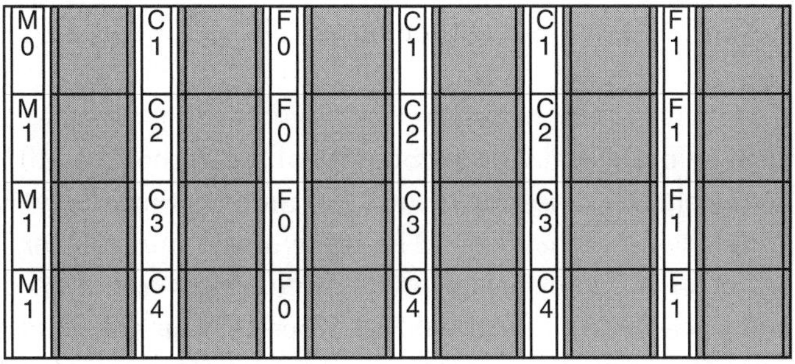

Figure 2-33 M12 frame comprised of four subframes and 48-bit payload fields.

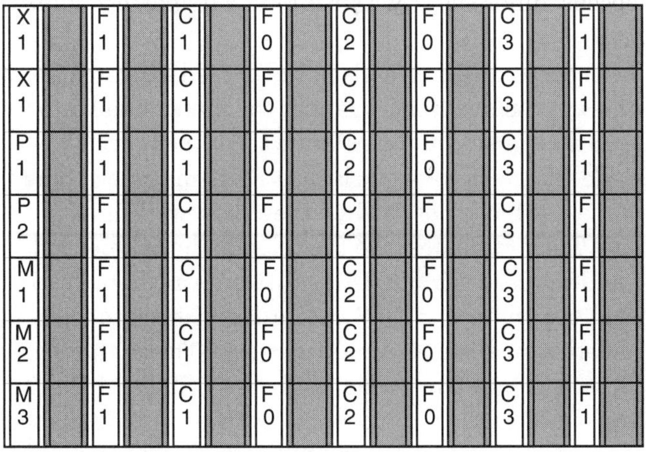

M0		C1		F0		C1		C1		F1	
M1		C2		F0		C2		C2		F1	
M1		C3		F0		C3		C3		F1	
M1		C4		F0		C4		C4		F1	

Figure 2-34 M13 frame comprised of seven subframes and 84-bit payload fields.

X1		F1		C1		F0		C2		F0		C3		F1	
X1		F1		C1		F0		C2		F0		C3		F1	
P1		F1		C1		F0		C2		F0		C3		F1	
P2		F1		C1		F0		C2		F0		C3		F1	
M1		F1		C1		F0		C2		F0		C3		F1	
M2		F1		C1		F0		C2		F0		C3		F1	
M3		F1		C1		F0		C2		F0		C3		F1	

[4]The process is similar for the E1 hierarchy.

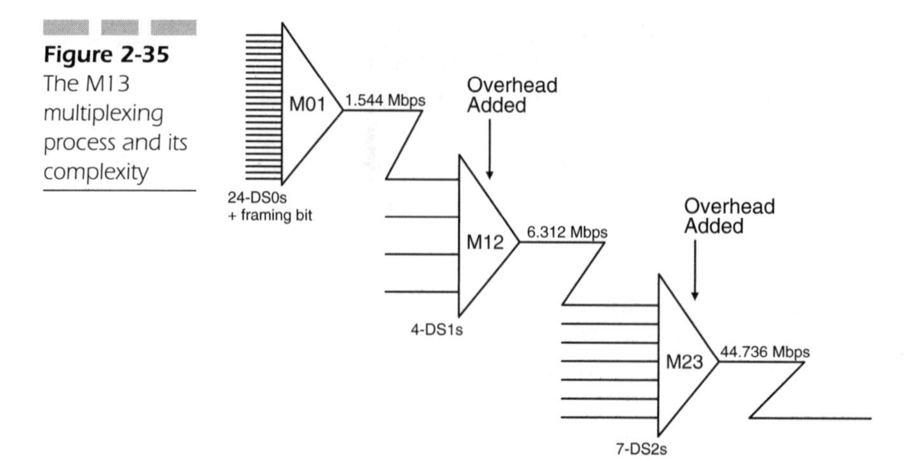

Figure 2-35 The M13 multiplexing process and its complexity

form an outbound 1.544 Mbps signal (I call this the M01 stage; that's my nomenclature, used for the sake of naming continuity). That signal then enters the intermediate M12 stage of the multiplexer, where it is combined (or bit interleaved) with three others and a good dollop of alignment overhead to form a 6.312 Mbps DS2 signal. That DS2 then enters the M23 stage of the mux, where it is bit interleaved with six others and another scoop of overhead to create a DS3 signal. At this point, we have a relatively high-bandwidth circuit that is ready to be moved across the WAN.

Of course, as our friends in the United Kingdom are wont to say, there is always the inevitable spanner that gets tossed into the works (those of us on the left side of the Atlantic call it a wrench). Keep in mind that the 28 (do the math) bit-interleaved DS1s may well come from 28 different sources—which means that they may well have 28 different destinations. This translates into the original digital hierarchy's greatest weakness. In order to drop a DS1 at its intermediate destination, we have to bring the composite DS3 into a set of back-to-back DS3 multiplexers (sometimes called M13 multiplexers). There, the ingress mux removes the second set of overhead, finds the DS2 in which the DS1 we have to drop out is carried, removes its overhead, finds the right DS1, drops it out, and then rebuilds the DS3 frame, including reconstruction of the overhead, before transmitting it on to its next destination. This process is complex, time consuming, and expensive. So what if we could come up with a method for adding and dropping signal components that eliminated the M13 process entirely? What if we could do it as simply as the process shown in Figure 2-36?

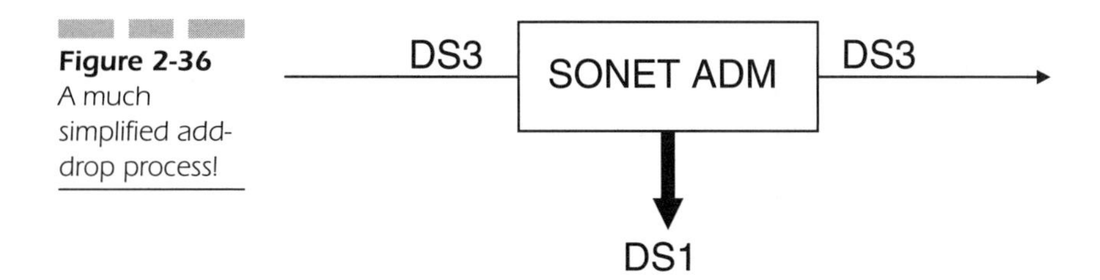

Figure 2-36 A much simplified add-drop process!

We have. It's called SONET in North America, SDH in the rest of the world, and it dramatically simplifies the world of high-speed transport.

SONET brings with it a subset of advantages that makes it stand above competitive technologies. These include interoperability; improved operations, administration, maintenance, and provisioning (OAM&P); support for multipoint circuit configurations; nonintrusive facility monitoring; and the ability to deploy a variety of new services. SONET (and its European equivalent, SDH) has become the mainstay of optical transport.

We now turn our attention to voice digitization.

From Analog to Digital

Digital signals, often called "square waves," comprise a very rich mixture of signal frequencies. Not to bring too much physics into the discussion, we must at least mention the Fourier series, which describes the makeup of a digital signal. The Fourier series is a mathematical representation of the behavior of waveforms. Among other things, it notes that if we start with a fundamental signal, such as that shown in Figure 2-37, and mathematically add to it its odd harmonics (a harmonic is defined as a wave whose frequency is a whole-number multiple of another wave), a rather remarkable thing happens: The waveform gets steeper on the sides and flatter on top. As we add more and more of the odd harmonics (there is, after all, an infinite series of them), the wave begins to look like the typical square wave. Now of course, there is no such thing as a true square wave; for our purposes, though, we'll accept the fact.

It should now be intuitive to the reader that digital signals comprise a mixture of low-, medium-, and high-frequency components, which means that they cannot be transmitted across the bandwidth-limited 4 KHz channels of the traditional telephone network. In digital carrier facilities, the equipment that restricts the individual transmission

Figure 2-37 Fourier transform of analog wave to digital wave

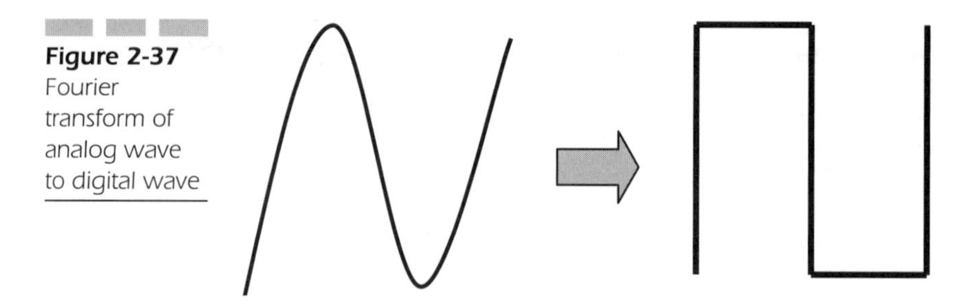

channels to 4 KHz "chunks" is eliminated, thus giving each user access to the full breadth of available spectrum across the shared physical medium. In frequency-division systems, we observed that we give users "*some* of the frequency *all* of the time;" in time-division systems, we turn that around and give users "*all* of the frequency *some* of the time." As a result, high-frequency digital signals can be transmitted without restriction.

Digitization brings with it a cadre of advantages, including improved voice and data transmission quality; better maintenance and troubleshooting capability, and therefore reliability; and dramatic improvements in configuration flexibility. In digital carrier systems, the time-division multiplexer is known as a "channel bank;" under normal circumstances, it allows either 24 or 30 circuits to share a single, four-wire facility. The 24-channel system is called T-carrier; the 30-channel system, used in most of the world, is called E-carrier. Originally designed in 1962 as a way to transport multiple channels of voice over expensive transmission facilities, they soon became useful as data-transmission networks as well. That, however, came later. For now, we focus on voice.

Voice Digitization

The process of converting analog voice to a digital representation in the modern network is a logical and straightforward process. It comprises four distinct steps: *pulse amplitude modulation* (PAM) sampling, in which the amplitude of the incoming analog wave is sampled every 125 microseconds; *companding*, during which the values are weighted toward those most receptive to the human ear; *quantization*, in which the weighted samples are given values on a nonlinear scale; and finally *encoding*, during which each value is assigned a distinct binary value. Each of these stages of PCM will now be discussed in detail.

Pulse Code Modulation (PCM)

We introduced PCM earlier, but here we will dig a little deeper into its functionality. Thanks to work performed by Harry Nyquist at Bell Laboratories in the 1920s, we know that to optimally represent an analog signal as a digitally encoded bitstream, the analog signal must be sampled at a rate that is equal to twice the bandwidth of the channel over which the signal is to be transmitted. Because each analog voice channel is allocated 4 KHz of bandwidth, it follows that each voice signal must be sampled at twice that rate, or 8,000 samples per second. In fact, that is precisely what happens in T-carrier systems, which we now use to illustrate our example. The standard T-carrier multiplexer accepts inputs from 24 analog channels as shown in Figure 2-38. Each channel is sampled in turn, every one eight-thousandth of a second in round-robin fashion, resulting in the generation of 8,000 pulse amplitude samples from each channel every second. The sampling rate is important. If the sampling rate is too high, too much information is transmitted and bandwidth is wasted; if the sampling rate is too low, then we run the risk of aliasing. *Aliasing* is the interpretation of the sample points as a false waveform, due to the paucity of samples.

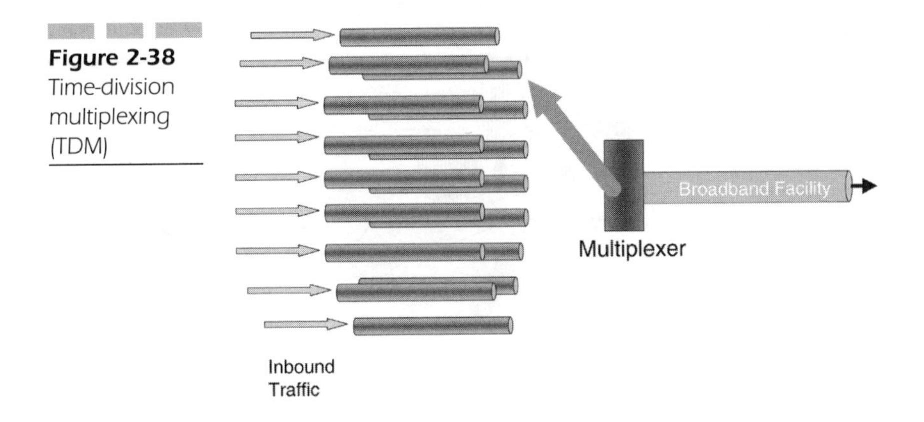

Figure 2-38 Time-division multiplexing (TDM)

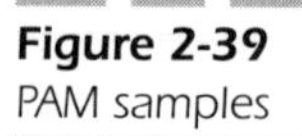

Figure 2-39 PAM samples

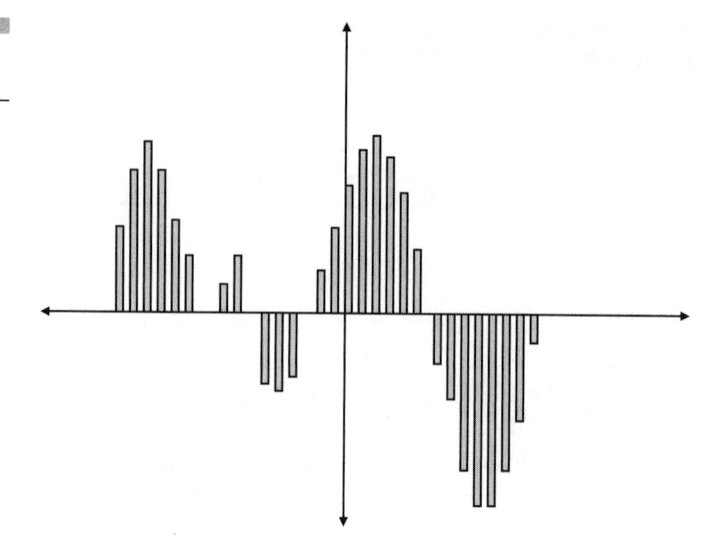

This PAM process represents the first stage of PCM, the process by which an analog baseband signal is converted to a digital signal for transmission across the T-carrier network. This first step is shown in Figure 2-39.

The second stage of PCM, shown in Figure 2-40, is called quantization. In quantization, we assign values to each sample within a constrained range. For illustration purposes, imagine what we now have before us. We have "replaced" the continuous analog waveform of the signal with a

Figure 2-40 Quantizing the samples

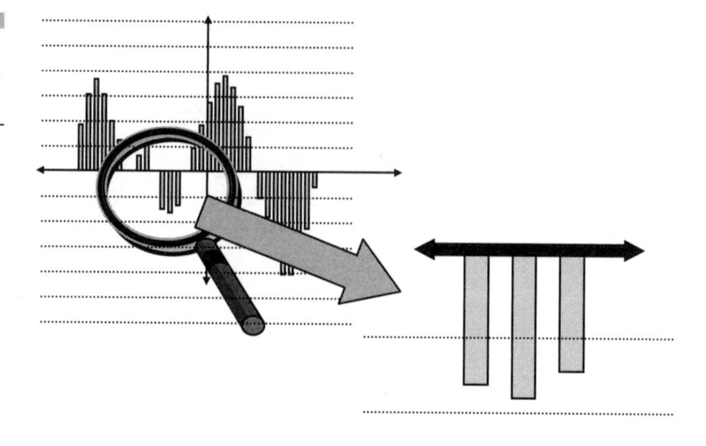

series of amplitude samples, which are close enough together that we can discern the shape of the original wave from their collective amplitudes. Imagine also that we have graphed these samples in such a way that the "wave" of sample points meanders above and below an established zero point on the x-axis, so that some of the samples have positive values and others are negative, as shown.

The amplitude levels allow us to assign values to each of the PAM samples, although a glaring problem with this technique should be obvious to the careful reader. Very few of the samples actually line up *exactly* with the amplitudes delineated by the graphing process. In fact, most of them fall *between* the values, as shown in the illustration. It doesn't take much of an intuitive leap to see that several of the samples will be assigned the same digital value by the *coder-decoder* (CODEC) that performs this function, yet they are clearly *not* the same amplitude. This inaccuracy in the measurement method results in a problem known as *quantizing noise* and is inevitable when linear measurement systems, such as the one suggested by the drawing, are employed in CODECs.

Needless to say, design engineers recognized this problem rather quickly and, equally quickly, came up with an adequate solution. It is a fairly well known fact among psycholinguists and speech therapists that the human ear is far more sensitive to discrete changes in amplitude at low volume levels than it is at high volume levels, a fact not missed by the network designers tasked with optimizing the performance of digital carrier systems intended for voice transport. Instead of using a linear scale for digitally encoding the PAM samples, they designed and employed a nonlinear scale that is weighted with much more granularity at low volume levels—that is, close to the zero line—than at the higher amplitude levels. In other words, the values are extremely close together near the x-axis and get farther and farther apart as they travel up and down the y-axis. This nonlinear approach keeps the quantizing noise to a minimum at the low amplitude levels where hearing sensitivity is the highest and allows it to creep up at the higher amplitudes, where the human ear is less sensitive to its presence. It turns out that this is not a problem, because the inherent shortcomings of the mechanical equipment (microphones, speakers, the circuit itself) introduce slight distortions at high amplitude levels that hide the effect of the nonlinear quantizing scale.

This technique of *compressing* the values of the PAM samples to make them fit the nonlinear quantizing scale results in a bandwidth savings of more than 30 percent. In fact, the actual process is called *companding,* because the sample is first compressed for transmission, and then expanded for reception at the far end, hence the term.

The actual graph scale is divided into 255 distinct values above and below the zero line. In North America and Japan, the encoding scheme is known as μ-Law (*Mu-Law*); the rest of the world relies on a slightly different standard known as *A-Law*.

There are eight segments above the line and eight below, one of which is the shared zero point. Each segment, in turn, is subdivided into 16 steps. A bit of binary mathematics now allows us to convert the quantized amplitude samples into an eight-bit value for transmission. For the sake of demonstration, let's consider a negative sample that falls into the thirteenth step in segment 5. The conversion would take on the following representation:

1 101 1101

Where the initial 0 indicates a negative sample, 101 indicates the fifth segment, and 1101 indicates the thirteenth step in the segment. We now have an eight-bit representation of an analog amplitude sample that can be transmitted across a digital network then reconstructed with its many counterparts as an accurate representation of the original analog waveform at the receiving end. This entire process is known as PCM, and the result of its efforts is often referred to as *toll-quality voice*.

Alternative Digitization Techniques

Although PCM is perhaps the best-known, high-quality voice digitization process, it is by no means the only one. Advances in coding schemes and improvements in the overall quality of the telephone network have made it possible for encoding schemes to be developed that use far less bandwidth than do traditional PCM. In this next section, we will consider some of these techniques.

Adaptive Differential Pulse Code Modulation (ADPCM)

ADPCM, referred to briefly in Chapter 1, is a technique that allows toll-quality voice signals to be encoded at half-rate (32 Kbps) for transmission. ADPCM relies on the predictability that is inherent in human speech to reduce the amount of information required. The technique still relies on PCM encoding but adds an additional step to carry out its task. The 64 Kbps PCM-encoded signal is fed into an ADPCM transcoder, which considers the *prior* behavior of the incoming stream to create a prediction of the behavior of the *next* sample. Here's where the magic happens. Instead of transmitting the actual value of the predicted sample, it encodes in four bits and transmits the *difference* between the actual and predicted samples. Because the difference from sample to sample is typically quite small, the results are generally considered to be very close to toll quality. This four-bit transcoding process, which is based on the known behavior characteristics of human voice, allows the system to transmit 8,000 four-bit samples per second, thus reducing the overall bandwidth requirement from 64 Kbps to 32 Kbps. It should be noted that ADPCM works well for voice, because the encoding and predictive algorithms are based upon its behavior characteristics. It does not, however, work as well for higher bit-rate data (above 4,800 bps), which has an entirely different set of behavior characteristics.

Continuously Variable Slope Delta (CVSD)

CVSD is a unique form of voice encoding that relies on the values of individual bits to predict the behavior of the incoming signal. Instead of transmitting the volume (height or y-value) of PAM samples, CVSD transmits information that measures the changing slope of the waveform. In other words, instead of transmitting the actual change itself, it transmits the *rate* of change.

To perform its task, CVSD uses a reference voltage to which it compares all incoming values. If the incoming signal value is *less* than the reference voltage, then the CVSD encoder reduces the slope of the curve to make its approximation better mirror the slope of the actual signal. If the incoming value is *more* than the reference value, then the encoder will increase the slope of the output signal, again causing it to approach and therefore mirror the slope of the actual signal. With each recurring sample and comparison, the step function can be increased or decreased as required. For example, if the signal is increasing rapidly, then the steps are increased one after the other in a form of step function by the encoding algorithm. Obviously, the reproduced signal is not a particularly exact representation of the input signal: In practice, it is pretty jagged. Filters, therefore, are used to smooth the transitions.

CVSD is typically implemented at 32 Kbps, although it can be implemented at rates as low as 9,600 bps. At 16 to 24 Kbps, recognizability is still possible. Down to 9,600, recognizability is seriously affected although intelligibility is not.

Linear Predictive Coding (LPC)

We mention LPC here only because it has carved out a niche for itself in certain voice-related applications such as voicemail systems, automobiles, aviation, and electronic games that speak to children. LPC is a complex process, implemented completely in silicon, which allows for voice to be encoded at rates as low as 2,400 bps. The resulting quality is far from toll quality, but it is certainly intelligible and its low bit-rate capability gives it a distinct advantage over other systems.

LPC relies on the fact that each sound created by the human voice has unique attributes, such as frequency range, resonance, and loudness. When voice samples are created in LPC, these attributes are used to generate *prediction coefficients*. These predictive coefficients represent linear combinations of previous samples, hence the name, *linear* predictive coding.

Prediction coefficients are created by taking advantage of the known *formants* of speech, which are the resonant characteristics of the mouth and throat that give speech its characteristic timbre and sound. This sound, referred to by speech pathologists as the *buzz*, can be described by both its pitch and its intensity. LPC, therefore, models the behavior of the vocal cords and the vocal tract itself.

To create the digitized voice samples, the buzz is passed through an inverse filter, which is selected based upon the value of the coefficients. The remaining signal, after the buzz has been removed, is called the *residue*.

In the most commonly used form of LPC, the residue is encoded as either a *voiced* or *unvoiced* sound. Voiced sounds are those that require vocal-cord vibration, such as the *g* in *glare*, the *b* in *boy*, the *d* and *g* in *dog*. Unvoiced sounds require no vocal-cord vibration, such as the *h* in *how*, the *sh* in *shoe*, the *f* in *frog*. The transmitter creates and sends the prediction coefficients, which include measures of pitch, intensity, and whatever voiced and unvoiced coefficients are required. The receiver undoes the process: It converts the voice residue, pitch, and intensity coefficients into a representation of the source signal using a filter similar to the one used by the transmitter to synthesize the original signal.

Digital Speech Interpolation (DSI)

Human speech has many measurable (and therefore predictable) characteristics, one of which is a tendency to have embedded pauses. As a rule, people do not spew out a series of uninterrupted sounds. They tend to pause for emphasis, to collect their thoughts, to reword a phrase while the other person listens quietly on the other end of the line. When speech technicians monitor these pauses, they discover that during considerably more than half of the total connect time, the line is silent.

DSI takes advantage of this characteristic silence to drastically reduce the bandwidth required for a single channel. Whereas 24 channels can be transported over a typical T1 facility, DSI allows for as many as 120 conversations to be carried over the same circuit. The format is proprietary and requires the setting aside of a certain amount of bandwidth for overhead.

A form of *statistical multiplexing* lies at the heart of DSI's functionality. Standard T-carrier is a time-division multiplexed scheme, in which channel ownership is assured. A user assigned to channel 3 will *always* own channel 3, regardless of whether he or she is actually using the line. In DSI, channels are not owned. Instead, large numbers of users share a pool of available channels. When a user starts to talk, the DSI system assigns an available timeslot to that user and notifies the receiving end of the assignment. This system works well when the number of users is large, because statistical probabilities are more accurate and indicative of behavior in larger populations than in smaller ones.

The downside to DSI, of course, comes in several forms. *Competitive clipping* occurs when more people start to talk than there are available channels, resulting in someone being unable to talk. *Connection clipping* occurs when the receiving end fails to learn what channel a conversation has been assigned to within a reasonable amount of time, resulting in signal loss.

Two approaches have been created to address these problems. In the case of competitive clipping, the system intentionally clips off the front end of the initial word of the second person who speaks. This technique is not optimal but prevents loss of the total conversation and obviates the problem of clipping out the middle of a conversation. The loss of an initial syllable or two can be mentally reconstructed far more easily than can the sounds in the middle of a sentence.

A second technique used to recover from clipping problems is to temporarily reduce the encoding rate. The typical encoding rate for DSI is 32 Kbps. In certain situations, the encoding rate may be reduced to 24 Kbps, thus freeing up significant bandwidth for additional channels. Both techniques are widely utilized in DSI systems.

Chapter Summary

In this chapter we have discussed the overall functionality of the traditional PSTN. We have looked at the access, transport, switching, signaling, and services components. We presented techniques for voice digitization and introduced the concept of voice compression used in many of the alternatives to PCM. This is particularly important as we turn our attention to VoIP. New compression algorithms combined with remarkably capable CODECs that perform the analog-to-digital conversion of voice signals in VoIP networks now make it possible to reduce the bandwidth required for a voice conversation from 64 Kbps to as little as 8 Kbps—an 8:1 reduction. In the next chapter, we introduce VoIP: how it works, what it is, and how its component parts work together to offer carrier-grade voice—and beyond.

CHAPTER 3

Beginnings: IP

Before we dig too deeply into voice over IP (VoIP), let's spend some time talking about the Internet Protocol (IP) and some of the corollary protocols that make it work as effectively as it does. Once we understand how the protocol functions, we'll overlay telephony and other services.

It would be irresponsible to publish this book without detailed coverage of the next generation of transport network that is currently emerging. Although frame relay and asynchronous transfer mode (ATM) continue to be important technologies for switched broadband transport, they are expensive, complex, and cumbersome to operate. And given the degree to which IP has taken the lead position in networking discussions today, it makes sense that IP should have its space, particularly given the fact that it is now being deployed in conjunction with *Multiprotocol Label Switching* (MPLS) as a carrier-grade, layer-three network infrastructure that offers truly granular quality of service (QoS).

We begin with a discussion of the remarkable protocol suite called Transmission Control Protocol (TCP)/IP and how it works.

IP History

The Internet as we know it today began as a Department of Defense (DoD) project (see Figure 3-1) designed to interconnect DoD research sites. In December of 1968, the government research agency known as the *Advanced Research Projects Agency* (ARPA) awarded a contract to Bolt Beranek and Newman to design and build the packet network that would ultimately become the Internet. It had a proposed transmission speed of 50 Kbps, and in September of 1969 the first node was installed at UCLA. Other nodes were installed on roughly a monthly basis at Stanford Research Institute, the University of California at Santa Bar-

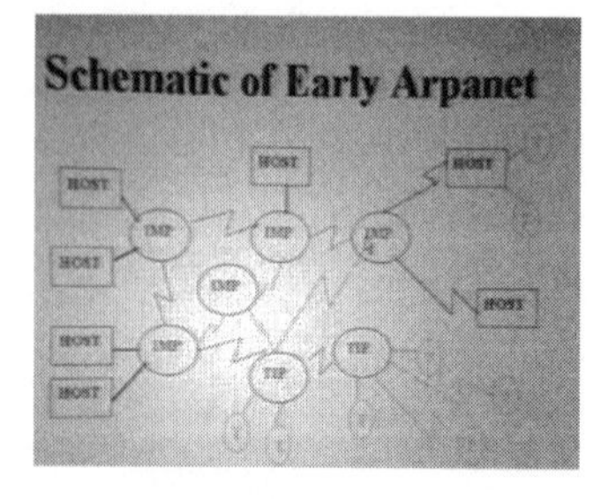

Figure 3-1 The original Arpanet diagram, which ultimately became the Internet (Image courtesy ISOC)

bara, and the University of Utah. The ARPANET spanned the continental United States by 1971 and had connections to research facilities in Europe by 1973.

The original protocol selected for the ARPANET was called the *Network Control Protocol* (NCP). It was designed to handle the emergent requirements of the low-volume architecture of the ARPANET network. As traffic grew, however, it proved to be inadequate to handle the load, and in 1974 a more robust protocol suite was implemented. This new *Transmission Control Protocol* (TCP) was an ironclad protocol designed for end-to-end network communications control. In 1978 a new design split the responsibilities for end-to-end versus node-to-node transmission among two protocols. The newly crafted IP was designed to route packets from device-to-device, and TCP was designed to offer reliable, end-to-end communications. Because TCP and IP were originally envisioned as a single protocol, they are now known as the *TCP/IP protocol suite*, a name that also incorporates a collection of protocols and applications that also handle routing, QoS, error control, and other functions.

One problem that occurred that ARPANET planners didn't envision when they sited their nodes at college campuses was visibility. Naturally, they placed the switches in the raised floor facilities of the computer science department, and we know what is also found there: *Undergraduus Nerdus,* the dreaded computer science (or worse yet, engineering) student. In a matter of weeks, the secret was out—ARPA's top secret network was top secret no longer. So, in 1983 the ARPANET was split into two networks. One half, still called ARPANET, continued to be used to interconnect research and academic sites. The other, called MILNET, was specifically used to carry military traffic and ultimately became part of the Defense Data Network.

That year was also a good year for TCP/IP. It was included as part of the communications kernel for the University of California's UNIX implementation, known as 4.2BSD (or Berkeley Software Distribution) UNIX.

Extension of the original ARPANET continued. In 1986, the *National Science Foundation* (NSF) built a backbone network to interconnect four NSF supercomputing centers and the National Center for Atmospheric Research. This network, known as NSFNET, was originally intended to serve as a backbone for other networks, not as a stand-alone interconnection mechanism. Additionally, the NSF's Appropriate Use Policy limited transported traffic to noncommercial traffic *only*. NSFNET continued to expand and eventually became what we know today as the

Internet. And although the original NSFNET applications were multiprotocol implementations, TCP/IP was used for overall interconnectivity.

In 1994, a structure was put in place to reduce the NSF's overall role on the Internet. The new structure consists of three principal components. The first of these was a small number of *network access points* (NAPs) where *Internet service providers* (ISPs) would interconnect to the Internet backbone. The NSF originally funded four NAPs in Chicago (operated by Ameritech, now part of SBC), New York (really Pensauken, New Jersey, operated by Sprint), San Francisco (operated by Pacific Bell, now part of SBC), and Washington, D.C. (MAE-East, operated by MFS, now a division of MCI).

The second component was the *very high-speed backbone network service*, a network that interconnected the NAPs and was operated by MCI. It was installed in 1995 and originally operated at OC-3 (155.52 Mbps), but was upgraded to OC-12 (622.08 Mbps) in 1997. The third component was the *routing arbiter*, designed to ensure that appropriate routing protocols for the Internet were available and properly deployed.

ISPs were given 5 years of diminishing funding to become commercially self-sustaining. The funding ended in 1998. Starting at roughly the same time, a significant number of additional NAPs have been launched. As a matter of control and management, three tiers of ISP have been identified. *Tier 1* refers to ISPs that have a national presence and connect to at least three of the original four NAPs. National ISPs include AT&T, Cable & Wireless, MCI, and Sprint. *Tier 2* refers to ISPs that have a primarily regional presence and connect to less than three of the original four NAPs. Regional ISPs include Adelphia.net, Verizon.net, and BellSouth.net. Finally, *Tier 3* refers to local ISPs, or those that do not connect to a NAP but offer services via the connections of another ISP.

Managing the Internet

The Internet is really a network of networks, all interconnected by high-bandwidth circuits leased (for the most part) from various telephone companies. It is something of an erroneous conclusion to think of the Internet as a stand-alone network; in fact, it is made up of the same network components that make up corporate networks and public telephone networks. It's simply a collection of leased lines interconnecting the routers that provide the sophisticated routing functions that make the

Internet so powerful. As such, it is owned by no one and owned by everyone, but more importantly it is *managed* by no one and *managed* by everyone! No single authority governs the Internet, because the Internet is not a single monolithic entity but rather a collection of entities. Yet it runs well, perhaps better than other more centrally managed services!

That being said, a number of organizations provide oversight, guidance, and a certain degree of management for the Internet community. These organizations, described in some detail in the following paragraphs, help to guide the developmental direction of the Internet with regard to such functions as communications standards, universal addressing systems, domain naming, protocol evolution, and so on. Please refer to Figure 3-2.

One of the longest standing organizations with Internet oversight authority is the *Internet Activities Board* (IAB), which governs administrative and technical activities on the Internet. It works closely with the *Internet Society* (ISOC), a nongovernmental organization established in 1992 that coordinates such Internet functions as internetworking and applications development. ISOC is also chartered to provide oversight and communications for the IAB.

The *Internet Engineering Task Force* (IETF) is one of the most well-known Internet organizations and is part of the IAB. The IETF establishes working groups responsible for technical activities involving the Internet such as writing technical specifications and protocols. Because

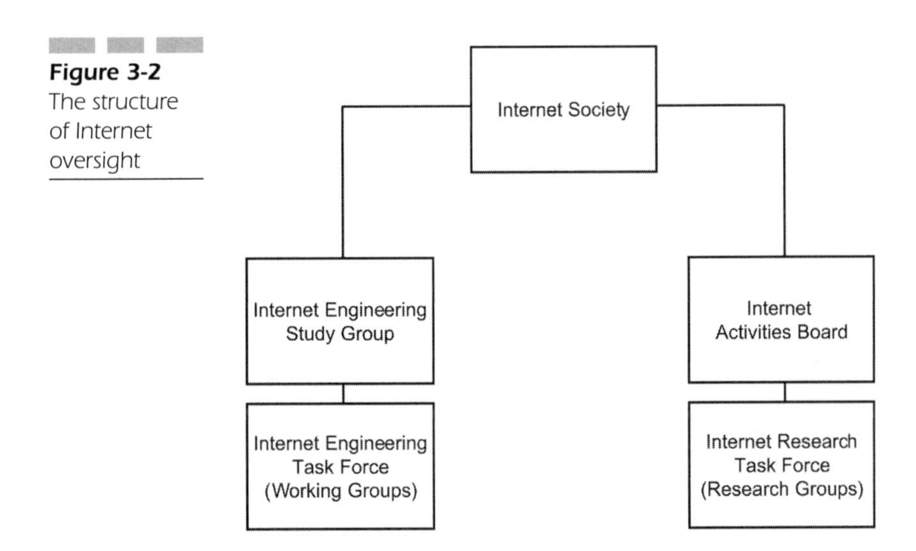

Figure 3-2 The structure of Internet oversight

of the organization's early, valuable commitment to the future of the Internet, the *International Organization for Standardization* (ISO) accredited the organization as an "official" international standards body at the end of 1994.

On a similar front, the *World Wide Web Consortium* (W3C) has no officially recognized role but has taken a leading role in the development of standard protocols for the World Wide Web to promote its evolution and ensure its interoperability. The organization has more than 400 member organizations and is currently leading the technical development of the Web.

Other smaller organizations play important roles as well. These include the *Internet Engineering Steering Group* (IESG), which provides strategic direction to the IETF and is part of the IAB; the *Internet Research Task Force* (IRTF), which engages in research that affects the evolution of the future Internet; the *Internet Engineering Planning Group* (IEPG), which coordinates worldwide Internet operations and helps ISPs achieve interoperability; and finally the *Forum of Incident Response and Security Teams,* which coordinates *computer emergency response teams* (CERTs) in various countries.

Together these organizations ensure that the Internet operates as a single, coordinated organism. Remarkable, isn't it? The least managed network in the world (and arguably the largest) is also the most well run!

Naming Conventions on the Internet

The Internet is based on a system of *domains,* that is, a hierarchy of names that uniquely identify the "operating regions" on the Internet. For example, "IBM.com" is a domain, as are "Verizon.net," "fcc.gov," and "ShepardComm.com." Domains help to guide traffic from one user to another by following a hierarchical addressing scheme that includes a top-level domain, one or more subdomains, and a host name. The postal service relies on a hierarchical addressing scheme, and it serves as a good analogy to understand how Internet addressing works. Consider the following address:

Samuel Cannon
168 Larsen Avenue
Blue Springs, IA
USA

In reality the address is upside down because a package addressed to Samuel must be read from the bottom up if it is to be delivered properly. To begin the routing process, the postal service will start at the bottom with USA, which narrows it down to a country. They will then go to Iowa, then on to Blue Springs, then to the address in Blue Springs, and finally to Samuel. Messages routed through the Internet are handled in much the same way. Routing arbiters deliver them first to a domain, then to a subdomain, and then on to a hostname, where they are ultimately delivered to the proper account.

The assignment of Internet IP addresses was historically handled by the *Internet Assigned Numbers Authority* (IANA). Whereas domain names were assigned by the *Internet Network Information Center* (InterNIC), which had overall responsibility for name dissemination. Regional NICs handled non-U.S. domains. The InterNIC was also responsible for the management of the domain name system (DNS), the massive, distributed database that reconciles host names and IP addresses throughout the Internet.

The InterNIC and its overall role have gone through a series of significant changes in the last decade. In 1993 Network Solutions, Inc. (NSI) was given the responsibility of operating the InterNIC registry by the NSF and had exclusive authority for the assignment of such domains as .com, .org, .net, and .edu. NSI's contract expired in April 1998 but was extended several times because no alternate agency existed to perform the task. In October 1998 NSI became the sole administrator for those domains, but a plan was created to allow users to register names in those domains with other companies. At roughly the same time, responsibility for IP address assignments was migrated to a newly created organization called the *American Registry for Internet Numbers* (ARIN). Shortly thereafter, in March 2000, NSI was acquired by VeriSign.

The most recent addition to the domain management process is the *Internet Corporation for Assigned Names and Numbers* (ICANN). Established in late 1998, ICANN was appointed by the U.S. *National Telecommunications and Information Administration* (NTIA) to manage the DNS.

Of course, it makes sense that sooner or later the most common top-level domains, which include such well-known suffixes as .com, .gov, .net, .org, and .net, would be deemed inadequate. And sure enough, in November 2000, the first new top-level domains, seven in all, were approved by ICANN. They are .aero (for the aviation industry), .biz (for businesses), .coop (for business cooperatives), .info (for general use), .museum (for, you guessed it), .name (for individuals), and .pro (for professionals).

The TCP/IP Protocol: What It Is and How It Works

TCP/IP has been around for a long time, and although we often tend to think of it as a single protocol that governs the Internet, it is in reality a fairly exhaustive collection of protocols that cover the functions of numerous layers of the protocol stack. And although open systems interconnection (OSI), discussed in the Appendix, has seven layers of protocol functionality, TCP/IP has only four, as shown in Figure 3-3. We will explore each layer in moderate detail in the sections that follow. For purposes of comparison, however, OSI and TCP/IP compare functionally as shown in Table 3-1.

The Network Interface Layer

TCP/IP was created for the Internet with the concept in mind that the Internet would not be a particularly well behaved network. In other words, designers of the protocol made the assumption that the Internet would become precisely what it has become—a network of networks. It uses a plethora of unrelated and often conflicting protocols, and transports traffic with widely varying QoS requirements. The fundamental building block of the protocol, the IP packet, is designed to deal with all of these disparities, whereas TCP (and other related protocols, discussed later) take care of the QoS issues.

Two network interface protocols are particularly important to TCP/IP. The *Serial Line Internet Protocol* (SLIP) and *Point-to-Point Protocol* (PPP) are used to provide data-link layer services in situations where no other data-link protocol is present, such as in leased-line or older dial-up environments. Most TCP/IP software packages for desktop applications include these two protocols, even though dial-up is rapidly fading into near-oblivion in the presence of growing levels of broadband access. With SLIP or PPP, a remote computer can attach directly to a host and connect to the Internet using IP rather than being limited to an asynchronous connection.

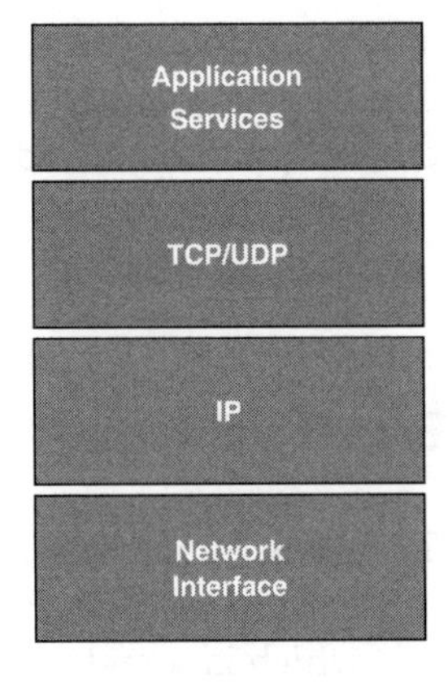

Figure 3-3 Simplified TCP/IP protocol stack

Table 3-1 TCP/IP vs. OSI Comparison Chart

TCP/IP Protocol Stack	OSI Reference Model
Network Interface Layer	Physical and Data Link Layers (L1 to L2)
Internet Layer	Network Layer (L3)
Transport Layer	Transport Layer (L4)
Application Services	Session, Presentation, Application Layers (L5 to L7)

The Point-to-Point Protocol (PPP)

PPP, as its name implies, was created for the governance of point-to-point links. It has the ability to manage a variety of functions at the moment of connection and verification, including password verification, IP address resolution, compression (where required), and encryption for privacy or security. It can also support multiple protocols over a single connection, an important capability for dial-up users who rely on IP or some other network-layer protocol for routing and congestion control. It also supports inverse multiplexing and dynamic bandwidth allocation via the *Multilink-PPP Protocol* (ML-PPP), commonly used in integrated services digital network (ISDN) environments where bandwidth supersets are required over the connection.

The PPP frame (see Figure 3-4) is similar to a typical high-level data link control (HDLC) frame, with delimiting flags, an address field, a protocol identification field, information and pad fields, and a frame-check sequence for error control.

The Internet Layer

IP is the heart and soul of the TCP/IP protocol suite and Internet itself—and perhaps the most talked about protocol in history. IP provides a connectionless service across the network, which is sometimes referred to as an *unreliable* service because the network does not guarantee delivery or packet sequencing. IP packets typically contain an entire message or a piece (fragment) of a message, which can be as large as 65,535 bytes in length. The protocol does *not* provide a flow-control mechanism.

IP packets, like all packets, have a header that contains routing and content information (see Figure 3-5). The bits in the packet are numbered from left to right starting at 0, and each row represents a single 32-bit word. An IP header must contain a *minimum* of five words.

IP Header Fields

The IP header contains *approximately* 15 unique fields. The *Version field* identifies the version of IP that is being used to encode the packet (IPv4 versus IPv6, for example). The *Internet Header Length (IHL) field* identifies the length of the header in 32-bit words. The maximum value of this field is 15, which means that the IP header has a maximum length of 60 octets.

Flag	Address	Control	Protocol	Data	CRC	Flag

Figure 3-4 PPP frame format

The *Type of Service (TOS) field* gives the transmitting system the ability to request different classes of service for the packets that it transmits into the network. The TOS field is not typically supported in IPv4 but can be used to specify a service priority (0 through 7) or route optimization.

The *Total Length field* indicates the length (in octets) of the entire packet, including both the header and the data within the packet. The maximum size of an IP packet is 64 KB (okay, 65,535 bytes).

When a packet is broken into smaller chunks (a process called *fragmentation*) during transmission, the *Identification field* is used by the transmitting host to ensure that all of the fragments from a single message can be reassociated at the receiving end to ensure message reassembly.

The *flags* also play a role in fragmentation and reassembly. The first bit is referred to as the *More Fragments* (MF) bit and is used to indicate to the receiving host that the last fragment of a packet has been received so that the receiver can reassemble the packet. The second bit is the *Don't Fragment* (DF) bit, which prevents packet fragmentation (for delay-sensitive applications, for example). The third bit is unused and is always set to 0.

The *Fragment Offset field* indicates the relative position of this particular fragment in the original packet that was broken up for transmission. The first packet of a fragmented message will carry an offset value of 0, while subsequent fragments will indicate the offset in multiples of eight bytes. Makes sense, no?

I love this field. A message is fragmented into a stream of packets, and the packets are all sent on their merry way to the destination system. Somewhere along the way, packet number 11 decides to take a detour and ends up in a routing loop on the far side of the world, trying in vain

Figure 3-5
IP header

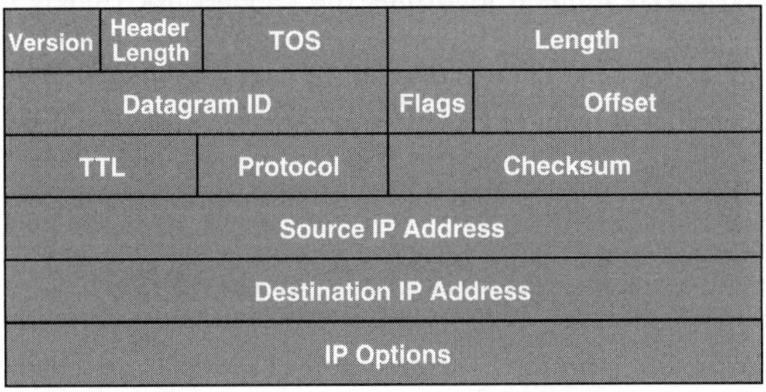

Version
Header Length
TOS
Length
Datagram ID
Flags
Offset
TTL
Protocol
Checksum
Source IP Address
Destination IP Address
IP Options

to reach its destination. To prevent this packet from living forever on the Internet, IP includes a *Time-to-Live* (TTL) field. This configurable field has a value between 0 and 255 and indicates the maximum number of hops that this packet is allowed to make before it is discarded by the network. Every time the packet enters a router, the router decrements the TTL value by one; when it reaches zero, the packet is discarded, and the receiving device will ultimately invoke error control to ask for a resend.

The *Protocol field* indicates the nature of the higher-layer protocol that is carried within the packet. Encoded options include values for Internet Control Message Protocol (ICMP; 1), TCP (6), User Datagram Protocol (UDP; 17), or Open Shortest Path First (OSPF; 89).

The *Header Checksum field* is similar to a frame-check sequence in HDLC and is used to ensure that the received IP header is free of errors. Keep in mind that IP is a connectionless protocol. This error check does not check the packet; it checks only the header.

When transmitting packets, it is always a good idea to have a *source address and a destination address.* You can figure out what they are for.

Understanding IP Addresses

As Figure 3-6 shows, IP addresses are 32 bits long. They are typically written as a sequence of four numbers, which represent the decimal value of each of the address bytes. These numbers are separated by periods ("dots" in telecom parlance), and the notation is referred to as *dotted decimal notation*. A typical address might be 168.152.20.10. These numbers are hierarchical. The hierarchy is described in the following paragraphs.

IP addresses are divided into two subfields. The *Network Identifier* (NET_ID) subfield identifies the subnetwork that is connected to the Internet. The NET_ID is most commonly used for routing traffic between networks. On the other hand, the *Host Identifier* (HOST_ID) subfield identifies the address of a system (host) within a subnetwork.

IP Address Classes

IP defines distinct *address classes,* which are used to discriminate between different size networks. Classes A, B, and C are used for host

192.168.1.1

Figure 3-6 32-bit IP address; each "segment" of the dotted decimal address (192,168,1,1) comprises eight bits.

addressing. The only difference between them is the length of the NET_ID subfield. A Class A address, for example, has an eight-bit NET_ID field and a 24-bit HOST_ID field. They are used for the identification of very large networks and can identify as many as 16,777,214 hosts in each network. To date, only about 90 Class A addresses have been assigned.

Class B addresses have 16-bit NET_ID and 16-bit HOST_ID fields. These are used to address medium-size networks and can identify as many as 65,534 hosts within each network.

Class C addresses, which are far and away the most common, have a 24-bit NET_ID field and an eight-bit HOST_ID field. These addresses are used for smaller networks and can identify no more than 254 devices within any given network. There are 2,097,152 possible Class C NET_IDs that are commonly assigned to corporations with fewer than 250 employees (or devices!).

Two additional address types are in use. Class D addresses are used for IP multicasting, such as transmitting a television signal to multiple recipients. Class E addresses are reserved for experimental use.

Some addresses are reserved for specific purposes. A HOST_ID of zero is reserved to identify an entire subnetwork. For example, the address 168.152.20.0 refers to a Class C address with a NET_ID of 168.152.20. A HOST_ID that consists of all ones (usually written "255" when referring to an all-ones byte, but also denoted as −1) is reserved as a broadcast address and is used to transmit a message to all hosts on a particular network.

Subnet Masking

One of the most valuable but least understood tools in IP management is called the *subnet mask.* Subnet masks are used to identify the portion of the address that specifies the network or the subnetwork for routing purposes. They are also used to divide a large address into subnetworks or to combine multiple smaller addresses to create a single, large domain. In the case of an organization subdividing its network, the address space is apportioned to identify multiple logical networks. This is accomplished by further dividing the HOST_ID subfield into a *subnetwork identifier* (SUBNET_ID) and a HOST_ID.

Adding to the Alphabet Soup: CIDR, DHCP, and Friends

As soon as the Internet became popular in the early 1990s, concerns began to arise about the eventual exhaustion of available IP addresses. For example, consider what happens when a small corporation of 11 employees purchases a Class C address. They now control more than 250 addresses, of which they may be using only 25. Clearly, this is a waste of a scarce resource.

One technique that has been accepted for address space conservation is called *classless interdomain routing* (CIDR). CIDR effectively limits the number of addresses assigned to a given organization, making the process of address assignment far more granular—and therefore efficient. Furthermore, CIDR has had a secondary yet equally important impact: It has dramatically reduced the size of the Internet routing tables because of the preallocation techniques used for address space management.

Other important protocols include *Network Address Translation* (NAT), which translates a private IP address that is being used to access the Web into a public IP address from an available pool of addresses, thus further conserving address space, and *Port Address Translation* (PAT) and *Network Address Port Translation* (NAPT), which allow multiple systems to share a single IP address by using different port numbers. Port numbers are used by transport-layer protocols to identify specific higher-layer applications.

Addressing in IP: The Domain Name System (DNS)

IP addresses are 32 bits long, and although not all that complicated, most Internet users don't bother to memorize the dotted decimal addresses of their systems. Instead, they use natural language host names. Most hosts, then, must maintain a comparative table of both numeric IP addresses and natural language names. From a host perspective, however, the names are worthless; the numeric identifiers must be used for routing purposes.

Because the Internet continues to grow at a rapid clip, a system was needed to manage the growing list of new Internet domains. Consider this: In January 1995, 4,852,000 recognizable hosts were on the Internet; in January 2005, 10 years later, there are 317,646,084! That naming system, the DNS, is a distributed database that stores host names and IP address information for all of the recognized domains found on the Internet. For every domain an *authoritative name server* exists that contains all DNS-related information about that domain, and every domain has at least one secondary name server that also contains the information. A total of 13 *root servers,* in turn, maintain a list of all of the authoritative name servers.

How does the DNS actually work? When a system needs a another system's IP address based upon its host name, the inquiring system issues a DNS request to a *local name server*, as shown in Figure 3-7. Depending on the contents of its local database, the local name server may be able to respond to the request. If not, it forwards the request to a root server. The root server, in turn, consults its own database and determines the most likely name server for the request and forwards it appropriately.

Early Address Resolution Schemes

When the Internet first came banging into the public psyche, most users were ultimately connected to the Internet via an Ethernet local area network (LAN). LANs use a local device address known as a *medium access*

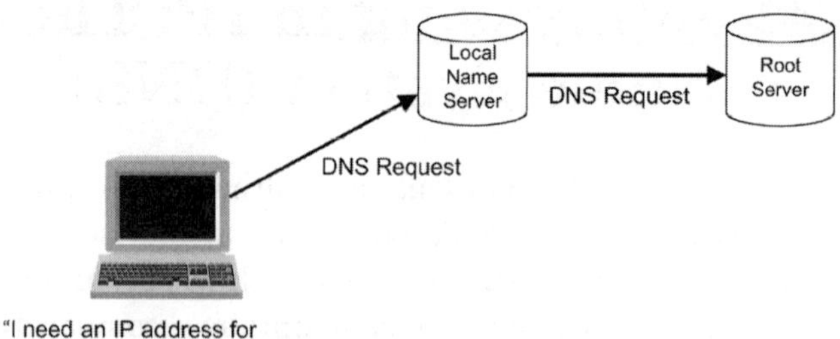

Figure 3-7 DNS service request

control (MAC) address, which is 48 bits long and nonhierarchical, meaning that it cannot be used in IP networks for routing.

To get around this disparity and to create a technique for relating MAC addresses to IP addresses, the *Address Resolution Protocol* (ARP) was created. ARP allows a host to determine a receiver's MAC address when it knows only the device's IP address. The process is simple: The host transmits an *ARP request packet* that contains the MAC broadcast address. The ARP request advertises the destination IP address and asks for the associated MAC address. Because every station on the LAN hears the broadcast, the station that recognizes its own IP address responds with an ARP message that contains its MAC address.

As ARP became popular, other address-management protocols came into play. *Reverse ARP* (RARP) gives a diskless workstation (a dumb terminal, for all intents and purposes) the ability to determine its own IP address, knowing only its own MAC address. *Inverse ARP* (InARP) maps are used in frame-relay installations to map IP addresses to frame-relay virtual-circuit identifiers. *ATMARP* and *ATMInARP* are used in ATM networks to map IP addresses to ATM virtual path/channel identifiers. And finally, *LAN Emulation ARP* (LEARP) maps a recipient's ATM address to its LAN emulation (LE) address, which is typically a MAC address.

Routing in IP Networks

Three routing protocols are the most commonly used in IP networks: the *Routing Information Protocol* (RIP), the OSPF protocol, and the *Border Gateway Protocol* (BGP).

OSPF and RIP are used primarily for intradomain routing—within a company's dedicated network, for example. They are sometimes referred to as *interior gateway protocols* (IGPs). RIP uses the hop count as the measure of a particular network path's cost; in RIP, it is limited to 16 hops.

When RIP broadcasts information to other routers about the current state of the network, it broadcasts its entire routing table, resulting in a flood of what may be unnecessary traffic. As the Internet has grown, RIP has become relatively inefficient because it does not scale as well as it should in a large network. As a consequence the OSPF protocol was introduced. OSPF is known as a *link state protocol.* It converges (spreads important information across the network) faster than RIP, requires less overall bandwidth, and scales well in larger networks. OSPF-based routers broadcast only changes in status rather than the entire routing table.

The BGP is referred to as an *exterior gateway protocol* (EGP) because it is used for routing traffic *between* Internet domains. Like RIP, BGP is a distance-vector protocol; unlike other distance-vector protocols, BGP stores the actual route to the destination.

The Not-So-Distant Future: IP Version 6 (IPv6)

Because of the largely unanticipated growth of the Internet since 1993 or so (when the general public became aware of its existence), it was roundly concluded that IP version 4 (IPv4) was inadequate for the emerging and burgeoning needs of Internet applications. In 1995 IP version 6 (IPv6) was introduced, designed to deal with the shortcomings of version 4. Changes included increased IP address space from 32 to 128 bits, improved support for differentiable QoS requirements, and improvements with regard to security, confidentiality, and content integrity.

Header Format

IPv6 differs substantially from IPv4. Whereas IPv4 has a variable-sized header depending on the included options, the IPv6 header is fixed at 40 bytes. This provides a substantial advantage: A fixed-length header can be rapidly read into hardware and interpreted, speeding up the processing of the header information. Options are dealt with through separate *extension headers*, which routers can ignore for the most part because they are designed to be processed by end-user devices.

Most of the IPv4 header comprises address information. Thirty-two bytes each are needed for the source and destination IP addresses. So in IPv6, a number of fields have been eliminated: Header Length fields, Identification fields, Flags and Fragment Offset fields, and Header Checksum fields. Elimination of the Checksum field accelerates header processing, and given that a checksum is calculated at the transport layer, there's no need for another one at layer three.

Fragmentation, long a function carried out by routers in the network, is no longer a major issue under IPv6. As a result the other fields mentioned in the prior paragraph disappear, further simplifying the process of packet analysis and treatment.

Of the remaining fields, the TTL field has evolved into a Hop Limit field, although its function remains precisely the same as in IPv4. The Protocol field is now called Next Header and serves to indicate whether or not an Extension Header follows. The Payload Length field no longer includes the length of the IP header, and the Type of Service field is now referred to as the Traffic Class field.

One new field has been added: the Flow Label field. This field allows groups of packets that belong to the same flow—that is, share the same source and destination—to be identified so that all packets can be processed together, thus reducing network overhead. Its primary function is to improve real-time traffic handling.

Extension Headers

IPv6 has added a collection of extension headers to the protocol model. They are listed here and *must* be ordered as shown:

- Hop-by-Hop Options header
- Destination Options header
- Routing header
- Fragment header
- Authentication header
- Encapsulating Security Payload header

These headers are processed only by the destination device.

The Fragment header allows end devices to fragment packets as required to satisfy maximum transmission unit (MTU) size limitations, whereas the Authentication and Encapsulation Security Payload (ESP) headers provide IP Security (IPSec) authentication for the IPv6 packet.

Addressing

IPv6 addresses are now 16 bytes (128 bits) long rather than 4 (32 bits), which means that additional address space must be allocated and devices transporting these packets must be able to handle the much-expanded address size. A number of other changes have taken place as well. For example, there is no longer a dedicated broadcast address. Instead, unicast, multicast, and any cast addresses have been provisioned.

The format of the content has changed as well. IPv6 addresses include a general routing prefix, a subnet ID, and an interface ID. This results in additional complexity, but will be handled easily by the resource-rich routers that now comprise the bulk of the network. Finally, the IPv6 address is expressed in hexadecimal, not decimal, so the old familiar IP address will now look rather—odd. The IPv6 packet structure is shown in Figure 3-8.

Transport-Layer Protocols

We turn our attention now to layer four, the Transport Layer. Two key protocols are found at this layer: TCP and UDP. TCP is an ironclad, absolutely guaranteed service delivery protocol, with all of the attendant protocol overhead you would expect from such a capable protocol. UDP, on the other hand, is a more lightweight protocol, used for delay-sensitive applications like VoIP. Its overhead component is relatively light.

In TCP and UDP messages, higher-layer applications are identified by *port identifiers*. The port identifier and IP address together form a *socket*, and the end-to-end communication between two or more systems is identified by a four-part complex address: the source port, the source address, the destination port, and the destination address. Commonly used port numbers are shown in Table 3-2.

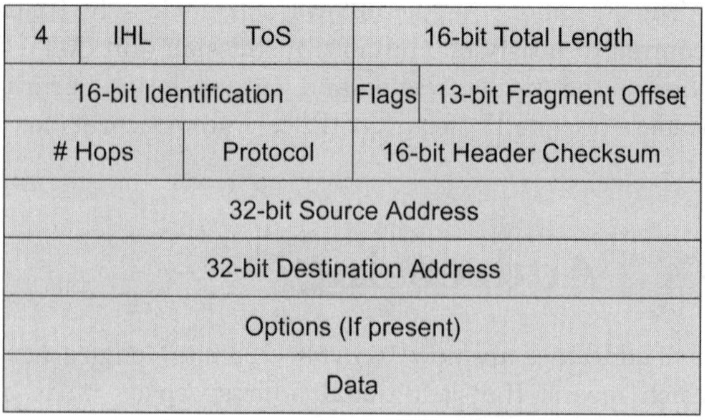

Figure 3-8 IPv6 packet format

4	IHL	ToS	16-bit Total Length	
16-bit Identification			Flags	13-bit Fragment Offset
# Hops		Protocol	16-bit Header Checksum	
32-bit Source Address				
32-bit Destination Address				
Options (If present)				
Data				

The Transmission Control Protocol (TCP)

TCP provides a connection-oriented communication service across the network. It stipulates rule sets for message formats, establishing virtual circuit establishment and termination, data sequencing, flow control, and error correction. Most applications designed to operate within the TCP/IP protocol suite use TCP's reliable, guaranteed delivery services.

In TCP, the transmitted data entity is referred to as a *segment* because TCP does not operate in message mode. It simply transmits blocks of data from a sender and receiver. The fields that make up the segment, shown in Figure 3-9, are described in the following paragraphs.

The *source port* and the *destination port* identify the originating and terminating connection points of the end-to-end connection, as well as the higher-layer application. The *sequence number* identifies this particular segment's first byte in the byte stream, and because the sequence number refers to a byte count rather than to a segment, the sequence numbers in sequential TCP segments are not, ironically, numbered sequentially.

The *acknowledgment number* is used by the sender to acknowledge to the transmitter that it has received the transmitted data. In practice, the field identifies the sequence number of the next byte that it expects from the receiver. The *Data Offset field* identifies the first byte in this particular segment; in effect it indicates the segment header length.

TCP relies on a collection of *control flags,* which do in fact control certain characteristics of the virtual connection. They include an *Urgent Pointer Field Significant* (URG) flag, which indicates that the current segment contains high-priority data and that the Urgent Pointer field

Table 3-2 Service Port Assignments

Port #	Protocol	Service	Port #	Protocol	Service
7	TCP	echo	80	TCP	http
9	TCP	discard	110	TCP	pop3
19	TCP	chargen	119	TCP	nntp
20	TCP	ftp-control	123	UDP	ntp
21	TCP	ftp-data	137	UDP	netbios-ns
23	TCP	telnet	138	UDP	netbios-dgm
25	TCP	smtp	139	TCP	netbios-ssn
37	UDP	time	143	TCP	imap
43	TCP	whois	161	UDP	snmp
53	TCP/UDP	dns	162	UDP	snmp-trap
67	UDP	bootps	179	TCP	bgp
68	UDP	bootpc	443	TCP	https
69	UDP	tftp	520	UDP	rip
79	TCP	finger	33434	UDP	traceroute

value is valid; an *Acknowledgment Field Significant* (ACK) flag, which indicates that the value contained in the Acknowledgment Number field is valid; a *Push Function* (PSH) flag, which is used by the transmitting application to force TCP to immediately transmit data that it currently has buffered without waiting for the buffer to fill; a *Reset Connection* (RST) flag, which is used to immediately terminate an end-to-end TCP connection; a *Synchronize Sequence Numbers* (SYN) flag, which is used to establish a connection and to indicate that the segments carry the proper initial sequence number; and finally a *Finish* (FIN) flag, which is set to request a normal termination of a TCP connection in whatever direction the segment is traveling.

The *Window* field is used for flow-control management. It contains the value of the permitted *receive window size,* the number of transmitted bytes that the sender of the segment is willing to accept from the

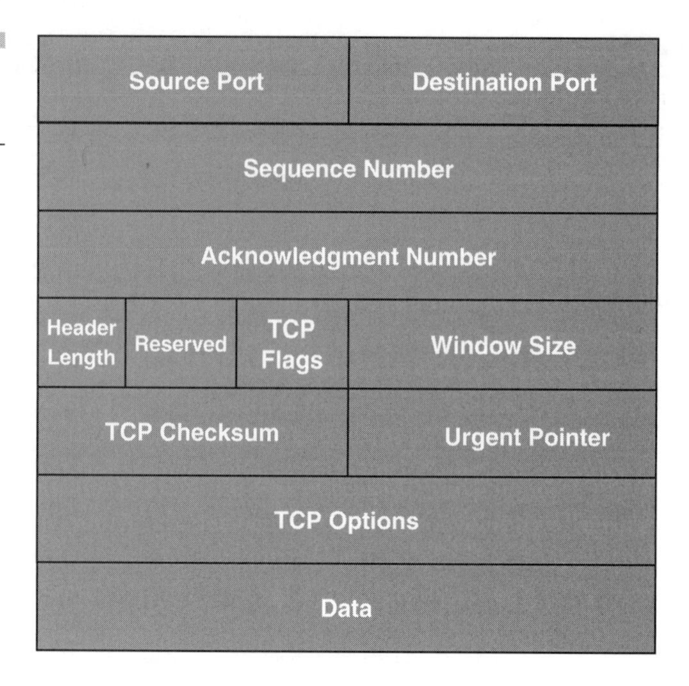

Figure 3-9 The TCP header

receiver. The *Checksum field* offers bit-level error detection for the entire segment, including both the header and the transmitted data.

The *Urgent Pointer field* is used for the management of high-priority traffic as identified by a higher-layer application. If so marked, the segment is typically allowed to bypass normal TCP buffering.

The last field is the *Options field.* At the time of the initial connection establishment, this field is used to negotiate such functions as maximum segment size and Selective Acknowledgement (SACK).

The User Datagram Protocol (UDP)

UDP provides connectionless service, and although "connectionless" often implies "unreliable," that is a bit of a misnomer. For applications that require nothing more than a simple query and response, UDP is ideal because it involves minimal protocol overhead. UDP's primary responsibility is to add a port number to the IP address to create a socket for the application. The fields of a UDP message (see Figure 3-10) are described in the following paragraphs.

The *Source Port field* identifies the UDP port used by the sender of the datagram, whereas the *Destination Port field* identifies the port used by the datagram receiver. The *Length field* indicates the overall length of the UDP datagram.

The *Checksum field* provides the same primitive bit error detection of the header and transported data as we saw with TCP.

The Internet Control Message Protocol (ICMP)

The ICMP is used as a diagnostic tool in IP networks to notify a sender that something unusual happened during transmission. It offers a healthy repertoire of messages, including *Destination Unreachable*, which indicates that delivery is not possible because the destination host cannot be reached; *Echo* and *Echo Reply*, which checks whether hosts are available; *Parameter Problem,* which indicates that a router encountered a header problem; *Redirect,* which makes the sending system aware that packets should be forwarded to another address; *Source Quench,* which indicates that a router is experiencing congestion and is about to begin

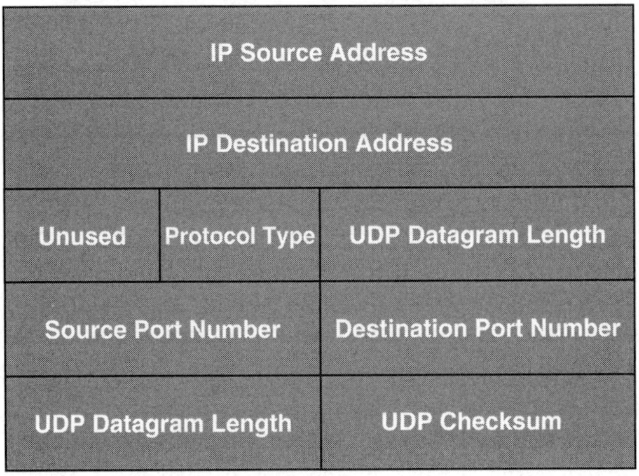

Figure 3-10 The UDP header

IP Source Address

IP Destination Address

Unused | Protocol Type | UDP Datagram Length

Source Port Number | Destination Port Number

UDP Datagram Length | UDP Checksum

discarding datagrams; *TTL Exceeded,* which indicates that a received datagram has been discarded because the Time-to-Live field (sounds like a soap opera for geeks, doesn't it?) reached zero; and finally *Timestamp* and *Timestamp Reply,* which are similar to Echo messages except that they timestamp the message, giving systems the ability to a measure how much time is required for remote systems to buffer and process datagrams.

The Application Layer

The TCP/IP Application-Layer protocols support the actual applications and utilities that make the Internet, well, useful. They include the BGP, the DNS, the *File Transfer Protocol* (FTP), the *Hypertext Transfer Protocol* (HTTP), OSPF, the *Packet Internetwork Groper* (Ping; how can you *not* love that name), the *Post Office Protocol* (POP), the *Simple Mail Transfer Protocol* (SMTP), the *Simple Network Management Protocol* (SNMP), the *Secure Sockets Layer Protocol* (SSL), and TELNET. This is a small sample of the many applications that are supported by the TCP/IP Application Layer.

A Close Relative: Multiprotocol Label Switching (MPLS)

When establishing connections over an IP network, it is critical to manage traffic queues to ensure the proper treatment of packets that come from delay-sensitive services such as voice and video. To do this, packets must be differentiable, that is, identifiable so that they can be classified properly. Routers, in turn, must be able to respond properly to delay-sensitive traffic by implementing queue management processes. This requires that routers establish both normal and high-priority queues and handle the traffic found in high-priority routing queues faster than the arrival rate of the traffic.

MPLS delivers QoS by establishing virtual circuits known as *Label Switched Paths* (LSPs), which are in turn built around traffic-specific

QoS requirements. An MPLS network, such as that shown in Figure 3-11, comprises *label switch routers* (LSRs) at the core of the network and *label edge routers* (LERs) at the edge. It is the responsibility of the LERs to set QoS requirements and pass them on to the LSRs, which are responsible for ensuring that the required levels of QoS are achieved. Thus, a router can establish LSPs with explicit QoS capabilities and route packets to those LSPs as required, guaranteeing the delay that a particular flow encounters on an end-to-end basis. It's interesting to note that some industry analysts have compared MPLS LSPs to the trunks established in the voice environment.

MPLS uses a two-part process for traffic differentiation and routing. First, it divides the packets into *Forwarding Equivalence Classes* (FECs) based on their QoS requirements and then maps the FECs to their next hop point. This process is performed at the point of ingress at the edge of the network. Each FEC is given a fixed-length "label" that accompanies each packet from hop to hop; at each router, the FEC label is examined and used to route the packet to the next hop point, where it is assigned a new label.

MPLS is a *shim* protocol that works closely with IP to help it deliver QoS guarantees. Its implementation allows for the eventual dismissal of ATM as a required layer in the multimedia network protocol stack. It offers a promising solution, but its universal deployment is still a ways off. However, it is growing in popularity and is rapidly becoming the preferred core routing technique in wide area IP networks.

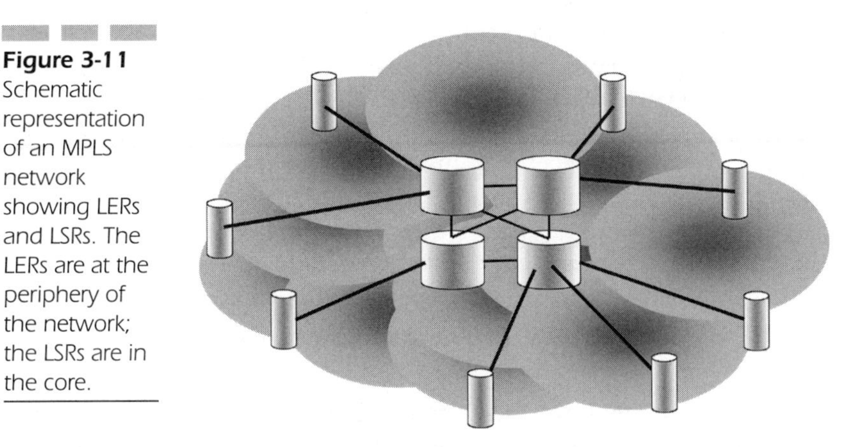

Figure 3-11 Schematic representation of an MPLS network showing LERs and LSRs. The LERs are at the periphery of the network; the LSRs are in the core.

Supporting Standards

In the final analysis, the VoIP network delivers a level of service and a consistent set of applications between users that are remarkably similar to those delivered by the public switched telephone network (PSTN). Voice and data are routinely handled by these systems. The real added value of VoIP becomes evident when it begins to deliver on the promise of convergence, and when it does so across the entirety of the network—from the LAN through the wide area network (WAN) and back to the LAN again.

Convergence means that the network becomes far more flexible and capable than it has ever been before. It means that network management can be performed from literally any point in the network, providing a level of capability and customer advocacy that until recently was unheard of. Of course, much of this functionality comes from the standards upon which VoIP systems are based; we turn our attention to them now.

Like Signaling System 7 (SS7) in the PSTN, VoIP networks must have their own set of protocols that manage call setup, maintenance, and teardown. They track the capabilities of the network and end-user devices and match them as required to ensure connectivity. We often say, "the nice thing about standards is that there are so many to choose from." That is certainly true in the IP domain. Two international standards bodies create and publish the standards that govern the transmission of voice and other services over IP networks: the International Telecommunications Union (ITU) and the IETF. Their approaches are somewhat different but have the same goal.

The ITU's primary standard is H.323; the IETF's is the Session Initiation Protocol (SIP). We'll describe each of them in turn.

VoIP networks work as well as they do because they are governed by a set of complex standards that all players—manufacturers and service providers alike—agree to abide by as they design components for the network. We begin our discussion with H.323.

H.323

H.323 started in 1996, an ITU-T standard for the transmission of multimedia content over ISDN. It was the first multimedia signaling standard and owes its provenance to H.320, the original standard for videoconferencing. Its original goal was to connect LAN-based multimedia systems to network-based multimedia systems. It originally defined a network architecture that included gatekeepers, which performed zone management and address conversion; endpoints, which were terminals and gateway devices; and multimedia control units, which served as bridges between multimedia types.

H.323 has now been rolled out in four phases. Phase one defined a three-stage call setup process: a precall step, which performed user registration, connection admission, and exchange of status messages required for call setup; the actual call setup process, which used messages similar to ISDN's Q.931; and finally a capability exchange stage, which established a logical communications channel between the communicating devices and identified conference management details.

Phase two allowed for the use of Real-Time Transport Protocol (RTP) over ATM, which eliminated the added redundancy of IP and also provided for privacy and authentication as well as greatly demanded telephony features such as call transfer and call forwarding. RTP has an added advantage: When errors result in packet loss, RTP does not request resends of those packets, thus providing for real-time processing of application-related content. No delays result from errors.

Phase three added the ability to transmit real-time fax after establishing a voice connection. And phase four, released in May 1999, added call connection over UDP, which significantly reduced call setup time, interzone communications, call hold, park, call pickup, and call and message-waiting features. This last phase bridged the considerable gap between IP voice, which was largely Internet-based catch-as-catch-can and carrier-grade IP telephony.

Several Internet telephony interoperability concerns are addressed by H.323. These include gateway-to-gateway interoperability, which ensures that telephony can be accomplished between different vendors' gateways; gatekeeper-to-gatekeeper interoperability, which does the same thing for different vendors' gatekeeper devices; and finally gateway-to-gatekeeper interoperability, which completes the interoperability picture.

H.323 offers a suite of functions for audio, video, and data conferencing, and it defines a set of functional modules similar to SIP. They include terminals, which are physical and soft phones; gateways, which provide the protocol conversion between packet and telephony; a gatekeeper, which performs address translation, connection admission control, and bandwidth management; and multipoint conferencing units (MCUs), which enable multiparty voice and videoconferencing. H.323 also supports a variety of collaborative applications, including screensharing and videoconferencing. Figure 3-12 shows a typical H.323-based network.

That being said, a number of downsides exist to H.323 that have resulted in its displacement by SIP as the preferred protocol for multimedia call management. Because the standard supports so many enhanced functions, the protocols are large and complex and therefore expensive to develop. Furthermore, H.323-compliant devices must host a wider range of protocols to be H.323 compliant and therefore must be more complex and expensive.

H.323 is an umbrella protocol, shown in Figure 3-13, meaning that it comprises a range of subtending standards that give it its functionality. Let's take a moment to walk through a call placed in an H.323 environment. When the user goes off-hook, the H.323-compliant phone relies on H.245 to select a channel to perform an endpoint capabilities exchange. When the exchange is complete, H.225.0 steps in and performs signaling and setup, after which the Registration-Admission-Status channel notifies the gatekeeper that a call is under way. If the call is to be transmitted off-net, the gateway must perform protocol conversion between the circuit and packet environments.

In reality H.323 is an umbrella standard that includes (among others) H.225 for call handling, H.245 for call control, G.711 and G.721 for coder/decoder (CODEC) definitions, and T.120 for data conferencing. Originally created as a technique for transporting multimedia traffic over a LAN, gatekeeper functions have been added that allow LAN traffic and LAN capacity to be monitored so that calls are established only if adequate capacity is available on the network. Later, the gatekeeper routed model was added, which allowed the gatekeeper to play an active

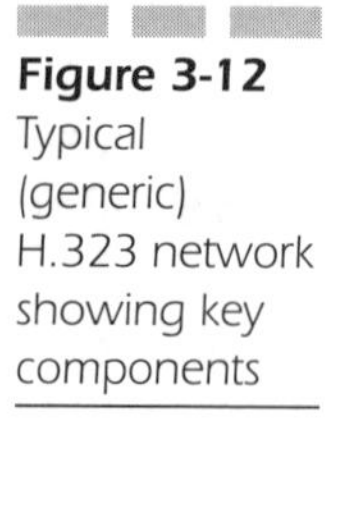

Figure 3-12 Typical (generic) H.323 network showing key components

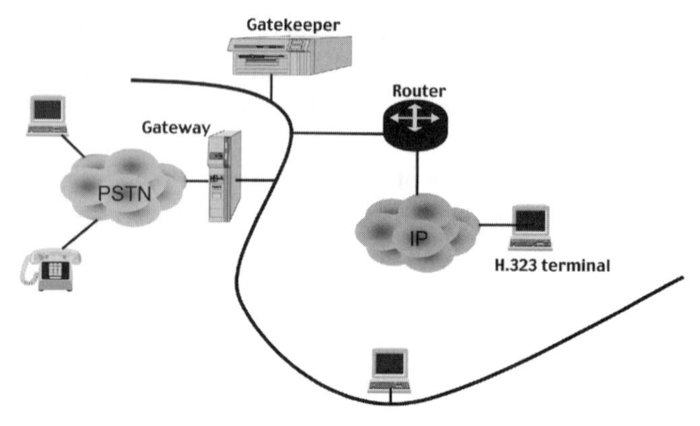

Figure 3-13 H.323 protocol components

Characteristic	H.323
Approved	1998
Network	No guaranteed bandwidth; packet switched networks, Ethernet
Video	H.261, H.263
Audio	G.711, G.722, G.728, G.723, G.729
Multiplexing	H.225.0
Control	H.245
Multipoint	H.323
Data	T.120
Comm. Protocol	TCP/IP

role in the actual call-setup process. This meant that H.323 had migrated from being purely a peer-to-peer protocol to having a more traditional, hierarchical design.

The greatest advantage that H.323 offers is maturity. It has been available for some time now, and while robust and full featured, it was not originally designed to serve as a peer-to-peer protocol. Its maturity, therefore, is not enough to carry it. It currently lacks a network-to-network interface and does not adequately support congestion control. This is not generally a problem for private networks, but it becomes problematic for service providers who wish to interconnect PSTNs and provide national service among a cluster of providers. As a result many service providers have chosen to deploy SIP instead of H.323 in their national networks.

The Session Initiation Protocol (SIP)

SIP was conceived of as a flexible and adaptable protocol that could serve as an alternative to its more mature cousin, H.323. Whereas H.323 was originally rolled out as a governance protocol to control the delivery of multimedia traffic on LANs, SIP was created specifically with VoIP in mind. SIP defines a set of standard objects and a message hierarchy for communicating among the various elements (described previously) that comprise the typical VoIP network.

SIP is made up of a collection of modules that work together to provide signaling functionality across the network. These modules, shown in Figure 3-14, include user agent clients; a location server, which relates a client device to a specific IP address; proxy servers, which are responsible to forward call requests from one server to another on behalf of SIP clients (hence the name proxy); and redirect servers, which transmit the called party's address back to the calling party so that the connection can be made.

These devices rely on a limited set of SIP messages that are used to perform basic signaling tasks—establishment, maintenance, and teardown. They include Invite, used to establish or join a call session; ACK, used to acknowledge an invitation; Register, used to register a user with a server; Options, which are used to petition for data about the capabilities of the server; Cancel, which cancels a previous request; and Bye, which ends a session. Figure 3-15 shows a typical exchange of information using the SIP protocol.

SIP was created by the IETF, the same organization responsible for many Internet-oriented standards. Its responsibility lies with session creation and destruction, while the actual in-progress session and connection details are left up to the communicating endpoint devices.

The protocol relies on a text-based command structure that uses the now universally familiar HTTP syntax and universal resource locator (URL) addressing, both ideal for delivering telephony over an IP network where the logical integration of applications such as voice, messaging, conferencing, and Web access can create an enhanced customer experience. Designed with the assumption that end-user devices have a

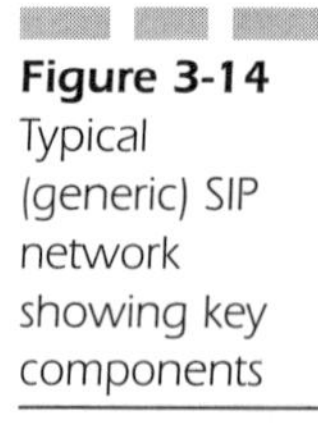

Figure 3-14 Typical (generic) SIP network showing key components

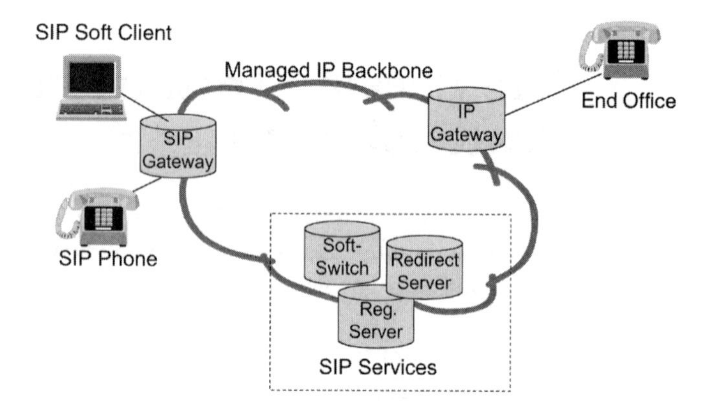

modicum of intelligence that can supplant a surfeit of intelligence in the core network, SIP is designed for communications among a collection of proxy and location servers. As a result it is immensely scalable, a solution for enterprise IT personnel who see growing user populations and unbounded demand as a major challenge to their ability to satisfy QoS demands.

SIP is designed to establish peer-to-peer sessions between Internet routers. The protocol defines a variety of server types, including feature servers, registration servers, and redirect servers. SIP supports fully distributed services that reside in the actual user devices, and because it is based on existing IETF protocols it provides a seamless integration path for voice/data integration.

Ultimately, telecommunications, like any industry, revolves around profitability. Any protocol that allows new services to be deployed inexpensively and quickly will immediately catch the eye of service providers. Like TCP/IP, SIP provides an open architecture that can be used by any vendor to develop products, thus ensuring multivendor interoperability. And because SIP has been adopted by such powerhouses as Lucent, Nortel, Cisco, Ericsson, and 3Com and is designed for use in large carrier networks with potentially millions of ports, its success is reasonably assured.

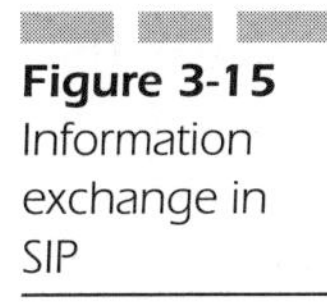

Figure 3-15 Information exchange in SIP

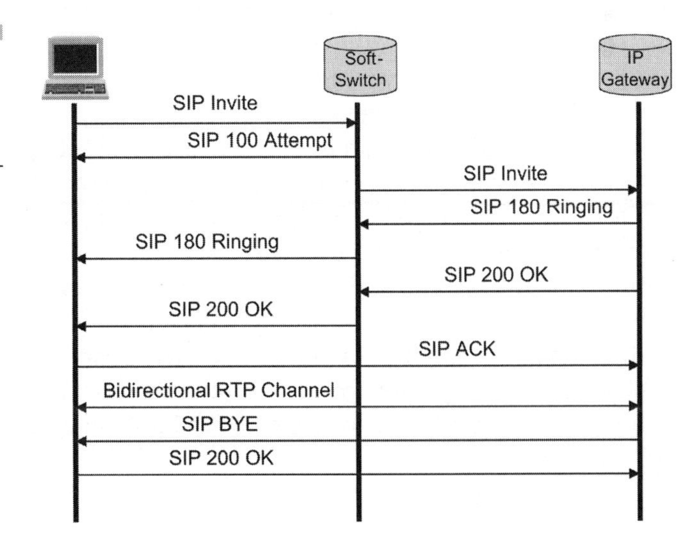

Originally, H.323 was to be the protocol of choice to make this possible. And although H.323 is clearly a capable suite of protocols and is indeed quite good for VoIP services that derive from ISDN implementations, it remains quite complex. As a result it has been relegated for use as a video-control protocol and for some gatekeeper-to-gatekeeper communications functions. Although H.323 continues to enjoy its share of supporters, it is slowly being edged out of the limelight. SIP supporters claim that H.323 is far too complex and rigid to serve as a standard for basic telephony setup requirements, arguing that SIP, which is architecturally simpler and imminently extensible, is a better choice.

The intense interest in moving voice to an IP infrastructure is driven by simple and understandable factors: cost of service and enhanced flexibility. However, in keeping with the "Jurassic Park Effect" (just because you *can* doesn't necessarily mean you *should*), it is critical to understand the differences that exist between simple voice and full-blown telephony with its many enhanced features. It is the feature set that gives voice its range of capability. A typical local switch such as Lucent Technologies' 5ESS offers more than 3,000 features, and more will certainly follow. Of course, the features and services are possible because of the protocols that have been developed to provide them across an IP infrastructure.

SIP in Action

SIP, detailed in RFC 2543, is part of the IETF's standards suite for managing the transmission and control of multimedia data. As noted earlier it is built on a client-server signaling model and is used almost exclusively in IP telephony networks.

Like other VoIP-related protocols, SIP is used to set up and tear down multimedia sessions between communicating endpoints. These multimedia sessions can include multiparty conferences, telephone calls, and distribution of multimedia content. These capabilities are central to the routine care and feeding of the enterprise, not to mention the enormous boon they represent for the enterprise call center.

SIP is a lightweight, text-based protocol transported via TCP or UDP and is designed to operate according to the ad hoc rules established for all Internet protocols: easy to implement, simple in operation, efficient, and scalable.

How SIP Works

SIP exchanges capabilities between communicating devices using "invitations" as a means to create *Session Description Protocol* (SDP) messages. Their role is to set up call-control-channel use and to ensure that call participants have the ability to negotiate a set of compatible media types.

Equally important is SIP's support of user portability. In an environment that is as fluid and dynamic as a college, a call center, or a research and development facility, it is crucial that network administrators have the ability to reconfigure the network as people move without incurring inordinately high move, add, and change (MAC) costs. SIP is a major part of the answer to this challenge. The protocol uses proxy and redirect services to match an incoming call to a user's *logical* location, and the logical location is then tied to a physical port. A user, for example, can inform the SIP server of her current physical location by transmitting a "registration message" to a software "registrar." Thus, a user can move from one location to another, take his phone with him, plug it in, and immediately be recognized and connected with appropriate access and security delivered on a user-by-user basis.

In proxy mode (see Figure 3-16), SIP clients (end-user devices) transmit server requests to the proxy server, and the proxy server either handles the service request directly or forwards the request to another SIP server. One advantage of proxy operation is that the proxy server can disguise users by transmitting a proxy rather than the original user-generated signaling message. As a result other users cannot "see" the original message. Instead they see a message from the proxy server.

In redirect mode (see Figure 3-17), the call setup request is transmitted to an SIP server. The server performs a lookup on the destination address, which it then returns to the originator of the call. The originator in turn signals the SIP client.

H.323 and SIP are two of the protocols that guide call setup, maintenance, and teardown in IP networks, but others exist. We'll discuss them now.

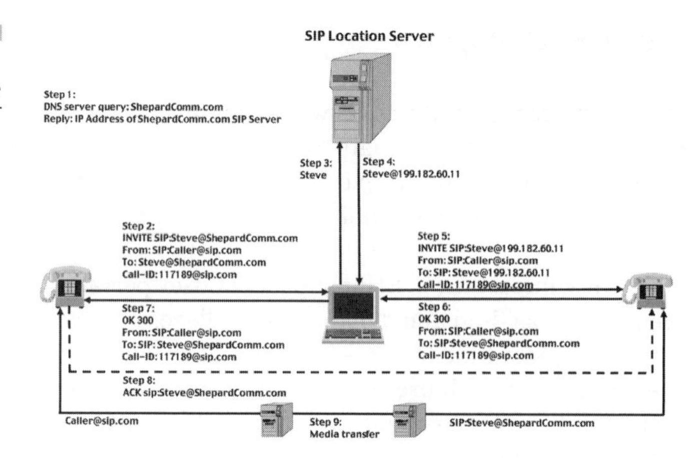

Figure 3-16 SIP proxy mode

Media Gateway Control Protocol (MGCP)—and Friends

Many of the protocols that guide the successful development of VoIP efforts today stem from work performed early by Level 3 and Telcordia, which together founded an organization called the International SoftSwitch Consortium. In 1998, Level 3 brought together a collection of

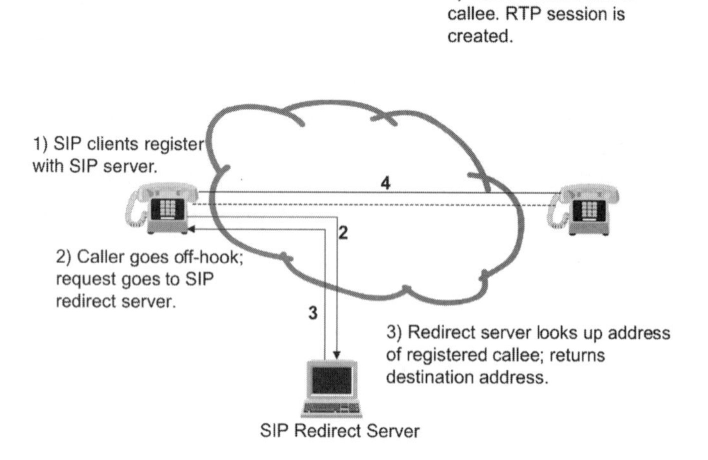

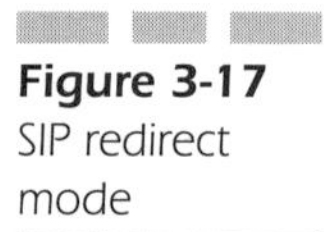

Figure 3-17 SIP redirect mode

vendors that collaboratively developed and released *Internet Protocol Device Control* (IPDC). At the same time, Telcordia created and released the *Simple Gateway Control Protocol* (SGCP). The two were later merged to form the *Media Gateway Control Protocol* (MGCP), discussed in detail in RFC 2705.

MGCP allows a network device responsible for establishing calls to control the devices that actually perform IP voice streaming. It permits software call agents and media gateway controllers to control streaming-media gateways at the edge of the network. These gateways can be cable modems, set-top boxes, private branch exchanges (PBXs), voice and telephony over ATM (VTOA) gateways, and VoIP gateways. Under this design, the gateways manage the circuit-switch-to-IP voice conversion, while the agents manage signaling and call processing.

MGCP makes the assumption that call control in the network is software based and resident in external intelligent devices that perform all call control functions. It also makes the assumption that these devices will communicate with one another in a primary-secondary arrangement, under which the call agents send instructions to the gateways for execution.

In effect, MGCP is designed to facilitate the effective management of the VoIP network by treating each gateway function (media gateway, media gateway controller, signaling gateway) as a separate functional unit. This facilitates the independent management of each VoIP gateway function as a separate logical device, thus facilitating enhanced network operations.

MGCP coordinates the responsibilities of media gateways in a primary-secondary relationship (see Figure 3-18). The media gateway controller, which manages signaling for call setup, is sometimes referred to as a *call agent*. It works closely with the media gateway, which has the responsibility to notify the call agent of service-oriented activities that require call setup or teardown. As you would expect, the media gateway controller notifies the media gateway to establish an RTP session between two communicating (or would-be communicating) endpoints.

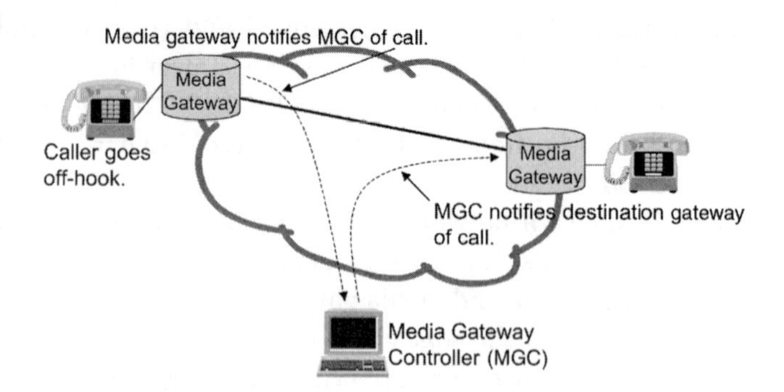

Figure 3-18 Call setup with MGCP

Media Device Control Protocol (MDCP) and MeGaCo

Meanwhile, Lucent created a new protocol called the *Media Device Control Protocol* (MDCP). The best features of the original three were combined to create a full-featured protocol called MeGaCo, also defined as H.248. In March 1999 the IETF and ITU met collaboratively and created a formal technical agreement between the two organizations, which resulted in a single protocol with two names. The IETF called it MeGaCo; the ITU called it H.GCP.

MeGaCo/H.GCP operates under the assumption that network intelligence is housed in the central office and therefore replaces the gatekeeper concept proposed by H.323. By managing multiple gateways within a single IP-equipped central office, MeGaCo minimizes the complexity of the telephone network. In other words, a corporation might be connected to an IP-capable central office, but because of the IP-capable switches in the central office, which have the ability to convert between circuit-switched and packet-switched voice, full telephony features are possible. Thus, the next generation switch converts between circuit and packet, whereas MeGaCo performs the signaling necessary to establish a call across an IP WAN. It effectively bridges the gap between legacy SS7 signaling and the new requirements of IP, and supports both connection-oriented and connectionless services.

MeGaCo/H.248, still in draft, is a cooperative standards effort between the IETF and the ITU. Similar to MGCP, it uniquely identifies individual media gateways (protocol conversion) and media gateway controllers (signaling). Like MGCP, MeGaCo defines a set of transactions under the control of a media gateway controller that establishes calls.

MeGaCo came about because of a desire on the part of standards bodies and manufacturers to create a single standard that would work with all devices, regardless of manufacturer—hence the involvement of both the ITU and the IETF. As you would expect, the players involved in the standard's design created a list of requirements (essentially benchmarks for success), including support for a broad range of telephones and related devices, support for both basic and evolving enhanced services for those devices, a standard that drives the cost of service to a level that the market will accept, and support for MeGaCo standards as defined in the documentation.

Real-Time Transport Protocol (RTP)

RTP is standardized through the IETF's RFC 1889 and RFC 1890, which describe the end-to-end delivery of data with real-time requirements such as interactive audio and video. Services covered by RTP include payload-type identification, sequence numbering, time stamping, and delivery monitoring.

RTP is a UDP-based application that is typically implemented in the media gateways; the protocol is used to deliver the voice traffic. The protocol's greatest contribution to an IP session is that it has the ability to monitor and manage timing, signal-loss detection, data security, content delivery, and the perpetuation of encoding schemes. An RTP session is identified by two IP addresses (the endpoint device addresses), and even though RTP rides on a connectionless UDP network, it supports a sequencing system that allows missing packets to be detected.

RTP is designed to manage the control requirements of real-time communication. Its structure is straightforward; its message format, shown in Figure 3-19, includes a Sequence Number field, which is incremented for each RTP packet, and a Time Stamp field, which records the sampling rate and the resultant message-playback rate.

One problem that RTP handles rather elegantly is packet loss. Rather than extrapolate by repeating the contents of a previous packet, it does

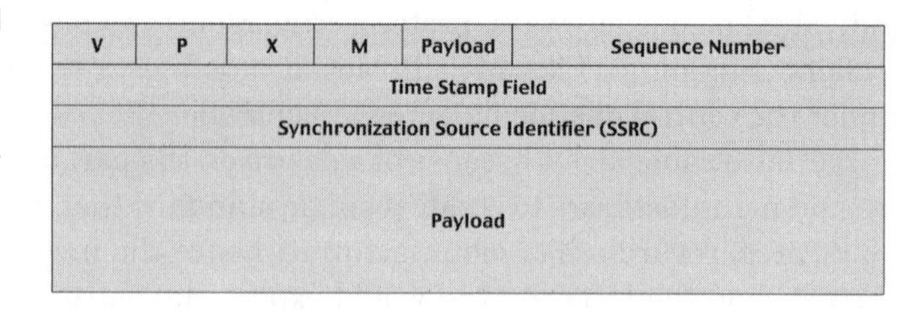

Figure 3-19 RTP message format

V: SIP Version
P: Encryption Padding
X: Extension Bit
M: More
Payload: Encoding methodology for video, other content
Sequence Number: Packet sequence number
Time Stamp: Compression algorithm-dependent
SSRC: Randomly assigned; identifies source

something different. To achieve high levels of QoS in the received audio signal, the sampling rate of the voice signal must be increased, leading to smaller packets. These smaller packets take less time to process than larger packets, but they create a different problem. Although the sampled content gets smaller, the header of each packet does not, leading to a strange form of data packet in which the header size approaches the size of the payload! RTP overcomes this inefficiency through a technique called *header compression*.

The RTP header compression technique comprises two stages. In the first stage, the system takes note of fields in the packet headers that do not change over the life of a flow. In the second stage, the system takes note of the fact that few flows exist at the "edge of the network" so that such information can be conveyed over the first hop by a single packet, after which they are identified by a short "connection identifier," which identifies the full state of the packet so that the first hop router can reconstruct the full packet. In RTP, certain fields change only a small amount from packet to packet, which means that the nonchanging information can be sent once and stored, then reused for multiple packets related to the same flow. If the router resets, the route changes, or the end system alters its state dramatically, an invalid checksum will be calculated, resulting in a reset of the stored state and an exchange of full-packet information necessary to recreate the current state. Under the right circumstances, RTP can result in a tenfold reduction in protocol overhead.

Because different media gateways rely on different CODECs with different payload-encoding schemes, the RTP Payload Type field identifies the encoding scheme that the originating media gateway uses to digitize the transported payload. The field is used to identify the RTP payload format and to guide its interpretation by the CODEC that is supported in the media gateway. Because payload types are somewhat limited in terms of their variability, profiles specify default static mappings of payload types to specific payload formats. These mappings support the ITU-G series of encoding schemes for CODECs.

Because of the variability among the various CODEC encoding schemes and packet generation rates, RTP packets vary in size and transmission interval. Network managers must therefore take these RTP characteristics into account when they plan for the addition and delivery of voice services. These characteristics dictate bandwidth consumption, a critical parameter given that voice will contribute greatly to overall network traffic.

Real-Time Transport Control Protocol (RTCP)

The RTCP serves as an optional adjunct to RTP. It's role, if implemented, is to collect network performance data and to provide feedback about the end-to-end quality control being performed by the network as it moves traffic from a source to a destination address. This function is central to RTCP's role; it relies on such factors as intermediate congestion and flow control in the network, both of which provide diagnostic data that a network administrator can use to manage QoS.

RTCP reports do not indicate the actual location of a problem in the network, but they do provide indicators. By collecting and analyzing reported data from multiple gateways in the network, an administrator can localize the source of a problem and take action to resolve it.

With RTCP, a network manager can monitor latency, jitter, end-to-end delay, and packet loss, each of which is indicative of different problem types in the network. RTCP collects and disseminates this information on a scheduled basis and distributes the data to both ends of a call on a per-call basis.

For the most part RTCP is used in enterprise installations, simply because residential customers would generally have no way of using the information delivered to them by the RTCP protocol.

Be Careful What You Wish For

One caveat should be mentioned with regard to the use of RTCP: Because it collects, processes, and then disseminates data on a periodic basis, it can become a resource hog if not carefully monitored. Network managers should pay close attention to the percentage of overall bandwidth consumption that the protocol is gobbling up and control it so that it delivers the information required, but not at the expense of network performance. The IETF RFC specifications recommend that the bandwidth allocated to RTCP be fixed at no more than 5 percent of the session's RTP traffic.

Enter VoIP

Now that we have presented an overview of IP as a protocol enabler, let's now turn our attention to the emerging VoIP network. The *external* goals of the two are one and the same: to deliver a set of services to the customer in a way that facilitates their ability to do business more effectively. *Internally,* however, the goals are slightly different. VoIP has the inherent ability to reduce the total cost of ownership of the network and to enhance the overall effectiveness of the network and its delivered services in the enterprise. Let's discuss, then, the inner workings of the VoIP network.

As we described in the PSTN discussion, the analog voice signal is sampled by the digital PSTN and converted to a digital bit stream. The signal can be compressed if desired, but either way the signal is transported through the network in digital format. The process is the same in a PBX environment with a few minor changes; the PBX is, after all, simply a remote switch of sorts connected to the PSTN via high-bandwidth facilities. A digital handset connected to a PBX is handled in very similar fashion.

In the IP environment, changes occur at this point. At this juncture the data signal must be packetized for transmission across the IP network as shown in Figure 3-20. It is important to note one key difference that exists between traditional PBXs and IP systems. Older PBX systems rely on proprietary signaling protocols, whereas IP PBXs employ internationally standardized signaling schemes.

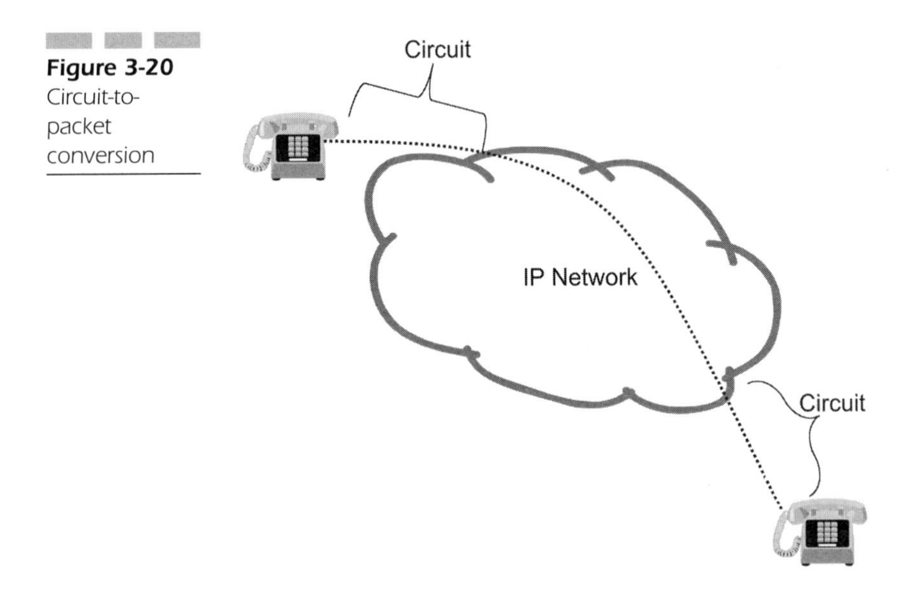

Figure 3-20 Circuit-to-packet conversion

VoIP Anatomy

The major components of the PSTN are the user access device (phone or PC), the local loop leading to the local switch or PBX, a corollary toll-switch array, and a separate signaling network. A VoIP network provides the same type of functionality (call establishment, maintenance, take-down, and associated accounting functions) but over a very different architecture. And although the devices all have corollaries to one another between the two networks, one component can be found in VoIP networks that is not found in the PSTN—a gateway device that bridges the gap between the PSTN and the VoIP infrastructure.

To establish a connection, the phone system must take action based on the setup signals received from the calling device. Most commonly this is done using software that manages the call setup process and the conversion that must take place between the standard ITU E.164 address (the one-plus-ten-digit telephone number) used in the PSTN and the IP addresses used in VoIP networks.

The key components of a converged IP-based voice network are the call-processing server, usually an IP-enabled PBX; one or more media gateways, sometimes called gatekeepers; an IP network over which calls will be placed; and endpoint devices that provide user functionality. We will describe each of these in turn; please refer to Figure 3-21.

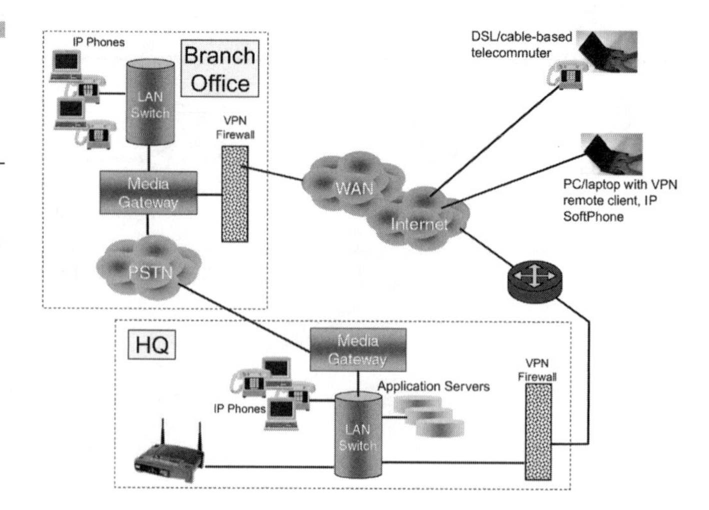

Figure 3-21 A typical VoIP enterprise network

SoftSwitch

SoftSwitch handles multiple tasks: recognition of an off-hook condition, dial-tone instigation, call routing, and call teardown. These functions are performed as part of the interaction that goes on between a telephone and a call server, functions that would naturally be performed by a traditional PBX.

SoftSwitch technology is designed to support next-generation networks that rely on packet-based voice, data, and video communications technologies and which can interface with a variety of transport technologies, including copper, wireless, and fiber. One goal of the SoftSwitch concept is to functionally separate network hardware from network software. In traditional circuit-switched networks, hardware and software are dependent on one another, resulting in what many believe to be an unnecessarily inextricable relationship. Circuit-switched networks rely on dedicated facilities and are designed primarily for delay-sensitive voice communications. The goal of SoftSwitch is to bring about dissolution of this interdependent relationship where appropriate.

In concert with the evolution of the overall network, the SoftSwitch concept has evolved to address a variety of network payload types. More and more, telecommunications networks are utilizing IP as a fundamental protocol, particularly in the network backbone. This backbone is also making VoIP services viable as an alternative to circuit-switched voice.

There is more to SoftSwitch, however, than simply separating the functional components of the network. Another key goal is to create an open-service, creation-development environment so that application developers can create universal products that can be implemented across an entire network. Part of the evolution will include the development of call control models that will seamlessly support data, voice, and multimedia services. The result of widespread SoftSwitch deployment will be a switching technique that does not have the same restrictions that plague circuit switches, such as intelligent network triggers, application invocation mechanisms, and complex service logic.

This functional distribution will result in faster and more targeted feature development and delivery, and significantly lower service-delivery costs. Softswitches will be architecturally simpler, operationally efficient, and less expensive to operate and maintain.

A number of corporations have now focused their efforts on the development of Softswitch products. Both Lucent and Nortel announced SoftSwitch products as early as 1999 in the form of the 7R/E (Lucent) and the Succession (Nortel). Following the collapse of the telecom bubble, they were all but forgotten, largely because the market simply wasn't ready to bridge the gap between the circuit-switched Bellheads and the IP bit-weenies. The good news is that with industry recovery has come a renewed interest in SoftSwitch technology, and both companies have come roaring back into the fray with Succession (Nortel) and iMerge (Lucent). Today the market is once again growing, and ferociously. Companies like Taqua Systems are beginning the process of reinvigorating the SoftSwitch marketplace. A key target sector is the small, rural telephone company. These companies are ideal SoftSwitch targets because they can take advantage of the incremental deployment nature of SoftSwitch and because they are one of the few segments that have money earmarked for capital growth.

SoftSwitch was originally intended as a local switch-replacement technology to facilitate the circuit-to-packet migration. When SoftSwitch originally rolled out, the economy was in turmoil and progress has not been as aggressive as the sector would like. In truth, SoftSwitches have been used to replace lines, not switches in a wholesale manner, and although the incumbents are not buying, they are shopping with RFPs and RFQs—a good indicator of later movement. To date the main applications that SoftSwitch solutions have addressed are enterprise applications such as PBX management, IP Centrex, VoIP, and wireless LANs.

The IP PBX

The call-processing server, or IP PBX, provides the central point of functionality for all VoIP calls. It has the responsibility to terminate all calls. Note that the IP PBX does not actually manage the call payload itself, which is typically a stream of RTP packets, which was described earlier. The call-processing server performs signaling and supervisory functions such as conferencing, transfer of calls from one call server to another, and providing ancillary services such as music-on-hold services. VoIP is a client-server architecture under which the IP PBX is the server and the user devices are clients, but that being said, all traffic flows peer to peer, terminal device to terminal device. The end devices manage traffic-flow requirements, and the call-processing servers interact with the network to negotiate the transmission parameters required for optimal QoS. One advantage of these networks compared to their PSTN predecessors is that the call processor or IP PBX is typically a software module that runs on one or more dedicated servers, giving the customer tremendous flexibility and the ability to provide enhanced levels of redundancy.

The Gateway

However, there's a piece missing: the interface that must exist between the IP domain and the PSTN, without which the chasm between the two would be impossible to bridge. This function is the responsibility of the gateway, which comprises three functional segments: the standard line or trunk interface for communication with traditional PSTN devices, the IP interfaces for communication with VoIP environments, and a gateway function in between that converts between the two and communicates with the SoftSwitch or call server to perform call setup. In many cases the two functions (gateway and SoftSwitch/call server) are integrated into a single device, but logically are stand-alone, interoperable components.

VoIP Gateways and Gatekeepers

Because of the ongoing need to interface the PSTN to the IP voice environment, a device is required that can perform the analog-to-digital conversion that must take place if traffic is to be seamlessly and transparently shared between the two networks. This device is the gateway, and in addition to its responsibilities for protocol conversion, it can also collect and report on network usage, perform echo cancellation, carry out silence suppression as required, and compress voice streams.

In essence, the gateway is the source of voice traffic. In most cases a voice conversation or session comprises an IP session that is transported inside a stream of RTP packets carried over UDP.

A word about terminology: "gateway" and "gatekeeper" have historically represented two different devices performing two distinct tasks. The gateway has always been the device described in the preceding paragraphs, while the gatekeeper's functions included bandwidth management, call admission, and call control—clearly gatekeeper responsibilities. As VoIP functionality has evolved, however, gatekeeper functions have been absorbed by the gateway, with both sets of functions now being performed by a single device.

Gateways incorporate a variety of functions that bridge the gap between the two communications environments and may come in a variety of forms. In some cases the gateway is a stand-alone rack-mounted device in a closet; in other cases it may be a PC that is set aside to run VoIP software and is beefed up with memory and a fast processor. The functions that a gateway performs include access gateways, which serve to connect a legacy PBX to the VoIP network; a trunk gateway, which provides for the interconnection of the PSTN to the VoIP network; a local or residential gateway, which is nothing more than a touch point for VoIP networks so that DSL, cable modems, and broadband wireless access services can reach the VoIP environment; and a media gateway, which offers an interface to the VoIP network for a digital PBX or "soft" (software-based) PBX.

User Devices

Whereas SoftSwitch takes care of switching and call-management functions and the gateway performs the protocol conversion between the PSTN and IP, the handsets provide user functionality. Most modern gateways are capable of supporting standard analog devices, but most new VoIP installations are designed to exclusively support IP phones. These phones come in two forms: a physical device that sits on the desktop and plugs into the Ethernet LAN or a logical softphone that runs on the desktop or laptop as an application. A softphone is nothing more than a software application that runs on a PC. It is typically designed to support mobile users who wish to avoid the need for a physical telephone device. The features are the same as those found on a typical enterprise telephone, and the application commonly runs in the background on the desktop so that it can be accessed at will and so that callers can reach the user on demand. Either way, the IP phone serves not only as the user-access device but also as the gateway, bridging the chasm between the PSTN and IP domains. "Telephone" may therefore be something of a misnomer under the circumstances; the device a user employs to access the network may be a physical device such as an IP telephone or a softphone running in a desktop or laptop PC.

Each VoIP telephone, regardless of whether it is a physical device or logical construct, is assigned an IP address for the subnet on which it is installed. These telephones therefore rely on the TCP/IP protocol suite for functionality but may also rely on other protocols for added capabilities such as directory, instant messenger (IM)/short message service (SMS), or screensharing applications. They rely on DHCP to auto configure. DHCP, the Dynamic Host Configuration Protocol, gives network administrators the ability to centrally manage and automate the assignment of IP addresses in their managed network. When an enterprise connects its users to the Internet, an IP address must be assigned to each machine. Prior to the arrival of DHCP, the IP address for each machine had to be entered manually, and if machines were moved from one location to another in a different part of the network, a new IP address had to be assigned—the famous MAC problem.

DHCP allows a network administrator to supervise and administer IP addresses from a central location and to automatically assign a new IP address whenever a computer is plugged into a different port on the network.

DHCP relies on a "leasing" technique that assigns a specific amount of time for which a given IP address can be "legally" used by a computer. The lease time varies depending on the nature of the Internet connection at a particular location and can be renewed automatically if desired. This technique is helpful on college campuses and in business parks that rely on "hoteling," that is, where users change their physical locations on a regular but often random basis. DHCP is a highly desirable and efficient addition to the network manager's toolkit. By using very short leases, DHCP can dynamically reconfigure a network in which more computers than available IP addresses are found!

DHCP also supports static addresses for computers serving as Web servers, which naturally need a permanent IP address. In practice, the

DHCP server tells the phone where to find the configuration server, which is typically collocated with or resident within the call-processing server.

"When we went through our feasibility study and vendor selection process, there were a number of questions that we wanted answers to," says Aaron Videtto. "We asked the vendors these kinds of things: Can the phone perform traffic classification to ensure that voice receives its proper level of treatment? What standard is used for signaling? Is there a secondary Ethernet port on the phone so that I can plug my laptop into it to gain access to the broadband network? Does the phone support Power over Ethernet (PoE) so that my users are protected in the event of a power failure? What is the interface like—is it intuitive and easy to use? Is it hard to install and configure prior to use? Even though we didn't necessarily plan to implement based on positive responses to 100 percent of these questions, they gave us benchmark data for comparing one vendor to another. And that was important."

The IP phones in common use today are extraordinarily flexible and support a wide range of software customization capabilities. Keep in mind that the phone is the user's "window on the services world;" it is essential that the view be as broad as it can possibly be.

VoIP Functionality: Integrating the PBX

An enormous installed base of legacy PBX equipment exists, and vendors did not enter the IP game enthusiastically. Early entrants arrived with enhancements to existing equipment that were proprietary and expensive, and they did very little to raise customer awareness or engender trust in the vendor's migration strategy. Over time, however, PBX manufacturers began to embrace the concept of convergence as their customers' demands for IP-based systems grew, and soon products began to appear. Most have heard the message delivered by the customer: Preserve the embedded base to the degree possible as a way of saving the existing investment, create a migration strategy that is seamless and as transparent as possible, and preserve the features and functions that customers are already familiar with to minimize churn.

Major vendors like Lucent Technologies and Nortel Networks have responded with products that do exactly that. They allow voice calls and faxes to be carried over LANs, WANs, the Internet, and corporate intranets. Their products function as both gateway and gatekeeper, providing circuit-to-packet conversion, security, and access to a wide variety of applications, including enhanced call features such as multiple-line appearances, hunt groups, multiparty conferencing, call forward, hold, call transfer, and speed dial. They also provide access to voicemail, computer telephony integration (CTI) applications, wireless interfaces, and call-center features.

The Private Branch Exchange (PBX)

Because the PBX is such a central component of the typical enterprise telephone environment, we will spend a few pages describing the typical PBX—what it is, how it works, and how it integrates into (and is also independent from) the PSTN. This discussion is particularly important for our later discussion of the IP PBX.

A PBX is really nothing more than a small, premises-based switch that is connected to the PSTN via a group of high-speed trunks. The number of trunks is determined by the size of the enterprise and the anticipated volume of calls that must pass to and from enterprise employees and the outside world. It is typically located on the premises of an enterprise facility.

Numerous advantages exist to having a PBX. First and foremost, it represents a telephony switch that can be custom configured for the enterprise, offering company-specific calling services, interlocation toll-free calling, and reduced dialing requirements. For example, a small office of several hundred people might have a dedicated PBX on the premises. Because all employees are logically located "within" the PBX, and because they all share the same prefix, they can call each other by dialing nothing more than the last four digits. For example, a corporate PBX may use prefix 823 for all numbers between 0000 and 9999. Therefore, the person with phone number 823-7677 can call the person with number 823-8119 by simply dialing 8119.

A larger PBX for an enterprise of several thousand people may host multiple prefixes, in which case employees must dial a single-digit prefix in addition to the four-digit telephone number.

If an employee wishes to call someone who is outside the PBX, the process is simple and straightforward. To gain access to an outside dial tone (that is, a dial tone delivered from the PSTN rather than from the PBX, sometimes referred to as a foreign exchange dial tone), the person must first (typically) dial a 9. By almost universal convention, this tells the PBX that the caller wishes to leave the domain of the PBX and connect to a number that is hosted elsewhere.

So in effect a PBX is really nothing more than a privately owned central office, hence the name "private branch exchange." An exchange, you will recall, is the old name for a telephone switching office.

PBX Anatomy

A typical PBX (see Figure 3-22) includes a management console that is used to perform administrative tasks on the intracorporate network; a switch matrix and call-processing capability used to establish, maintain, and tear down calls; trunk interfaces that facilitate interconnection with the PSTN; line interfaces that provide connectivity to the feeder cables, which deliver service to users throughout the enterprise location; desktop cabling plant that feeds the actual telephones, fax machines, and other devices; call management and accounting software; and a collection of enterprise-selected enhanced services.

PBX Connectivity

Two primary connections exist on the PBX, both of which were mentioned earlier. The first is the customer premises equipment (CPE) line card, which connects the desktop telephone to the PBX. Each phone is typically connected to a port on a line card in the PBX, although multiline phones are naturally connected to multiple ports.

Depending on the nature of the telephones being deployed throughout the enterprise, the line cards will be either analog or digital, matching the nature of the device to which they grant access to the network.

The back side of the PBX offers connectivity for the trunk interfaces, which provide connectivity between the PBX and the PSTN. These trunks can be configured in a wide variety of ways using an equally wide array of technology options that includes ISDN, T1, E1, and SONET/SDH.

In addition to PSTN connectivity, the trunk interfaces also provide access to a range of enhanced feature sets. Most of these features are based on caller ID functionality and include such capabilities as basic caller ID and *direct inward dialing* (DID). In the case of caller ID, two options are available. The first is *multiple data message format* (MDMF),

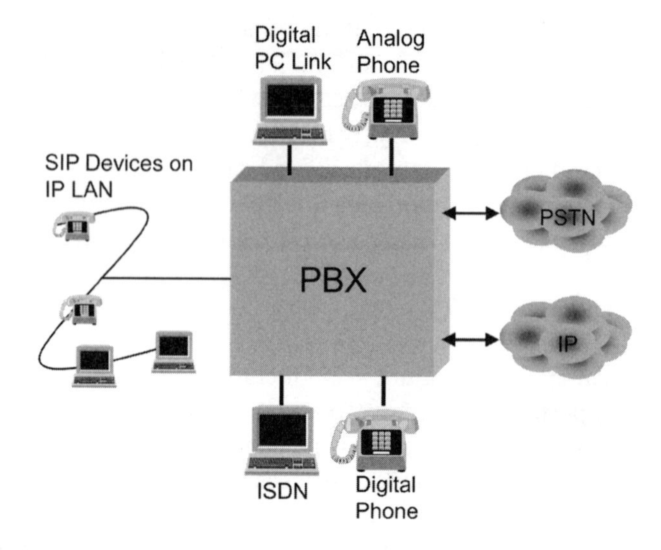

Figure 3-22 Typical PBX interfaces

which makes the caller's name and number visible to the called party, and *single data message format* (SDMF), which displays the calling number only. For purposes of completeness, we should note that caller ID information can also be delivered through a PBX environment using *automatic number identification* (ANI) and *dialed number identification service* (DNIS). Both are available as enhanced services.

DID allows outside callers to reach an individual behind the PBX without having to first speak with an attendant or go through an automated call director. DID setup is fairly straightforward; when an enterprise such as Champlain College wants to deploy DID trunks, the telephone company assigns blocks of twenty 10-digit numbers to the customer and assigns them to dedicated DID trunks connected to the PBX. When an outside caller places a call to a DID number, the PSTN strips off all but the last three or four digits of the 10 dialed digits during the call-setup process and forwards the reduced digits to the PBX. The PBX in turn routes the call to the extension associated with the dialed number.

It is important to note that analog DID trunks cannot be configured as two-way trunks—they are inbound only. For enterprise installations that have multiple locations or multiple PBXs that must be interconnected, the interconnection is typically done using one or more multichannel carrier (T1 or similar) connections connected to specific interfaces on the PBX. These interfaces were designed to connect to and interoperate with large central office switches. This means that at least one of the PBXs in the enterprise must perform the equivalent of local central office signaling to ensure interoperability between the two (or more) communicating devices. When IP is introduced into the mix, a similar interoperability exercise must be undertaken to ensure that the legacy equipment and the IP gateway devices are able to communicate properly.

PBX Feature Sets

Because PBX users were trained to expect certain things from their telephone system before they were served out of a PBX, the PBX must offer a similar set of capabilities if it is to be viewed as carrier grade. Some of these features offer nothing more than convenience and time saving for the user; others are considered mission critical for enterprise users. The most common of them includes last number redial, call transfer, three-way calling, mute, speakerphone, call forward, call pickup, and an alphanumeric display. A number of so-called value-added features enjoy widespread uptake, including voicemail, automated attendant, CTI connectivity, and conference bridging. To a large extent these features are integrated within the PBX, but not always. In some cases (and in fact, in growing numbers), they are delivered to the PBX via high-speed interfaces between the PBX and a series of application-specific adjunct devices. This model serves to create something of a building-block approach to service and application delivery, but can also complicate the overall architectural design and add substantively to the cost of implementation.

We will discuss the automated attendant feature in greater detail because it has a powerful impact on customer-service enhancement if implemented properly.

The Automated Attendant Feature

The question of whether or not to go with an automated attendant feature was a matter of significant discussion at Champlain College. "Our customer-service folks were very reluctant to go with an electronic voice as opposed to a human on the line, because they were worried about the perception it would create among the customers who called in," says Aaron Videtto. "So we did data collection at the main number of the switchboard for several months to verify the situation regarding calls placed versus calls that were actually answered or dropped. What we found was interesting and was in fact what we expected: About half the time the switchboard was helping people that could have easily helped themselves through a well-designed IVR [interactive voice response] interface! So after we walked them through the process and showed them how the customer interaction would actually work and after we showed them the level of response they could reasonably expect from the system, management agreed to let us put in the IVR. We tracked it like a hawk: In fact, I ran a report starting about 5 months into the process, and it demonstrated that the IVR was very effective in terms of handling requests and reducing the number of hang-ups. It was a good idea. However, we spent a *lot* of time designing it to make sure it delivered what we wanted it to deliver."

The auto-attendant feature provides a mechanism for callers to manage their own destiny—to route themselves to the information they seek or to a particular person they would like to talk to. Auto-attendant relies on in-band dual-tone multifrequency (DTMF) signaling to control intrasystem routing.

Auto-Attendant Design

The auto-attendant feature is a powerful adjunct to any phone system, but it can be problematic. One of its key advantages is its ability to reduce dependency on human operators, which clearly reduces expense in the enterprise. However, it can also become a distancing mechanism and at worst can cause serious frustration on the part of callers who have to navigate an exhaustively complex and deep system of routing commands. At Champlain, Videtto took pains to work with vendors to ensure that the menu design was complete, but not so exhaustively complex that customers calling in for information had to spend too much time navigating their way through the system.

"We created a list of questions that we wanted answers to before we put the thing online," says Videtto. "We wanted a complete system, but we didn't want too many menu levels, because we didn't want people getting lost or *feeling* lost in the system. We wanted to be able to customize the menus according to the time of the year. Because admissions is our busiest time, and incoming freshmen and their parents always have questions that they won't ask during other times of the academic year, we wanted to be able to modify the information to best accommodate the needs of the customer.

"We were also concerned about how customers could most effectively escape from the system to seek refuge with an operator if they needed to," he added. "We wanted system bypass capability for regular callers, an easy-to-use directory, and the ability for the customer to get back to the same person they spoke with previously. The good news is that we were able to do that and it works very well. But I caution you: It took a lot of time and a lot of planning effort. However, we wouldn't do it any other way. These are customers we're talking about, after all."

Because the legacy PBX relies on a wholly centralized architecture, the degree to which it lends itself to be well managed and is flexible is somewhat limited. Customers are often heard to say, "Sure, my PBX is designed for adaptation. I have to adapt my business operations to it." The traditional PBX relies on centralized intelligence, the end result of which is that management and control functions are carried out from a single console. The console is typically connected directly to the PBX and does not allow remote access, and as a result system management or changes require maintenance and support personnel to be on site. For single-location enterprises this may not present a problem. But for multisite corporations it rapidly becomes prohibitively complex and expensive. *Moves, adds, and changes* (MACs), which should not be particularly complicated, become inordinately costly and complicated because they require highly trained personnel who are often either contractors or resources provided by an outsourcing agent. Either way they increase the overall cost of PBX operations.

The complexity does not end with the PBX. Most legacy digital telephones designed for use with a PBX use proprietary communications protocols, which lock the customer into a single vendor and ultimately drive up the price of the overall system. A company with four locations and four different PBXs may discover that a simple employee move becomes a logistical nightmare as it may require interaction with a plethora of noninteroperable components and interfaces.

Finally, we have to consider the issue of scalability. PBXs are traditionally configured and built for enterprise customers that fall within a specific size range. In the event that the enterprise should add personnel or grow the business, it will find itself in a position where it may have to forklift upgrade to a higher capacity switch, the cost of which may be more than a small- to medium-size business can handle.

Flexibility, Thy Name Is IP PBX

The obvious benefits of an IP-based PBX are lowered *total cost of ownership* (TCO), greater flexibility, and more effective manageability. With an IP PBX and the converged infrastructure that results, voice and data share the same network infrastructure. This offers the potential result of a significant reduction in operating expenditures (OPEX) and some capital expenditures (CAPEX). And because IP PBX environments rely on standards-based systems that are commonly available and therefore inexpensive, the cost of ownership drops.

According to most network administrators, MAC activity represents one of the most significant recurring cost elements they have to deal with. In an IP PBX system, MAC cost is roughly a third of the cost of engaging in the same activity under a legacy PBX regime. One reason for this is the virtual nature of a system that is based on the IP protocol: There is no inviolable physical relationship between the telephone number and the local loop or device. If a user wants to move from one building, campus, or country to another, he or she simply unplugs his or her phone, moves, and plugs it back in. The system immediately recognizes and registers his or her new logical location.

"Our PBX is so simple to look at, it almost looks like a toy, yet it delivers amazing capabilities to us," says Professor Dave Whitmore. Dave is one of Champlain's telecommunications wizards, and he walked me through the school's equipment rooms. Here's the PBX," he says, gesturing to a small rack of equipment (see Figure 3-23). "That's all there is to it. The bottom racks are our analog DIDs [Direct Inward Dialing] and failsafe equipment, along with the E911 server just above them. Then we have the Net6 push server that takes care of delivering the Web content to all 750 or so campus phones, and just above that is the speech-to-text server that allows callers to say the name of the person or department they are looking for and the call director will connect them. We then have the call director server, the voicemail server, the call center publisher, and the remote keyboard video mouse (KVM) switch for moving between functions. That's it! All those functions you're writing about are done by that little pile of servers."

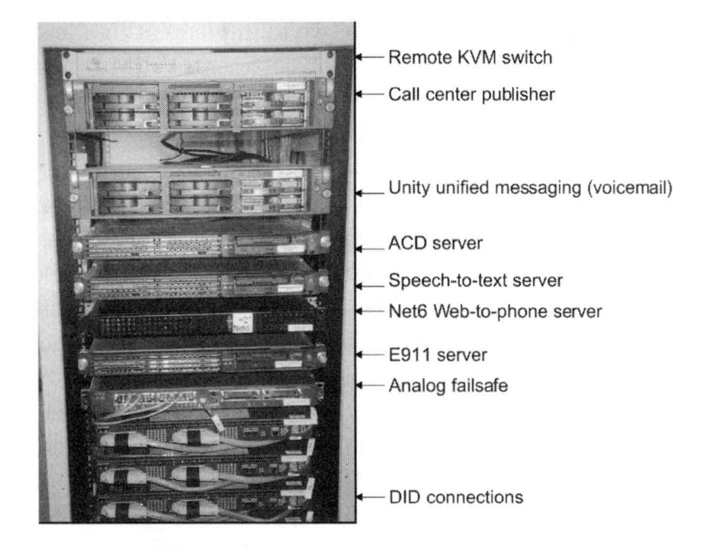

Figure 3-23 The Champlain College IP PBX

PBX-to-IP PBX Transition Challenges

The physical and logical interconnection of the IP PBX into the legacy world (or vice versa, depending on perspective) is relatively straightforward. More complex is the integration of legacy applications into the IP-enabled environment. A critical consideration is how the IP PBX will support the preexisting voicemail system, if in fact that is an issue that requires consideration. Two approaches have emerged. The first is to keep the legacy voicemail system and the IP voicemail systems completely separate (assuming that both legacy and IP PBXs will coexist), using standard interchange protocols to pass messages between the two as required. These protocols include the *Audio Messaging Interchange Specification* (AMIS) and the *Voice Profile for Internet Messaging* (VPIM). VPIM emulates the DTMF signals typically generated by the phone to call-forward voicemails from one AMIS-compliant system to another, using trunks established between the two systems for that purpose. VPIM, on the other hand, provides a standard *multipurpose Internet mail extensions* (MIME) encoding scheme that allows voicemail messages to be sent as multimedia attachments over the LAN or WAN.

AMIS is the simpler of the two to implement in a legacy environment and requires significantly less investment. VPIM is a more recent standard but is typically offered as an add-on and is therefore more costly (and complex) to implement than is AMIS.

The alternate technique to the intersystem gateway described previously is to either preserve the existing voicemail system on the PBX or to migrate all users to voicemail on the IP PBX.

Regardless of the system that is ultimately chosen, certain key functions must be preserved. For example, the message-waiting light must be activated and deactivated. Legacy telephone sets should indicate the presence of voicemail messages, regardless of whether the device is attached to a legacy or IP PBX. The *simplified message desk interface* (SMDI) was specifically developed for serial connections between voicemail systems and PBXs. SMDI indicates the presence of voicemail for a particular extension to the management interface of the legacy PBX. The PBX uses this information to notify the CPE that its message-waiting light should be illuminated. Once the message has been retrieved, the lamp should be extinguished using the same mechanism.

So, how does the VoIP architecture work to provide calling services? The same way the PSTN does, in essence. The three major components of the converged VoIP network—the SoftSwitch, the media server, and the IP telephones—intercommunicate with one another to perform call setup. When the phone goes off-hook, it transmits a state change message to the SoftSwitch, which then communicates with the phone to collect the dialed digits (the address), select a route for the call, enter the appropriate state information into a table, and proceed with the steps necessary to set up the actual call. The SoftSwitch easily converts back and forth between the two native address modes and establishes the call as requested by the caller. Once the call is established, the SoftSwitch bows out of the equation, its job done—for the moment.

Assuming that the call is placed to an off-net number (outside the enterprise IP network), the gateway performs the appropriate conversion of the IP packets to make them palatable to the PSTN. If, on the other hand, the call is on-net and is to be routed to another IP phone, then no circuit-to-packet conversion is required, although the packets may be handled by more than one SoftSwitch.

So how is a telephone call carried across an IP network? In very much the same way that a call traverses the PSTN. A customer begins the call, often using a traditional telephone. The call is carried across the PSTN to an IP telephony gateway, which is nothing more than a special-purpose router designed to interface between the PSTN and a packet-based IP network. As soon as the gateway receives the call, it interrogates an associated gatekeeper device, which provides information about billing, authorization, authentication, and call routing. When the gatekeeper has delivered the information to the gateway, the gateway transmits the call across the IP network to another gateway, which in turn hands the call off to the local PSTN at the receiving end, completing the call.

Let's also address one important misconception. At the risk of sounding Aristotelian, "All Internet telephony is VoIP, but all VoIP is not Internet telephony." Got it? Let me explain. It is quite possible today to make telephone calls across the Internet, which by definition are IP-based calls simply because IP is the fundamental operational protocol of the Internet. However, IP-based calls can be made without involving the Internet at all. For example, a corporation may interconnect multiple locations using dedicated private-line facilities, and they may transport IP-based phone calls across that dedicated network facility. This is clearly VoIP but is *not* Internet telephony. There's a big difference.

VoIP Services Evolution

There was a time when VoIP was only for the brave—those intrepid explorers among us who were willing to try Internet-based telephony just to prove that it could be done, with little regard for QoS, enhanced services, and the like. In many ways it was reminiscent of the 1970s, when we all drove around in our cars and talked to each other on CB radios. The quality was about that good, and it was experimental, and new, and somewhat exciting. It faded in and out of the public's awareness but didn't really catch on in a big way until late 2002 or so, when the level of broadband penetration reached a point where adequate access bandwidth was available and CODEC technology far enough advanced to make VoIP not only possible but actually quite good. Everyone toyed with the technology. Cisco, Avaya, Lucent, and Nortel, not to mention Ericsson and Siemens, all built VoIP products to one degree or another, because they knew that sooner or later the technology would become mainstream. But as always happens, it took *early introducers* (not early adopters) to make the transition from experiment to service. And although companies like Telus in Canada played enormous roles in the ongoing development of VoIP (they were the first company to install a systemwide IP backbone, and one of the first innovators to offer true VoIP services), it took a couple of somewhat off-the-wall companies to really push it to the front of the public consciousness. The first of these was Vonage; the second, Skype.

Vonage

Vonage, "the Broadband Phone Company," was founded in January 2001 in Edison, New Jersey. With 600 employees, Vonage (www.vonage.com) completes more than five million calls per week and has more than 400,000 active lines in service. To date they have completed more than 200 million calls over their network, which is Internet-based VoIP.

Signing up for the service could not be easier, because it is all done at the Vonage Web site—no human is needed in the loop. Would-be customers can read about the service and get answers to their questions. And if they decide they want to sign up, they give Vonage a credit card via a secure connection, and the card is billed each month for the service. They then enter their area code (NPA), after which Vonage assigns them a phone number. It is important to note that the number is virtual—that is, although I live in Vermont, I don't *have* to have a Vermont area code (there's only one for the entire state). For example, a good friend of mine lives in New Jersey, but his parents live in southern California. To prevent them from having to pay long-distance charges when they call him, he has a Los Angeles area code. I have even heard about professional people asking for a 212 area code so that they would appear as if they were in New York City for prestige reasons!

Once the phone number has been assigned, a user can then select a wide variety of supplementary services (caller ID and so on) at no extra charge. One thing the user has to do is configure his or her voicemail service, which is included as part of the low-cost package. There are two options; one is the traditional model, where a subscriber dials an access number, enters a PIN code, and then listens to his messages. Alternatively, subscribers can set it up (all done at the Web site) so that received voicemail messages are converted to .wav files and sent to them as e-mail attachments. Yet another example of unified messaging and converged services.

Two options exist for connecting to the Vonage network. One is to use a small terminal adapter that Vonage will mail to you; it is connected to the broadband network, and a phone and laptop are in turn plugged into the box (see Figure 3-24). Alternatively, subscribers can download a SoftPhone (see Figure 3-25) and make calls from their laptop. The SoftPhone boots at startup and simply sits on the desktop. Once the user has logged in he or she can make and receive calls with the SoftPhone application.

I have been a Vonage subscriber for quite some time, and the service could not be easier to use. When I travel I simply plug my laptop into the broadband connection in my hotel room. The Vonage SoftPhone automatically boots up when I start the machine, and as soon as it's online I can start making (and more importantly, receiving) telephone calls. Think about it: *I can be in a hotel room in Mumbai and if someone dials my Vermont-based Vonage telephone number, the phone will ring in India —and the caller has no idea where I am, nor do they need to.* Users beware: Be sure to shut down your laptop when traveling internationally, because the phone will inevitably ring at 3:30 AM because the caller doesn't realize you're traveling abroad!

Vonage offers remarkable service for a remarkably low price, and most of the time it is indistinguishable from traditional circuit-switched voice.

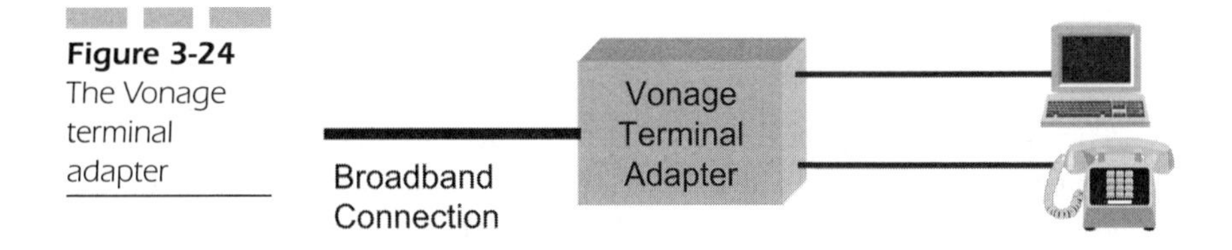

Figure 3-24 The Vonage terminal adapter

Figure 3-25 The Vonage SoftPhone

Their service is so good that other companies have come into the marketplace, including AT&T (CallVantage) and Verizon (VoiceWing). The number of players grows daily.

To date Vonage has approximately 700,000 customers. Let's now turn our attention to another purveyor of VoIP, which has a few more—*roughly 100 million subscribers.*

Skype

> I knew it was over when I downloaded Skype. When the inventors of KaZaA are distributing for free a little program that you can use to talk to anybody else, and the quality is fantastic, and it's free—it's over. The world will change now inevitably.
>
> — Michael Powell, Chairman, FCC

> The idea of charging for calls belongs to the last century. Skype software gives people new power to affordably stay in touch with their friends and family by taking advantage of their technology and connectivity investments.
>
> — Niklas Zennström, CEO and cofounder of Skype

Many readers will recall KaZaA, the file-sharing program that became well known during the turbulent Napster controversy over the unauthorized sharing of commercial music. The same two programmers who created KaZaA, Niklas Zennström and Janus Friis (see Figure 3-26), put their heads together and concluded that although peer-to-peer music works well, it is fraught with legal ramifications. So what else could be shared in a peer-to-peer fashion? *Voice.*

To date, more than 100 million people have downloaded the free Skype installer (www.skype.com/download), and as I sit here writing this section in mid-April 2005, there are 2.2 million people online (see Figure 3-27). I have used Skype for over a year, and the service is excellent. The only downside is that the original, free Skype is a walled garden technology—users can talk only with other Skype users. But with over two million people online at any time, and with the application's ability to search for other users by city and country, *you'll find someone to talk with.* Skype also supports messaging, voicemail, and videoconferencing.

Of course, the walled garden concept is limiting, so in early 2004 Skype announced SkypeOut and SkypeIn, paid services that allow Skype

Figure 3-26 From left, Niklas Zennström and Janus Friis, the founders of Skype

users to place calls to and receive calls from regular (non-Skype) telephones in about 20 countries (and growing) for about $.02 per minute.

Vonage and Skype are but two of the growing herd of public VoIP providers that are redefining the world of voice services. In the enterprise space, VoIP is having an equally dramatic impact. Let's examine the various ways in which enterprise VoIP can be deployed to the benefit of customers.

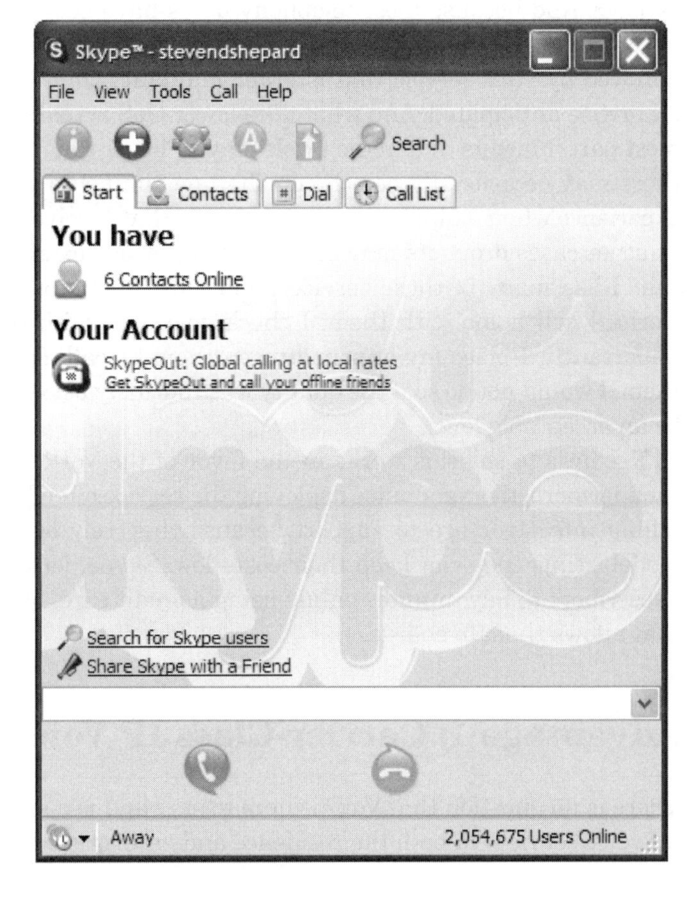

Figure 3-27 The Skype user interface. Note the number of users online (bottom right corner).

VoIP Implementation

IP voice has become a reasonable technological alternative to traditional PCM voice. VoIP gateways are in the late stages of development with regard to reliability, features, and manageability. Consequently, service providers wishing to deploy VoIP solutions have a number of options available to them. One is to accept the current state of the technology and deploy it today while waiting for enhanced versions to arrive, knowing that although the product may not be 100 percent carrier grade, it will certainly be an acceptable solution.

A second option is a variation of the first: Implement the technology that is available today, but make no guarantees of service quality. This approach is currently being used in Western Europe and in some parts of the United States as a way to provide low-cost long-distance service. The service is actually quite good, and although it is not toll quality, it is inexpensive. And because most telephony users have become inured to the lower-than-toll-quality of cellular service, they are less inclined to be annoyed by VoIP service that is lower in quality than they might have otherwise anticipated. And what does lower QoS actually mean? For the most part it means that audio levels may be lower than expected, or that there may occasionally be echo on the line similar to what we used to experience when making international calls that involved a satellite hop. In other cases dropouts may occur, although they are rare. The truth is that I use many of these services, and I use them daily—all over the world. I call home with them, I check voicemail with them, but most importantly, I place my imprimatur on them by calling *customers* with them. I would not do so if the quality was routinely below what I deem to be an acceptable level.

Needless to say this works in the favor of the VoIP service provider. Furthermore, the companies deploying the service often have no complex billing infrastructure to support because they rely on prepaid billing models, thus they can keep their costs low. Skype, for example, allows subscribers to buy minutes online using a credit card, one of the factors that allows them to charge.

Advantage 1: Carrier-Class IP Voice

There is no question that VoIP is here to stay and is a serious contender for voice services in both the residence and enterprise markets. The dollars don't lie: Domestic revenues for VoIP will grow exponentially between 2004 and 2009, from $600 million today to $3 billion.

From a customer's perspective, the principal advantages of VoIP include consolidated voice, data, and multimedia transport; the elimination of parallel systems; the ability to exercise PSTN fallback in the event that the IP network becomes overly congested; and the reduction of long-distance voice and facsimile costs.

For an ISP or competitive local exchange carrier (CLEC), the advantages are different, but no less dramatic, and include the efficient use of network investments due to traffic consolidation, new revenue sources from existing clients because of demand for service-oriented applications that benefit from being offered over an IP infrastructure, and the option of transaction-based billing. These advantages can collectively be reduced to the general category of customer service, which service providers such as ISPs and CLECs should be focused on. The challenge these service providers face will be to prove that the service quality they provide over their IP networks will be identical to that provided over traditional circuit-switched technology.

Major carriers are voting on IP with their own wallets, a sure sign of confidence in the technology.

As we observed earlier, the key to IP's success in the voice-provisioning arena lies with its invisibility. If done correctly, service providers can add IP to their networks, maintaining service quality while dramatically improving their overall efficiency. IP voice (not to be confused with Internet voice) will be implemented by carriers and corporations as a way to reduce costs and move to a multiservice network fabric. Virtually every major equipment manufacturer, including such notables as Lucent, Nortel, and Cisco, have added SS7 and IP voice capability to their router products and access devices. These manufacturers recognize that their primary customers are looking for IP solutions. They have recognized that interdependencies exist among SS7, SIP, H.323, and other signaling protocols, and they have embraced them.

Advantage 2: IP-Enabled Call Centers

Ultimately, well-run call centers are nothing more than enormous routers. They receive incoming data delivered using a variety of media (phone calls, e-mail messages, facsimile transmissions, mail order) and make decisions about handling them based on information contained in each message. One challenge that has always faced call-center management is the ability to integrate message types and route them to a particular agent based on specified decision-making criteria such as name, address, telephone number, e-mail address, automatic number identification (ANI) triggers, product purchase history, the customer's geographic location, or language preference. This has resulted in the development of a technical philosophy called *unified messaging.* With unified messaging, all incoming messages for a particular agent, regardless of the media over which they are delivered, are housed centrally and clustered under a single account identifier. When the agent logs into the network, his PC lists all of the messages that have been received for him, giving him the ability to much more effectively manage the information contained in those messages and respond to service requests on a highly personal level.

Today, unified messaging systems also support road warriors. A traveling employee can dial into a message gateway and download all messages—voice, fax, e-mail—from that one central location, thus simplifying the process of staying connected while away from the office.

Call centers are undergoing tremendous change as the IP juggernaut hits them. The first of these is a redefinition of the market they serve. For the longest time, call centers have primarily served large corporations because they are expensive to deploy. With the arrival of IP, however, the cost is dropping precipitously, and all major corporations are moving toward an IP-enabled call-center model because of the ability to create unified application sets and to introduce enormous flexibility into their calling models.

Chapter Summary

IP is here to stay and is profoundly changing the nature of telecommunications at its most fundamental levels. Applications for IP range from carrier-class voice that is indistinguishable from that provided by traditional circuit-switching platforms to best-effort services that carry no service-quality guarantees. The interesting thing is that all of the various capability levels made possible by the incursion of IP have an application in modern telecommunications and are being implemented at rapid-fire pace. A tremendous amount of hype is still associated with IP services as they edge their way into the protected fiefdoms of legacy technologies, and implementers and customers alike must be wary of "brochureware" solutions and the downside of the Jurassic Park Effect, which warns that just because you *can* implement an IP telephony solution doesn't necessarily mean you *should.* Hearkening back once again to the old telephone company adage that observes, "If it ain't broke, don't fix it," buyers and implementers alike must be cautious as they plan their IP migration strategies. The technology offers tremendous opportunities to consolidate services, to make networks and the companies that operate them more efficient, to save costs and pass savings on to the customer. Ultimately, however, IP's promise lies in its ubiquity, and its ability to tie services together and make them part of a unified delivery strategy. The name of the game is service, and IP provides the bridge that allows service providers to jump from a technology-centric focus to a renewed focus on the services that customers care about.

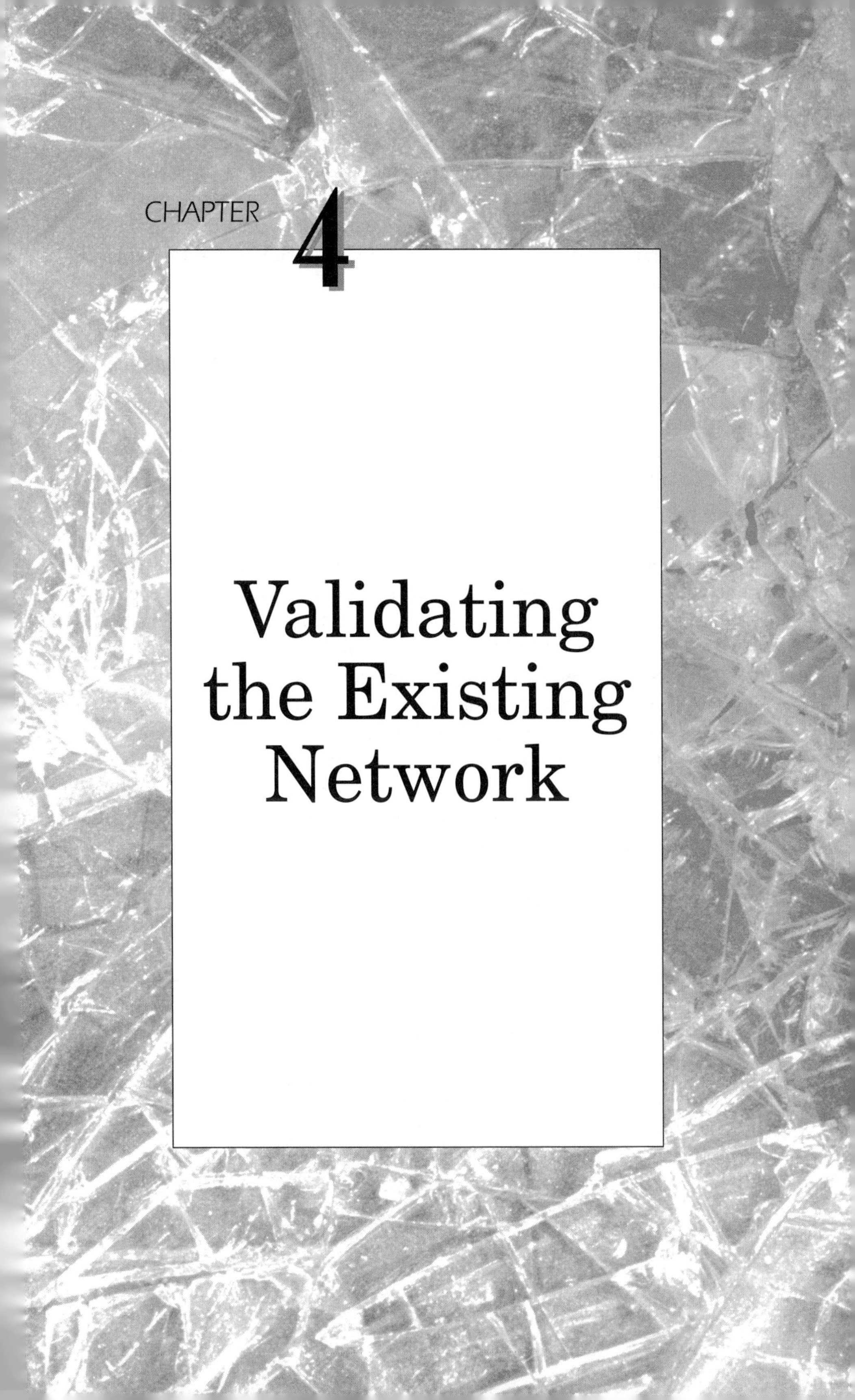

CHAPTER 4

Validating the Existing Network

Most enterprises have a substantial investment in their preexisting communications infrastructure and are therefore loath to lose the value of that investment if they can help it. The evolution to a converged Internet Protocol (IP) architecture does not necessarily imply that the existing installed base must be replaced in toto. Many components can probably be reused, and there is no question that a carefully crafted migration strategy can result in a true migration in which the legacy infrastructure is eased out when (and only when) it makes business sense to do so.

Before the Request for Information (RFI) or Request for Proposal (RFP) process is undertaken with vendors, it makes sense to perform an asset and capability inventory of the existing network. As with any well-run project, a company should avoid the *Alice in Wonderland* approach ("If you don't know where you're going, any road will take you there") by developing a clear understanding of the current network environment. Convergence is a good thing, but before making a decision to implement it, it makes sense to know why. What capabilities are required that the current network is incapable of delivering? What are the tangible and intangible costs that can be mined out as a result of implementation? Is the existing backbone adequate for current and future needs? Fundamentally, three questions emerge: What are my most important and critical enterprise applications? Among them, which are latency sensitive? And among *those*, how many prioritized traffic flows will I have to create to ensure that they receive the treatment they deserve?

Many companies make the mistake of jumping on the IP bandwagon without going through a thorough analysis of the current state of affairs, the result of which is a complex and expensive effort with minimal return. Four areas should be examined as part of the pre-assessment phase: the physical network, the logical network, network utilization, and usage practices and policies.

The Physical Network

The first question that should be asked is this: How old is the equipment that is potentially to be replaced or upgraded? In the case of Champlain College, the private branch exchange (PBX) had reached the end of its useful life and was in fact actively consuming itself. "The PBX had gotten so old and unstable that it was randomly sending spurious current into itself and shorting out the line cards," said Aaron Videtto. "It got to the point that we couldn't rely on it, so for us the decision was an easy one." Of course, in most cases the decision is more complex than that. The issue isn't simply one of dealing with failed or extraordinarily old and nonfunctional network components; it is one of using equipment that is halfway depreciated or of an older vintage of technology that will become obsolete sooner rather than later. It will soon reach a stage where its ability to support emerging applications will be seriously jeopardized by its age and technological capabilities. This falls under the "Only a fool brings a knife to a gunfight" rule: There comes a time when a decision must be made. It is important therefore to aggressively examine the network and determine which components must be replaced and which can simply be upgraded.

It is also critical to look at the investment in inventory. Many large corporations keep a ready supply of plug-ins and other components on hand, and they can represent a substantial amount of money. Like the equipment into which it may be installed, the inventory must also be examined in terms of its own vintage and future value; this must be part of the assessment equation.

Another key question has to do with environmental support. How many vendors are currently involved in maintaining and running the network? Is it a single-vendor environment, or are there multiple vendors involved in the ongoing operation and maintenance of the infrastructure? Are any of the applications that currently run on the network in any way proprietary, and are they therefore so hardware platform dependent that migration to an alternate vendor could result in significant application porting costs (assuming that it is even possible)?

The applications themselves, and the way in which they interact with each other, can also present concerns that must be investigated. As convergence occurs and quality of service (QoS) becomes more visible and important, questions of traffic management come to the forefront and cannot be ignored. Which kinds of traffic policies are already in place that govern such factors as traffic shaping, connection admission control, flow management, and prioritization? How is traffic analysis performed, and how must it be done in the future network? Are packets inspected at wire speed, or is there substantial buffering? Finally, how are security requirements enforced? What policies are in place and must be continued or augmented in the new network, including such functions as authentication, signaling, authorization, access auditing, and nonrepudiation? Is stateful packet filtering performed? *Stateful* typically refers to a process carried out by the firewall, which spends a great deal of its time monitoring packet information at Layers One through Four. However, it also has the ability to monitor goings-on at higher layers such as the Application Layer, carefully monitoring the activities of inbound data. If an inbound packet being monitored matches an existing firewall rule, the packet is allowed to pass through the firewall, and its arrival is duly noted in the state table. That initial packet "sets the rules" for other packets that follow. Because they follow the same set of rules, they are allowed access without examination. Only the IP (address) and Transmission Control Protocol/User Datagram Protocol (TCP/UDP; port number) data is inspected and verified against the data stored in the state table. Stateful management improves firewall performance and is a better technique for packet inspection than proxy-based techniques that must examine every inbound packet in excruciating detail.

The Logical Network

Equally as important as the examination of the physical network is a careful audit of the logical infrastructure. This includes an examination of the physical layout of the network itself, a view of the topology deployed over the physical network. (For example, a point-to-point logical circuit can be deployed over a physical ring, as shown in Figure 4-1.) An infrastructure audit would also include a study of the traffic patterns analyzed by time of day, regionality, work group, and whatever other parameters make sense. The applications themselves should be examined to determine how they contribute to specific traffic-pattern devel-

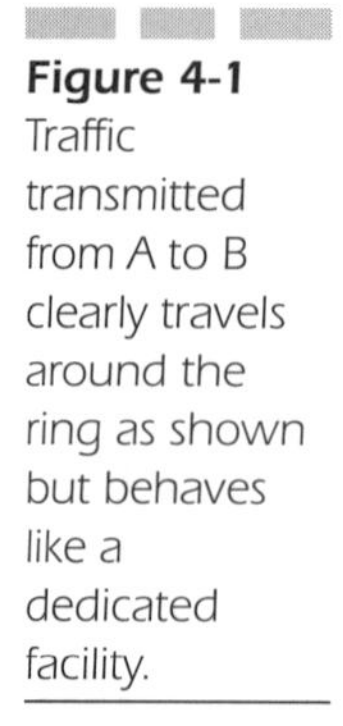

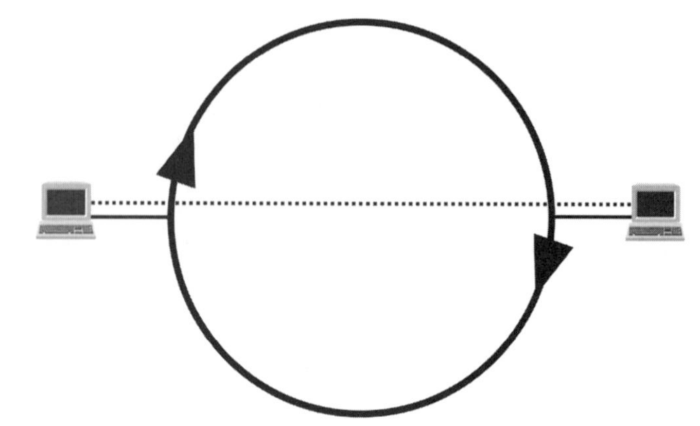

Figure 4-1 Traffic transmitted from A to B clearly travels around the ring as shown but behaves like a dedicated facility.

opment based on the nature of the manner in which they interact with the user community and with each other. For example, a multicast application generates very different traffic flows than a unicast application.

Network Utilization

The point of having a network is very simple: to provide an access mechanism between applications running on one or more systems and the users (both fleshware and software) that depend on them. One of the most important functions is capacity planning. Keep in mind that packet infrastructures tend to be inordinately "bursty" in terms of the manner in which they consume bandwidth. It is therefore prudent to consider such issues as current utilization on an application and workgroup basis, taking into account not only traffic generation related to transaction processing, but also storage, report generation, and routine voice traffic that will soon find itself on the converged network. Some networks permit oversubscription to take place. If that is the case with your network, how will that play out following the move to a converged infrastructure?

Closely linked to capacity planning is network design. If policies and procedures are already in place for network and application design, are they portable to the new environment? Will they be *appropriate* in the new environment, or will other policies need to be written?

Tightly linked to capacity planning and network design are a number of somewhat less tangible issues related to QoS, security, and directory services. QoS is maintained in a variety of ways, most of which are related to either infrastructure control or expectation management. Networks experience QoS problems when points of congestion are mismanaged or when usage policies are improperly implemented. For example, servers that host applications that experience high usage at certain times of the day can become points of contention due to users' inability to gain access to those applications. Traffic prioritization, if not managed properly, can result in the indiscriminate treatment of diverse traffic types, such as voice packets receiving the same priority as asynchronous data traffic.

QoS can also be directly affected by improperly thought-out security policies. In many companies, security enforcement is still in many ways an ad hoc process done on a work group-by-work group basis rather than as the result of a strong, centralized policy. As part of the evolution, particularly because of the unique vulnerabilities that IP presents as a voice/data hybrid, it is crucial that companies monitor intrusion detection, authentication, connection admission control, and other security-related policies.

Project Posting and Vendor Identification

Once the needs of the evolving business have been identified, the strengths and weaknesses of the existing network categorized, and an assessment of policies and procedures completed, the project can be posted for consideration by a group of bidding vendors. The vendor selection criteria must be carefully crafted to not only ensure that the vendor's product is up to the task of addressing the business needs implicit in the convergence evolution, but that the vendor is in a position to provide ongoing, long-term support for the newly installed system. This is an important and often misunderstood part of the process, and in fact companies have been known to select a vendor based solely on the viability of their technology when in fact the technology is perhaps the least important of all the criteria. Some companies, for example, have a policy that says that if their company's purchases represent more than 20 percent of a vendor's annual revenues, they won't buy from that vendor. It doesn't seem to make logical sense until you think about the decision strategically. If your company represents 20 percent of that company's revenues and you have a bad year and cut back on expenditures, you could put the vendor in financial jeopardy, which would make it impossible for them to provide the level of support you require in your enterprise. It is important therefore to consider vendor selection from multiple, seemingly unrelated (and oftentimes irrelevant) points of view. Vendors are typically evaluated in four key areas: overall cost, breadth and richness of services, financial viability, and overall solution functionality.

"We won't buy from you if we represent more than 20 percent of your annual revenues."

Cost

Overall cost is a measure of flexibility and operational effectiveness, and the manner in which the vendor presents their solution is a measure of how well they understand your "pain points" and business drivers. For example, does the vendor go to significant lengths to understand your current situation and take whatever steps are available to protect any preexisting investments that you, as the customer, would like to protect? Does the vendor offer a variety of solutions or packages that provide "silver, gold, and platinum" options if you request them? Does the vendor offer anything in the way of modeling tools that will help you make a buying decision?

Services

Services are represented not only by the capabilities of the vendor's solution, but also by the added value that the vendor overlays on top of and after the sale. For example, a converged network infrastructure solution is a complex sale and an even more complex implementation. Is your expectation that the vendor will work in lockstep with you before, during, and after the implementation? Will they rely entirely on their own resources for conversion or will they also rely on contracted capability? What are the qualifications of the people who will be your primary implementation resources? What kind of post-sales support will you receive? Will there be resources on site for a period of time following project completion or will you be dependent on remote resources that must be dispatched if they are needed?

Financial Viability

Mentioned earlier, this particular criterion is of utmost importance, particularly in large, long-term installations that are deemed "mission critical." If the vendor does not strike you as being viable, don't consider them in the final running. Keep in mind that the implementation of a complex converged network overlay does not end the day the system is turned up. In essence, it never ends and requires various levels of vendor support throughout its life cycle. Critical questions to ask, then, should revolve around the vendor's business model, survivability, cash position, vision, and commitment to product longevity. For example, to what degree is the vendor committed to building products around open standards? What does the vendor's future vision look like for themselves and their products? Can they tell a compelling and logical story about where they see themselves in three to five years? What customer testimonials will the vendor make available to you? Who are their customers, and what sector do they hail from? Are they exclusively in one sector or another, or do they span the range from enterprise, to carrier, to consumer? Do they work with channels, and if so, what's the nature of the relationship? How long have they been in business? What is their cash position?

Solution Functionality

Related to these basic questions are questions around the offered product, service, or solution. How well does the vendor understand your business, where you are, and where you want to go? How broad are the considerations it addresses? Does the proposal go beyond pure technology to include economic, human capital, end-user, application, and competitive concerns? How effectively does the vendor "push back" in project discussions to raise concerns about potential issues? Do they ask as many questions as they attempt to answer? Have they created and offered a technique for assessing the overall effectiveness of the implementation and distinct, well-defined intervals? Similarly, to what degree has the vendor identified management concerns?

Project Assessment

When filling out the paperwork in the old Bell System for a work-related accident, regardless of how minor, one of the last questions asked on the form was, "How could this accident have been prevented?" If a converged solution is to be put in place, and it is to deliver on its promises of cost reduction, efficiency, and enhanced workplace effectiveness, project managers must create an assessment methodology. Such a methodology should monitor project progress, compare it to one's expectations, and perform a qualitative analysis of the results relative to those expectations. There is no such thing as a perfect, problem-free project, but there *is* such a thing as a well-managed project that experiences minimal problems, typically because of a well-designed assessment and intervention process.

Elements of such a process should include, at a minimum, such qualities as cost, identified benchmarks, a clear list of anticipated (desirable) outcomes, solid, extensive project documentation, and well-crafted flowcharts that identify the interactivity among all logical and physical elements of the project. It should also include a well-designed training program to ensure that the personnel responsible for operating, maintaining, and using the new system can do so with maximum levels of capability. There should also be an audit process in place to track the things that went well, as well as those things that didn't, and track discussions about alternative paths that could or should have been taken. Before any of this is done, however, the project must be handed off to a small collection of vendors in the form of an RFP.

The Request for Proposal (RFP) Process

When formulating the contents of the RFP, it is important to precisely describe the desired system. This is the document upon which vendors will base their detail design and pricing, so if an error is made in the RFP —a module or feature is omitted—the vendor's response to the RFP will not accurately reflect the expectations. On the other hand, one should take care not to "overengineer" the system. The inclusion of features that are unnecessary or that are to be implemented at a later time will add to the cost and reflect badly on cost-benefit analyses. The features should be clearly identified.

When Aaron Videtto and Paul Dusini went through the RFP process with their vendors, they were surprised at the diverse levels of knowledge among them. Although all of them were clearly well informed about convergence technologies and their impact on enterprise operations, they were not equally informed about all the aspects of the environment that Champlain wanted to implement. Some knew more about one aspect, whereas others were more informed about another. Some vendor representatives had worked extensively with educational institutions, whereas others were more experienced with enterprise installations.

It is not prudent to assume any particular level of knowledge or experience when working with vendors. They will clearly all have the knowledge necessary to create a solution and craft it in response to the specifics of the RFP, but the specific knowledge required to most effectively work with a particular industry may be less than anticipated. Be prepared to educate the vendor if necessary.

Components of the RFP

Keep in mind what an RFP is: It is a detailed request, extended to one or more vendors, to offer up a proposal to resolve a specific customer business problem. It should include feature and option definitions, infrastructure requirements, the level of offered pre- and post-installation support, space and power requirements, training, and other important data necessary to make an informed decision. Examples of the types of data that should be included are as follows.

Features and Options

This section should comprise a complete list of both required (primary) and desired (secondary) options. The list should display the nature of the feature or option, an indication of the vendor's compliance or noncompliance with the feature, and a place for vendor comments related to the former. See Table 4-1 for an example.

Table 4-1

A Typical Features and Options Table

Feature/Option	Compliance?	Exception?	Vendor Comments
Power over Ethernet (PoE) according to 802.3af	X		This option has been available since product was first announced.
Redundant power supply		X	This option will be available for upgrade in 3Q05.

Cabling and Wiring

This section should stipulate how the vendor intends to handle the installation or modification of existing inside and outside wiring plant. The RFP should make it very clear that all cabling, material, and work must comply with national electrical code requirements and all local building codes. It should detail whether the distribution plant will be copper or fiber, how it will be installed, and the degree of redundancy that will be taken into consideration during the installation process.

E911 Compliance

Today all states require compliance with E911 regulatory mandates. Enhanced 911, often referred to as E911, service is a North America-specific telephony feature that automatically associates a caller's physical address with his or her telephone number. This association is performed using a reverse telephone directory and provides emergency responders with the location of an emergency without the person who is calling for help having to provide it. E911 has been deployed in most metropolitan areas. Privacy legislation allows emergency responders to obtain the caller's information, even if the caller's number is blocked for caller ID purposes.

A second phase of E911 will allow a wireless or mobile telephone to be located geographically using radio triangulation from the cellular radio network or through a global positioning system (GPS). Other proposed features will allow callers within corporate networks to be located down to a specific office on a particular floor of a building.

Vendors should address their plans for compliance in their responses. "This is one of the main reasons we went with Cisco," says Paul Dusini. "E911 was a huge consideration for us. Because of the nature of the IP network, you can unplug your phone and move from building to building if you want—which is great for a network administrator. We haven't had to do a MAC [media access control] in a year and a half! It's beautiful! Now we still use our analog copper plant for modems, fax machines, and credit card machines, and we do have some copper gateways as well. But here's the real issue: With traditional 911, if you move your phone, you move your physical address, and that's a problem. The PSAP [public service access point] doesn't get updated automatically.

"It turns out that Cisco has a great solution to this called Emergency Responder, which is an enhancement to the Cisco CallManager. Basically, we buy a separate set of DIDs [Direct Inward Dialing] and assign a DID to a port on a Cisco switch in the basement of each building. The emergency responder has a database of IP addresses for every port on those switches; it also has a database of DIDs tied to a location name (physical address). So we assign a DID to a location and then we assign the location to a port. We build our call-routing tables so that whenever 911 is called, the network is instructed to send the call to Emergency Responder rather than out the trunks, and Emergency Responder calls the PSAP with the exact location. Every hour Emergency Responder uses SNMP [Simple Network Management Protocol] to poll all of the campus switches to verify what phones and numbers are on what ports. So it performs a physical/logical association every hour. Our faculty move constantly, but I no longer care as much as I used to since the PSAP doesn't need to be updated.

"Now the relationship between the *real* address and a *fake* address is only made when a call is placed. It doesn't maintain it; we have this set of 911 DIDs. If you call one of those numbers, you'll get a busy signal. If I see a call coming in on a 911 DID, I'll just send it to the Emergency Responder. So I only have to update PSAP when we add locations. And actually, we have a security computer dedicated for 911 calls. It blinks the screen when a 911 call is placed.

"So, in summary, here's what happens. We assign a DID to a port on a switch within the Cisco Emergency Responder (CER). The CER maintains a database of all switch IP addresses, port numbers, location names, and DID numbers. When a 911 call is made, the Call Manager forwards that call to the Emergency Responder, and it takes over call processing. CER then looks in its database to see which DID is associated with the physical port that the call was made from, and then transmits *the DID of the switch port* to the PSAP, *not* the DID of the phone. This model works beautifully. In fact, the only thing we have to be careful about is to ensure that no one changes the location of a cable in the switch room. That would result in database inaccuracies that could be dangerous.

"The only problem with this whole system—and it's really a nonproblem, considering what it does for us—is that I have 400 DIDs that I can only use for this system and they cost 18 bucks per month or so."

Telephone Company Interfaces

Depending on the nature of the installation, the RFP should indicate the degree of service provider connectivity and interoperability that the customer requires. For example, if the customer wishes to be able to pass signaling information back and forth between the IP PBX and the public telephone network, then that needs to be stipulated in the RFP.

Power

If the converged installation is to deliver carrier-grade service, then uninterruptible power must be part of the delivered package and must be covered in the details of the RFP documentation. Details should include the breadth and scope of uninterruptible power and the expected duration of battery backup that the system will provide in the event of a failure of commercial power.

Redundancy

Although uninterruptible power is a form of system redundancy, other factors play a part as well. It is critical to give serious consideration to the level of physical and logical redundancy that is required in the system before issuing the RFP. The document should describe in precise detail the nature and extent of redundancy required. This should include (but not be limited to) network redundancy, spares inventory, repair turnaround, and failover mechanisms.

System Backup

Vendors responding to the RFP should cover this particular feature in detail. It is critical that systems be backed up on a regular basis, and the vendor's documentation should cover this as a matter of course. Their response should include such issues as preferred methodology and medium for performing backups, recommended frequency for archival and incremental backups, and offsite storage.

Security

Security should cover not only logical and physical security considerations, but system administration as well. Vendor documentation should address such basics as virtual private network (VPN) and virtual local area network (VLAN) configuration, user authentication, and resource management.

The RFP should also include information about the following:

- **Removal of outmoded hardware.** If upgrade activity results in the wholesale replacement of hardware components that are no longer serviceable (such as Champlain's terminally ill PBX), then the vendor should be asked to remove the old hardware as part of the package.
- **Process for choosing subcontractors.** It is common practice for vendors to hire trusted subcontractors to perform installation, testing, and maintenance procedures during a large-scale project.
- **Documentation and system manuals.** Needless to say, the new system should come with a full suite of user manuals. However, are they paper, electronic, or both? Are updates available online? How often is documentation updated?
- **Customer training.** Careful: As the customer, you should expect extensive training on all aspects of the new system, and you have the right to dictate whether follow-up "refresher" sessions will be made available and, if so, when and how often.
- **Project timeline details.** You must work closely with the vendor to establish project benchmarks, critical success factors, and measures of completion. These serve as mutually beneficial tools to keep both sides honest with regard to project progress.
- **Physical space requirements.** Real estate is expensive. A distributed system like a voice over IP (VoIP) installation will require distributed real estate. You must stipulate any requirements that you have *up front* about space limitations that the vendor must work within, as this can affect the overall design and layout of installed hardware.
- **Environmental requirements.** What kind of cooling, power, and air filtration (if any) are required? This information is crucial to a successful and long-term system installation.
- **Service and repair requirements.** (including times and days per week that the vendor will respond to trouble calls and anticipated

response times). From the time the call is made until the vendor shows up on site, what is the interval? Furthermore, what is the average mean time between failures (MTBF) and the average mean time to repair (MTTR) for the hardware that is to be installed?

- **Emergency response.** In the event of a major failure, what extraordinary measures can the customer expect from the vendor in terms of prioritization, technician dispatch, and spares' availability? What escalation procedures should be put into place to ensure the proper response to a catastrophic event?
- **Location of nearest maintenance personnel and spare parts.** This is critical. Many companies have entered into agreements with vendors without asking this question, only to discover that their East Coast installation is supported by technicians based in California—or worse. It is also a valid and reasonable request to ask that an inventory of critical spares be kept on site.

System Options

As we mentioned earlier, there will more than likely be a collection of options "for future consideration" identified in the RFP but clearly labeled as options. Vendors should understand that they should provide information about the optional selections listed and that the customer may bid on their inclusion.

References

At the time of RFP publication, you should ask bidding vendors to include a list of some customer references you can contact to seek a better understanding of the vendor's responsiveness, price competitiveness, and other considerations that could affect the vendor selection process.

Selecting System Partners

Ultimately, Champlain College selected Cisco as its hardware platform provider and Net6 (now part of Citrix Corporation) as its partner to develop system applications. "We wanted to put into place a collection of IP solutions that would give us better ways to communicate with the students in the dorms as well as with the faculty—although student contact was our primary driver," says Videtto. "We have about 700 residential students scattered across a handful of dorms [see Figure 4-2], so we ended up buying large screen phones and installing them in the dorm rooms [see Figure 4-3]. We also entered into an agreement with Net6, a

Figure 4-2 A typical Champlain College dormitory

Figure 4-3 One of the IP phones installed throughout Champlain's facilities. Note the large screen on the device for displaying data.

software company out in California, which has a very capable push server that allows us to send messages out to all the phones in the system. We set up a screensaver on all the phones that sends out a new message every 15 seconds. I have limited the number of people who have the authority to insert messages to about 15, but they include the head of security, the head of residential life, the people who run the physical plant, the folks responsible for student activities, and the head resident of each dorm. Putting up the messages is actually quite easy—it doesn't require any technical skill. It's just a Web page that they navigate to and insert messages that they feel are important. If a call comes in, whatever's on the screen goes away so that it doesn't interfere with call processing. Every four hours the server synchs up with the Cisco CallManager database, which is all SQL [Structured Query Language] and says, 'Give me the IP addresses of all your phones.' I then created groups on the push server, with names like 'Summit Hall, Room 1' to create specific groups that can be individually targeted. We also have an all-student and all-admin group. And while it's pretty simple, we had to think hard about this before we actually implemented it."

Doing the Vendor Dance

A critical component of the overall process is the selection of the vendor or vendors that will provide the hardware and software required to effect the conversion. This is a difficult process that requires significant up-front effort, but a well-thought-out effort will result in a far less painful evolution and a quicker timeline to success.

Aaron Videtto remembers the selection process that he and Paul Dusini went through. "By the time we got to the vendor selection process, we had already done our homework and we knew that we would most likely go with a converged environment, although the letter of intent that we sent out did not explicitly say that. We went to the market with a Request for Information (RFI), which we sent to a collection of vendors. Based on their data we narrowed the list down to three, and we sent them a Request for Proposal (RFP). Their responses were remarkably different. One vendor very quickly returned a one-page response to us, telling us that our requirements would cost $1 million. Needless to say, they were out, since our initial research showed us that we could do the project for half that or less. A second vendor had a good product, and their pricing was within bounds, but they had no local support, and since this was a new installation using relatively new technology in an application that is critical to the operation of the college, we decided that they were not a viable choice either.

"You have to consider where we were at the time that we were making our selections. We knew that the industry was going to VoIP, and we were teetering on the edge. This was three years ago and the technology was really new. Few organizations had installed it, it wasn't fully baked, and given the magnitude of the investment that were about to make—and the potential impact that it could have if it went south—did we really want to go with the 'bleeding edge?'

"So we decided that the plan was good. We revamped the original RFP, adding an addendum that specifically requested an IP solution."

Videtto smiles as he thinks back. "It was amazing how different the proposals were as they came in. One vendor's original solution was part TDM [time division multiplexing], part VoIP; it was actually very interesting. In fact, our final decision came down to that vendor and Cisco. Before we made a final vendor decision, we visited another college that had recently completed the installation of that same vendor's hybrid solution, and while it was very good, it was basically a cabinet and card-based product. So we made a decision: We wanted a server-based design, so ultimately we went with Cisco. The process was tough, however. Paul and I were all over each other because I wanted the other vendor and Paul wanted Cisco. It took us four months of arguing back and forth to make a decision, and in the final analysis it came down to company financials. The vendor I wanted was a private company; Cisco was public. And since this was an investment that would be in use for a very long time, we began to think about concerns over long-term support. It has turned out to be a very good decision."

Paul Dusini presents another reason why the Cisco solution was the best for the campus community. "There's another aspect that we had to take into account as well," he explains. "This is, after all, a telephone system that we're talking about. We were seriously concerned about survivability and redundancy, all the things that make a network carrier-grade. When we looked at Cisco, we realized that the way they do redundancy is unheard of in the VoIP industry. Their call manager is unique and solid, and the only way you can provide true redundancy in this kind of a network.

"Here's how it works—and why we like it. We have two redundant call managers that run under a proprietary version of Windows 2000. What Cisco has done is strip the kernel (the central component of the operating system that provides basic services for all other operating system functions, including memory, process, file, and I/O [input/output] management) of everything that is not telephony related. When a phone 'boots up,' it receives a DHCP [Dynamic Host Configuration Protocol] address and a TFTP [Trivial File Transfer Protocol] server IP address, which is the IP address of the server."

Note that TFTP is interesting. When a device boots, it queries the DHCP server and asks for network configuration information. The server responds with an IP address for the device, a subnet mask, a default gateway, a domain name server (DNS) address, and a TFTP server address. The device then requests a configuration file from the TFTP server, which searches both the primary and alternate paths for the requested configuration file. If the server finds it, it sends the file to the requesting device. If the device receives the Cisco CallManager name, it resolves the name using DNS and opens a connection with the Cisco CallManager. If the device does *not* receive an IP address or name, it defaults to the default server name. The server, one function of which is to serve as a publisher, stores the MAC addresses and the settings of all the active phones in the system. When the phone boots, it talks to the server, and the server builds a record.

"Included in the TFTP file from the server are IP addresses of both of the redundant call managers—the publisher and the subscriber," Paul Dusini continues. "That information is received from the DHCP server before it talks to the call managers. For protection purposes, our call managers are on opposite sides of the campus. Now every 30 seconds the phone talks to the publisher and asks, "are you there?" If it fails to get a response after 30 seconds, it immediately fails over to the second IP address—the other server—and the users never know it happens."

Videtto agrees. "One reason that the Cisco design is nice is because we run Windows 2000, which means that I can easily patch the servers when it's required. And because they are fully redundant and designed to support each other, I can turn off all the services on one server, upgrade it, then turn all the services back on, and repeat the process for the other server, all without service interruption. They just fail back and forth transparently. I can lose the call manager and calls aren't affected. Only Cisco does it this way. It's true failover redundancy, and it's one of the primary reasons we chose them as our system vendor."

Other Vendor Issues

Champlain College's choice of Cisco as their primary vendor was based on a wide-ranging set of questions that were not only technological in nature, but also had a great deal to do with postsales support, interoperability, the ability of the platform to adapt to evolving applications, and support for the constantly changing needs of the business. These questions must be asked regardless of the vendor that is selected for a project of any magnitude.

Vendors should be able to speak intelligently about the components of their solutions and about the relative advantages that their designs offer. A typical converged network is comprised of customer premises equipment (CPE) devices, often referred to as clients; communications servers, which deliver VoIP; application servers, which deliver enhanced services that are above and beyond the voice application; and gateways, which provide for interworking with the public switched telephone network (PSTN). The chosen vendor must have a clear and compelling support story for each of these device types. Clients, which may include both analog and digital telephones, desktop PCs, laptops, a plethora of personal digital assistant (PDAs), and SoftPhones, are extensively used throughout most enterprises and their ability to access services on the new converged network must be available.

Communications servers must be fully supported and upgradeable. Originally intended as the device that delivers voice, these servers now offer a range of media services as well that integrate with and enrich basic voice.

Application servers further enrich the customer experience by offering alternative communication modalities to callers. These include support for short message system/instant message (SMS/IM), unified messaging, audio and videoconferencing, a collection of Web-based customer contact applications, and collaborative applications enabled with the Session Initiation Protocol (SIP) such as screen-sharing and follow-me services.

Finally, gateways facilitate access to the PSTN, the cellular network, and other access and transport options. They also perform protocol conversion as required to ensure that legacy installations can continue to be used, thus protecting whatever investment remains that is not yet fully amortized.

The End State: Network Convergence

In the final analysis, convergence is a redefinition of the manner in which services are delivered throughout the enterprise as an enhanced mechanism for delivering services to end users and customers. No "magic" is involved in the conversion; it is simply a different collection of protocols and procedures that make it easier for an enterprise to do business with its customers.

The IP "core" network (a bit of an oxymoron, because everything happens at the edge) is comprised of routers, local area network (LAN) switches, wireline and wireless transmission facilities, robust firewalls, and support for Packet over Ethernet (PoE). As convergence within the network continues, we will begin to see the emergence of other components, including gateways, which provide protocol conversion between the IP network and legacy devices such as telephones and fax machines on the user side and T-1 facilities on the network side. They will support next-generation voice and video-signaling protocols, such as SIP and H.323, described earlier, as well as "corollary protocols" that facilitate the delivery of a broad swath of services. These include the Lightweight Directory Access Protocol (LDAP), SNMP, and Java Database Connectivity (JDBC). LDAP is a set of procedures for accessing information directories. Although it has not yet been widely implemented, LDAP will ultimately make it possible for virtually any application running on any computer to obtain directory information such as e-mail addresses and public keys. Because LDAP is an open protocol, applications have no reason to concern themselves with the nature of the server on which the data is hosted.

SNMP is a standard for gathering statistical data about network traffic and about the behavior of network components. It relies on management information bases (MIBs) to define the information available from any manageable device on the network.

Finally, JDBC technology is an application-programmer interface (API), included in both the Java 2.0 Standard Edition (J2SE) and Java 2 Enterprise Edition (J2EE) releases, that provides connectivity to a wide range of SQL databases and access to other data sources such as spreadsheets or flat files. With a JDBC-enabled driver, systems can gain access to all corporate databases regardless of the nature of the hosting server.

Convergence and Performance

Looking at the process of convergence in functional stages, we find IP at the lowermost layer of the "protocol stack." These networks are application independent and now offer differentiable QoS, which is a far cry from their original capabilities through which all traffic was treated equally poorly. There was no sense of QoS.

Making the Call

Most carriers agree that a collection of factors affects the decision to choose an enterprise IP telephony solution. They include availability, QoS (as in quality of the audio signal), the degree to which the selected system is built around a collection of known standards and protocols, and the degree to which the selected product facilitates the evolution to a converged network infrastructure and a converged set of communications applications. In a nutshell, these are the factors to consider before deploying a VoIP network solution.

Availability

Availability is a measure of the likelihood that the service being measured will be available for use whenever it's needed. Sometimes referred to as "carrier-class service," availability is measured as a function of the number of minutes of downtime the network experiences in a given year. The most commonly known number, "five nines," simply means that the network's delivered resources will be available 99.999 percent of the time. That works out to just over 5¼ minutes of downtime per year (365 days/year × 24 hours/day × 60 minutes/hour × 0.0001), a fairly respectable performance number, particularly considering all of the factors that can affect the network's ability to deliver on that performance promise. Vendors of IP telephony solutions should be able to discuss in some detail the ability of their hardware and software to meet the carrier-grade availability challenge.

One of the factors that guarantees carrier-grade service delivery is uninterruptible power. In the PSTN, power is delivered over the copper local loop, guaranteeing that a customer can call the power company in the event of a failure of commercial power. Under normal circumstances, a VoIP network does not deliver power over its access infrastructure. However, with the advent of the Institute of Electrical and Electronics Engineers (IEEE) standard 802.3af (data terminal equipment power via media-dependent interface) in 2003, power can now be delivered to the desktop device in a way that emulates the PSTN.

802.3S1haf

IEEE 802.3af, sometimes called *PoE*, defines a technique to deliver power to terminals from other Ethernet devices in the network. The standard calls for the delivery of 48 volts of AC power to the device over unshielded twisted-pair (UTP) wiring. One great advantage of the standard is that it works over existing cable plant (including Category 3, 5, 5e, and 6), horizontal and patch cables, outlets, and connecting hardware without any modification to the existing plant. Another advantage is that 802.3af obviates the need for installed outlets and the labor cost required to install them. Finally, legacy Ethernet devices and newer LAN-powered devices can seamlessly be combined on the same network.

The standard came about because network devices such as laptops, cameras, wireless hot spots, and IP telephones required both a power connection and an Ethernet connection to function. Inasmuch as the LAN has become the most likely connectivity option in the enterprise today, it makes sense that a technique for physically combining the two would be developed in short order.

802.11af supports point-to-multipoint power distribution that is parallel to the data network. This allows IT staff to rely on a single uninterruptible power supply (UPS) installation (also similar to the PSTN) for multidevice power backup. Using SNMP commands, the standard supports remote access management.

The node current is limited to 350 milliamps. The total amount of continuous power that can be delivered to each node, assuming a certain amount of drop over the cable run, is 12.95 watts. Inasmuch as most devices rarely require more than 10 watts, this is more than adequate.

Because of safety standards and cable limitations, the 802.3af standard includes a mechanism to detect the presence of noncompliant devices so that they can be left out of the power-up sequence, thus avoiding the potential for fire. The only terminals that will receive power in a standards-compliant 802.3af environment are those that bear the valid PoE "signature."

The IEEE 802.3af standard defines two techniques to deliver line-based power to IP phones. The first is called *end span*. In end-span implementations, Ethernet switches are replaced with alternate devices that use current delivered over pairs 1/2 and 3/6 (on the RJ-45 jacks) that are typically used for data. This technique works well in new buildings or for major network upgrades. Because it requires replacement of the existing Ethernet switches, it can be costly.

The second technique is called *mid-span.* In mid-span implementations, a two-port adjunct device is added between the Ethernet switch and the powered end-user device. It delivers power to the CPE over pairs 4/5 and 7/8, which are not used in the RJ-45 jack for Ethernet implementations. Because it does not require a hardware replacement, it is significantly less expensive than the end-span option.

The RJ-45 eight-conductor data cable comprises four pairs of wire. Each pair consists of a solid-colored wire and a white wire with a stripe of the same color. The pairs are twisted together within the cable. To ensure transmission reliability, the pairs should not be untwisted more than approximately 1 cm.

RJ-45 has two wiring standards: T-568A and T-568B. They differ only in connection sequence; their color code is identical. Figure 4-4 illustrates the T-568B standard. The pairs designated for 10BaseT Ethernet are orange and green. The other two pairs, brown and blue, can be used for a second Ethernet line or for phone connections.

Figure 4-4 RJ-45 wiring scheme

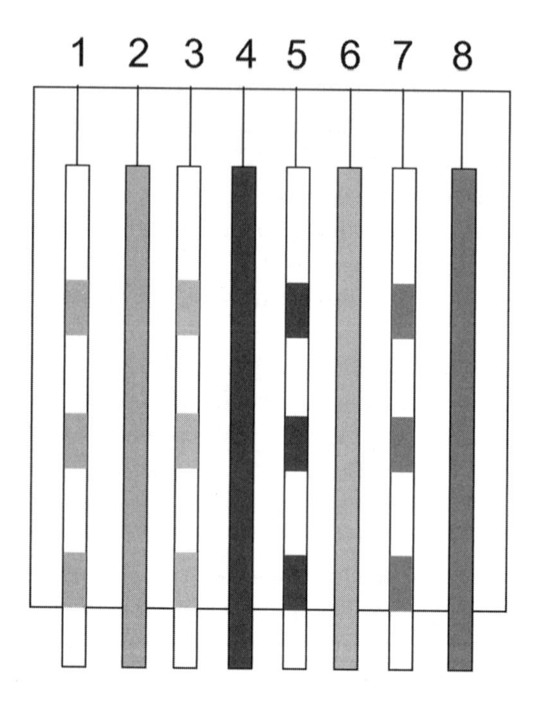

Table 4-2

RJ-45 Pin Numbers' Colors and Information

Pin	Color	Pair	Name
1	White/orange	2	Transmit data (+)
2	Orange	2	Transmit data (-)
3	White/green	3	Receive data (+)
4	Blue	1	
5	White/blue	1	
6	Green	3	Receive data (-)
7	White/brown	4	
8	Brown	4	

Note that the blue pair is on the center conductors and corresponds to the red and green pair in a normal phone line. The connections shown are specifically for an RJ-45 connector. The wall jack may be wired in a different sequence because the wires are actually crossed inside the jack. The jack should include a wiring diagram or should include designated pin numbers that can be matched to the color code shown in Table 4-2.

Under a naming convention that is oddly reminiscent of the Synchronous Optical Network (SONET), 802.3af defines two types of power sources: end-span devices and mid-span devices, described earlier. An end-span device is simply an Ethernet switch that has embedded PoE technology. These devices deliver both power and data over transmission pairs 1/2 and 3/6.

Mid-span devices, on the other hand, are similar to patch panels and normally have 6 to 24 channels. They are typically located between legacy switches and the devices that are to be powered. Each port has an RJ-45 data *input* connector as well as an RJ-45 data and power *output* connector. Mid-span systems rely on pairs 4/5 and 7/8 to carry power, while data is carried on the other pairs.

In new installations, end-span Ethernet switches are most commonly selected. Mid-span devices, on the other hand, work well to upgrade a network without replacing the switch and in situations where there is low port density. Nevertheless, because of the proliferation of edge devices, most installations succumb to both because the end-span switch connects to IP telephones, wireless LAN hot spots, and other devices.

And although Ethernet-powered end-user devices are a terrific addition to the enterprise network, servers and switches should be connected to a UPS as well. Between the capabilities of 802.3af-compliant devices and UPS-backed servers, network administrators are guaranteed that no single point of failure (at least from a power delivery perspective) exists on the network.

Audio Quality

Because of the fundamental way in which IP works, IP networks are not overly concerned with the bandwidth assigned to each conversation. As a result, so-called "wideband phones," recently introduced in the IP voice market, actually improve the quality of audio carried at frequencies higher than 3.3 KHz (the lower limit of the so-called voice band) and as high as 7 KHz. In fact, audio quality is directly related to the sampling rate, that is, the frequency at which audio samples are collected from an incoming analog audio signal. DVD-quality audio, for example, is sampled at 192 KHz. On the other hand, CD-quality audio is sampled at 44.1 KHz, and classic telephony is sampled at 8 KHz. Higher sampling rates enable the network to capture harmonics that provide tone and richness, facilitating better perceived quality for applications such as "music on hold." For people with hearing difficulties, wideband phones may improve productivity and enhance the customer's experience using the telephone.

The standards for wideband audio include G.711 Adaptive Differential Pulse Code Modulation (ADPCM), discussed earlier, and G.729 compression. To improve application performance and to provide remote users with greater end-to-end control of the session, compression control should most effectively reside in the IP phone rather than at a central shared router.

A significant number of factors determine the level of perceived audio quality, including network design, physical plant limitations, operational considerations, and the nature of the communicating end devices themselves. In today's networks, the physical integrity of the plant itself is considered to be a given; it's just assumed that the network will be of the highest possible quality. And although this isn't always the case, it's largely the case that the network infrastructure is good. That being said, it stands to reason that the resolution of issues related to the physical plant is a straightforward process, and indeed it is compared to other issues that manifest themselves in interesting and challenging ways. Wiring problems, equipment failures, and distance-related issues can be logically resolved in short order. More difficult, however, are problems related to network operations, architecture, and overall design. These include such things as the presence (or lack of) echo cancellation mechanisms, the distance between the two (or more) communicating devices, packet prioritization settings in the LAN switch, and VLAN voice traffic configuration.

VLAN considerations are critical components of the typical VoIP enterprise network. Before we discuss the VLAN's role in the VoIP network, however, let's take a few pages to describe what Ethernet LANs are, how they work, and how they in turn relate to VLANs. (Note: This is one of those places where readers who are not particularly interested in the technological details can skip forward. See you on the other end.)

A Technical Overview of Ethernet

In the 20 years since Ethernet first arrived on the scene as a low-cost, high-bandwidth 10 Mbps LAN service for office automation applications, it has carved a niche for itself that transcends its original purpose. Today it is used in LANs, metropolitan area networks (MANs), and wide area networks (WANs) as both an access and transport technology. Its bandwidth capabilities have evolved to the point that the original 10 Mbps version seems almost embarrassing; a 10 Gbps version of the technology now exists.

Today's Ethernet product is based on the IEEE 802.3 standard and its many variants, which specify the Carrier Sense Multiple Access with Collision Detection (CSMA/CD) access scheme (explained in a moment) and a handful of other details. It is perhaps important to note that Ethernet is a product name, originally invented and named by 3Com founder Bob Metcalfe. CSMA/CD is the more correct name for the technology that underlies Ethernet's functionality and the 802.3 standard. The term Ethernet, however, has become the de facto name for the technology.

Ethernet evolved in stages, all based upon the original IEEE 802.3 standard, which passed through various evolutionary phases of capability including 10 (Ethernet), 100 (Fast Ethernet), 1,000 (Gigabit Ethernet), and 10,000 Mbps (10 Gbps Ethernet) versions. All are still in use, which allows for the design and implementation of tiered corporate Ethernet networks. A corporation, for example, could build a gigabit backbone, connected to Fast Ethernet access links, connected in turn to 10

Mbps switches and hubs that communicate with desktop devices at the same speed.

In its original form, Ethernet supported point-to-point connectivity as far as 0.062 miles over UTP, up to 1.2 miles over multimode fiber, and up to 3.1 miles over single-mode fiber (SMF) at a range of speeds. Today 10 Mbps is inadequate for many enterprise applications, so more robust versions of Ethernet have evolved and become standardized.

How Ethernet Works

Ethernet is called a *contention-based LAN.* In contention-based LANs, devices attached to the network vie for access using the technological equivalent of gladiatorial combat. If a station on the LAN wants to transmit, it does so, knowing that the transmitted signal *could* collide with the signal generated by another station that transmitted at the same time, because even though the transmissions are occurring on a LAN, still some delay still exists between the time both stations transmit and the time they both realize that someone else has transmitted. This collision results in the destruction of both transmitted messages. If a collision occurs, both stations stop transmitting, wait a random amount of time, and try again. This technique is called *truncated binary exponential backoff* and will be explained later.

Ultimately, each station *will* get a turn to transmit, although how long they have to wait is based on how busy the LAN is. Contention-based systems are characterized by *unbounded delay*, because no upward limit exists on how long a station may wait to use the shared medium. As the LAN gets busier and traffic volume increases, the number of stations vying for access to the shared medium, which only allows a single station at a time to use it, also goes up, which results in more collisions. Collisions result in wasted bandwidth so LANs do everything they can to avoid them.

The protocol employed in traditional Ethernet is called CSMA/CD. In CSMA/CD, a station observes the following guidelines when attempting to transmit. Remember the old western movies when the tracker would lie down and put his ear against the railroad tracks to determine whether a train is coming? Well, CSMA/CD is a variation on that theme. First, it listens to the shared medium to determine whether it is in use. If the LAN is available (not in use), the station begins to transmit but continues to listen while it is transmitting, knowing that another station could transmit simultaneously. If a collision is detected, both stations back off and try again.

LAN Switching

To reduce contention in CSMA/CD LANs, LAN switching was developed, a topology in which each station is individually connected to a high-speed switch, often with Asynchronous Transfer Mode (ATM) on the backplane. The LAN switch implements a full-duplex transmission at each port, reduces throughput delay, and offers per-port rate adjustment. The first 10 Mbps Ethernet LAN switches emerged in 1993, followed closely by Fast Ethernet (100 Mbps) versions in 1995 and Gigabit Ethernet (1,000 Mbps) switches in 1997. Today they are considered the standard for LAN architectures, and because they use star wiring topologies they provide high levels of management and security to network administrators. VLANs followed; we'll describe them a bit later.

Ethernet has been around for a long time and is understood, easy to use, and trusted. It is a low-cost solution that offers scalability, rapid and easy provisioning, granular bandwidth management, and simplified network management due to its single protocol nature. As its popularity has increased and it has edged into the wide area network (WAN) environment, it now supports such applications as large-scale backup, streaming media server access, Web hosting, application service provider (ASP) connectivity, storage area networks (SANs), video, and disaster recovery.

Another important service is the VLAN. Often characterized as a VPN made up of a securely interconnected collection of LANs, a number of vendors now offer VLANs as a service, including Lantern Communications, which claims to be able to support thousands of VLANs on a single metro ring. Similarly, transparent LAN (TLAN) services have gained the attention of the industry; they allow multiple metro LANs to be interconnected in a point-to-point fashion.

Many vendors are now offering what is known as an Ethernet private line (EPL), an alternative to a legacy dedicated private line. Implemented over SONET/Synchronous Digital Hierarchy (SDH), the service combines SONET/SDH's reliability, security, and wide presence with Ethernet's rapid, flexible provisioning ability.

Of course, Ethernet does have its disadvantages as well. These include the potential for frame (and therefore packet) loss and the QoS issues that result. However, two standards have emerged that help to eliminate this problem. The IEEE 802.17 Resilient Packet Ring standard is designed to give the ironclad protection and reliability offered by carrier-grade SONET/SDH networks to Ethernet, using a Data Link Layer protocol[1] ideal for packet services across not only LANs, but MANs and WANs as well. Primarily targeted at voice, video, and IP traffic, it restores a breached ring within 50 ms. The second standard, International Telecommunication Union (ITU) X.86, maps Ethernet (OSI layer two) frames into SONET/SDH physical layer (OSI layer one) frames for high-speed transmission across the broad base of legacy networks.

A Brief History of Ethernet

Ethernet was first developed in the early 1970s by Bob Metcalfe and David Boggs at the Xerox Palo Alto Research Center (Xerox PARC). Originally designed as an interconnection technology for the PARC's many disparate minicomputers and high-speed printers, the original version operated at a whopping 2.9 Mbps. Interestingly, this line speed was chosen because it was a multiple of the clock speed of the original Alto computer. Created by Xerox Corporation in 1972 as a personal computer targeted at research, Alto's name came from the Xerox Palo Alto Research Center, where it was originally created. The Alto was the result of developmental work performed by Ed McCreight, Chuck Thacker, Butler Lampson, Bob Sproull, and Dave Boggs, engineers and computer scientists attempting to create a computer small enough to fit in an office but powerful enough to support a multifunctional operating system and powerful graphics display. In 1978, Xerox donated 50 Alto computers to Stanford, MIT, and Carnegie Mellon University. These machines were quickly adopted by campus user communities and became the benchmark for the creation of other personal computers. The Alto comprised a beautiful (for the time) graphics display, a keyboard, a mouse that con-

[1]Because it is so central to an understanding of how Ethernet and other systems work, readers not familiar with the OSI Reference Model for protocols should refer to the appendix for an overview.

trolled the graphics, and a cabinet that contained the central processing unit (CPU) and hard drive. At $32,000, the thing was a steal.

The concept of employing a visual interface (what you see is what you get [WYSIWYG]) began in the mid-70s at Xerox PARC, where a graphical interface for the Xerox Star system was introduced in early 1981. The Star was not a commercially successful product, but its creation led to the development of the Apple Lisa in 1983 and later the 1984 Macintosh.

In July 1976, Metcalfe and Boggs published "Ethernet: Distributed Packet Switching for Local Computer Networks" in *Communications of the Association for Computing Machinery*. A patent followed with the mouthful of a name, "Multipoint Data Communications System with Collision Detection," issued to Xerox in December 1977.

The Roles of Xerox and DEC

In 1979, DEC, Intel, and Xerox Corporations standardized on an Ethernet version that any company could use anywhere in the world. In September 1980, they released Version 1.0 of the "DIX standard," so named because of the initials of the three founding companies. It defined the so-called "Thicknet" version of Ethernet (10Base5), which offered a 10 Mbps CSMA/CD protocol. It was called "thick" Ethernet (or Thicknet) because it relied on relatively thick coaxial cable for interconnections among the various devices on the shared network. The first DIX-based controller boards became available in 1982, and the second (and last) version of DIX was released in late 1982.

In 1983, the IEEE released the first Ethernet standard. Created by the organization's 802.3[2] Committee, *IEEE 802.3 Carrier Sense Multiple Access with Collision Detection (CSMA/CD) Access Method and Physical Layer Specifications* was basically a rework of the original DIX standard with changes made in a number of areas, most notably to the frame format. However, because of the plethora of legacy equipment based on the original DIX specification, 802.3 permitted the two standards to interoperate across the same Ethernet LAN.

---

[2]A bit of history: The IEEE 802 committee is so-named because it was founded in February (2) 1980 (80).

The ongoing development and tightening of the standard continued over the next few years. In 1985, IEEE released 802.3a, which defined "thin Ethernet," sometimes called "cheapernet" (officially 10Base2). It relied on a thinner, less expensive coaxial cable for the shared medium that made installation significantly less complex. And although both media offered satisfactory network performance, they relied on a bus topology that made change management problematic.

Two additional standards appeared in 1987. 802.3d specified and defined what came to be known as the Fiber Optic Interrepeater Link (FOIRL), a distance extension technique that relied on two fiber optic cables to extend the operational distance between repeaters in an Ethernet network to 1,000 meters. IEEE 802.3e defined a 1 Mbps transmission standard over twisted pair, and although the medium was innovative it never caught on.

1990 proved to be a banner year for Ethernet's evolution. The introduction of 802.3i 10BaseT standard by the IEEE made it possible to transmit Ethernet traffic at 10 Mbps over Category 3 UTP cable. This represented a major advance for the protocol because of the use of UTP in office buildings. The result was an upswing in the demand for 10BaseT.

With the expanded use of Ethernet came changes in the wiring schemes over which it was deployed. Soon network designers were deploying Ethernet using a star wiring scheme (Figure 4-5) instead of the

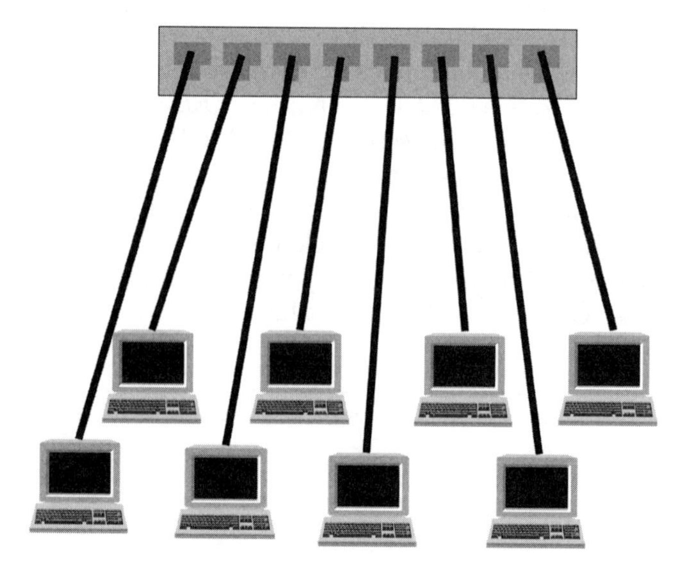

Figure 4-5 A star-wired arrangement; all devices "home" back to the hub, shown at top.

more common bus architecture, which was significantly more difficult to manage, install, and troubleshoot.

In 1993, the IEEE released 802.3j (10BaseF), a technological leap for Ethernet that extended the transmission distance significantly by stipulating the use of a pair of optical fibers instead of copper facilities. With this technique, transmission distances could be extended to 2,000 meters, a distance that dramatically expanded Ethernet's utility in office parks and in multitenant units (MTUs). 802.3j represented an augmentation and update of FOIRL, mentioned earlier.

Fast Ethernet

In 1995, the IEEE announced the 100 Mbps 802.3u 100BaseT standard. Commonly known as "Fast Ethernet," the standard actually comprises three disparate "substandards," which in turn support three different transport media. These include 100BaseTX, which operates over two pairs of CAT 5 twisted pair; 100BaseT4, which operates over *four* pairs of CAT 3 twisted pair; and 100BaseFX, which operates over a pair of multimode optical fibers. It was deemed necessary to offer a version for each of these media types because of the evolving customer environment; more on this later.

Two-Way Ethernet

"Due to popular demand" as they say, the IEEE announced 802.3x in 1997, which provided a standard that governed full-duplex Ethernet. By allowing stations on an Ethernet LAN to simultaneously transmit and receive data, the need for a medium contention mechanism such as CSMA/CD was eliminated, allowing two stations to transmit simultaneously over a shared point-to-point facility. The standard was written to accommodate all versions of Ethernet, including Fast Ethernet. The IEEE also released the 802.3y 100BaseT2 standard in 1997, which provided a mechanism for transmitting at 100 Mbps over two pairs of CAT 3 balanced wire.

Gigabit Ethernet

In 1998, Gigabit Ethernet arrived on the commercial scene. Like Fast Ethernet, the 802.3z 1000BaseX standard supports three different transmission mechanisms with three substandards. 1000BaseSX relies on an 850-nanometer laser over multimode fiber, 1000BaseLX uses a 1,300-nanometer laser over both single- and multimode fiber, and 1000BaseCX is designed for operation over short-haul copper (sometimes called twinax) shielded twisted pair (STP). The IEEE also released 802.3ac in 1998, a standard that defines the extensions required to support VLAN tagging on Ethernet networks, mentioned earlier in this section.

The advances continued. In 1999, 802.3ab 1000BaseT hit the market, supporting 1 Gbps transmission over four pairs of CAT 5 UTP.

The Ethernet Frame

Ethernet is a layer two Data Link standard—in effect, a switching technique. The standard transmission entity at layer two is the frame. The frame is nothing more than a "carriage" mechanism for the user's data, which is usually a packet. The frame, then, has a field for user data, as well as a series of fields designed to carry out the responsibilities of the Data Link Layer. These responsibilities include framing, addressing, transparency control, and bit-level error detection and correction. The Ethernet frame, is designed to accomplish its task well. Each field of the frame is described in detail in Figure 4-6.

The *preamble* is a seven-byte sequence of alternating ones and zeroes, designed to signal the beginning of the frame and to provide a degree of synchronization, as receiving devices detect the presence of a signal and prepare to process it correctly.

The *start frame delimiter* comprises a specific sequence of eight bits with the 10101011. This field signifies the functional beginning of the frame of data.

The *destination MAC address* identifies the station that is to receive the transmitted frame. *The source MAC address,* on the other hand, identifies the transmitting station. In 802.3, these address fields can be

Figure 4-6 Ethernet frame

Preamble (7 bytes)	Start frame delimiter (1 byte)	Dest. MAC address (6 bytes)	Source MAC address (6 bytes)	Length/ type (2 bytes)	MAC client data (0-n bytes)	Pad (0-p bytes)	Frame check sequence (4 bytes)

either two bytes or six bytes in length, although most networks today use the six-byte option. The destination address field can specify either an individual address targeted at a single station or a multicast address targeted at a group of stations. If the destination address consists of all one bits, then the address is a broadcast address and will be received by all stations on the LAN.

The *length/type field* is used to indicate either the number of bytes of data contained in the MAC client data field or the type of protocol carried there. If the value of the field is less than or equal to 1,500, then it indicates the number of bytes in the MAC client data field. If it is greater than or equal to 1,536, then the field indicates the nature of the transported protocol.

The *MAC client data field* contains the user data transmitted from the source to the destination. The maximum size of this field is 1,500 octets. If it is less than 64 bytes, then the pad field that follows it in the frame is used to bring the frame to its required minimum length.

The *pad field* allows a "runt frame" to be augmented so that it meets the minimum frame size requirements. When used, extra bytes are added in this field. The minimum Ethernet frame size is 64 bytes, measured from the destination MAC address field through the frame check sequence.

The *frame check sequence field* provides a four-byte cyclical redundancy check (CRC) calculation for error checking the content of the frame. As with all layer two error control schemes, the originating station creates the MAC frame and then calculates the CRC before transmitting the frame across the network. The calculation is performed on all the fields in the frame from the destination MAC address through the pad fields, ignoring the preamble, start frame delimiter, and, of course, the frame check sequence. When the transmitted frame arrives at the destination device, the receiving system calculates the CRC and compares the two. If they do not match, the station assumes an error has occurred and discards the frame.

The original standards for Ethernet defined the minimum frame size to be 64 octets and the maximum size to be 1,518 octets. This size includes everything from the destination MAC address field through the FCS. In 1998, the IEEE issued 802.3ac, which extended the maximum frame size to 1,522 octets so that a VLAN tag could be included for evolving VLAN applications. The VLAN protocol allows an identifier or tag to be inserted in the frame so that the frame can be associated with the VLAN to which it pertains, so that logical traffic groups can be created for management purposes. IEEE 802.1Q defines the VLAN protocol.

Let's take a side trip to examine VLANs in some detail.

Understanding VLANs

As we just described, Ethernet LANs are typically categorized as *broadcast domains*. The typical enterprise LAN comprises hubs, bridges, switches, and other devices that serve to provide the connectivity infrastructure for communicating endpoint devices. Devices on the same LAN can communicate directly with one another, while devices located on different LANs must communicate through a router, as shown in Figure 4-7.

In Figure 4-8, the LANs are separated from each other by a router. The LANs (and therefore distinct broadcast domains) are separated by dotted lines and numbered a through e.

As enterprise networks grow, additional routers are required to separate "communities of users" into what are known as broadcast and collision domains and as a way to more effectively provide connectivity between disparate LANs. In Figure 4-9, LANs D and E illustrate how a router groups users within a single building into multiple broadcast domains.

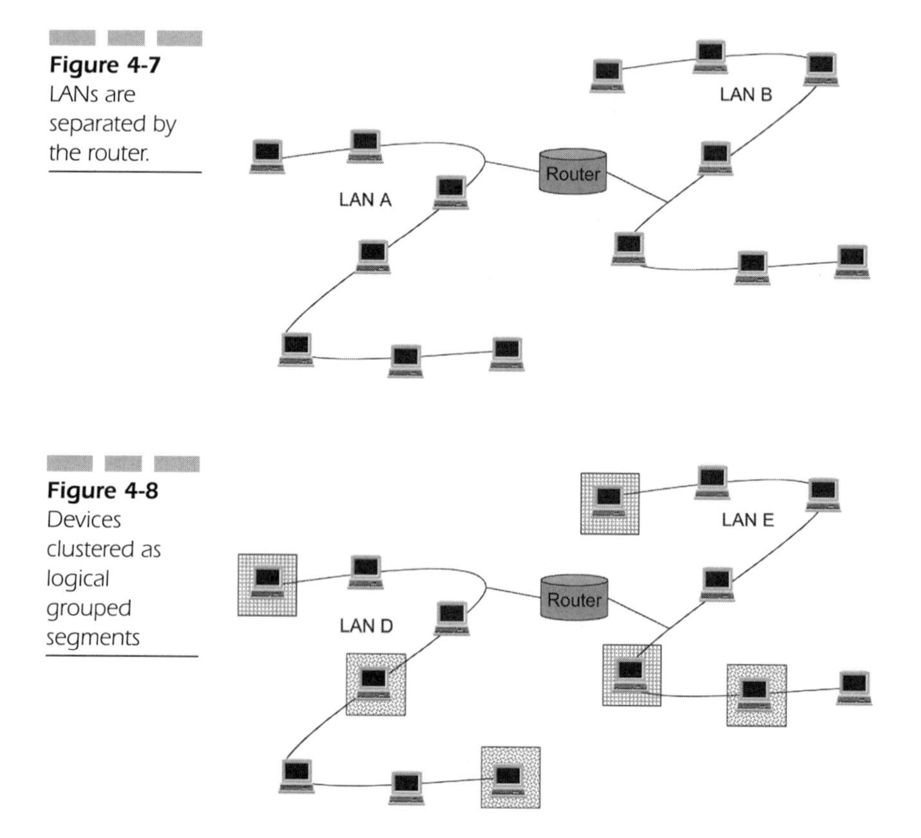

Figure 4-7 LANs are separated by the router.

Figure 4-8 Devices clustered as logical grouped segments

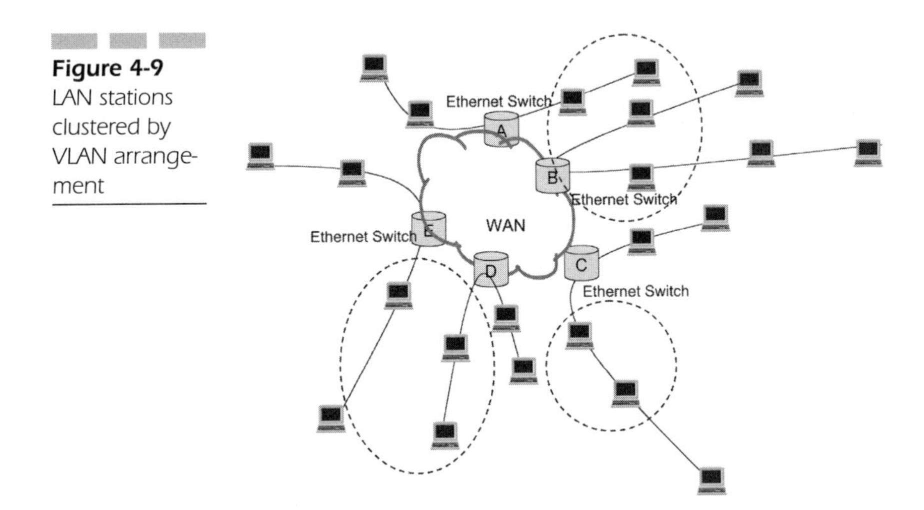

Figure 4-9 LAN stations clustered by VLAN arrangement

One drawback to the creation of router-based collision domains is that routers add latency, which affects overall transmission efficiency. This is the result of packet processing within the router that is required to move packets from one domain to another—in effect, the interdomain routing process.

A VLAN is actually a cluster of devices located on different physical LAN segments but that communicate with one another as if they were on the same physical LAN segment. They provide a number of benefits but require a modified network topology if they are to provide maximum benefit.

The switched network shown in Figure 4-10 provides the same connectivity as Figure 4-11. Although the transport network enjoys speed

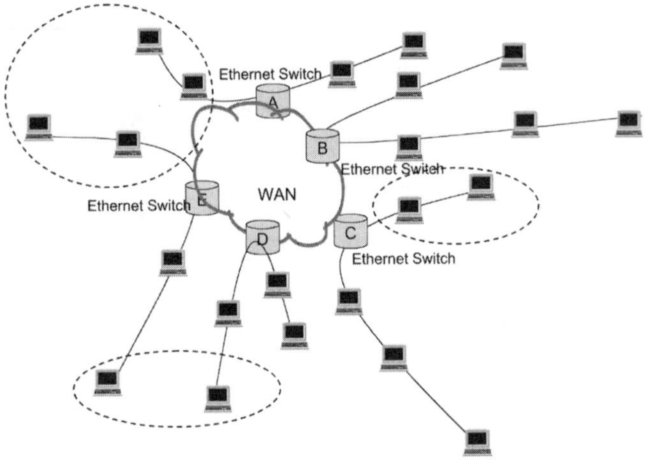

Figure 4-10 LAN stations with more granular clustering

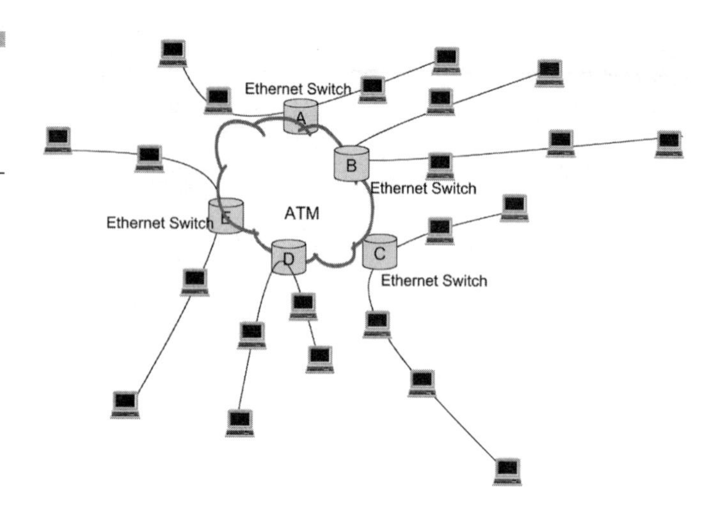

Figure 4-11 LANs are separated by an ATM switch.

and latency advantages over the network shown in Figure 4-12, it is by no means an ideal solution. The single greatest drawback to this design is that all the stations are now within the same broadcast domain, the greatest downside of which is massive traffic that every station must deal with, even though the bulk of that traffic has nothing to do with most stations. As a result, increased broadcast traffic (typical of broadcast LANs) runs the risk of flooding the network and bringing it to its knees.

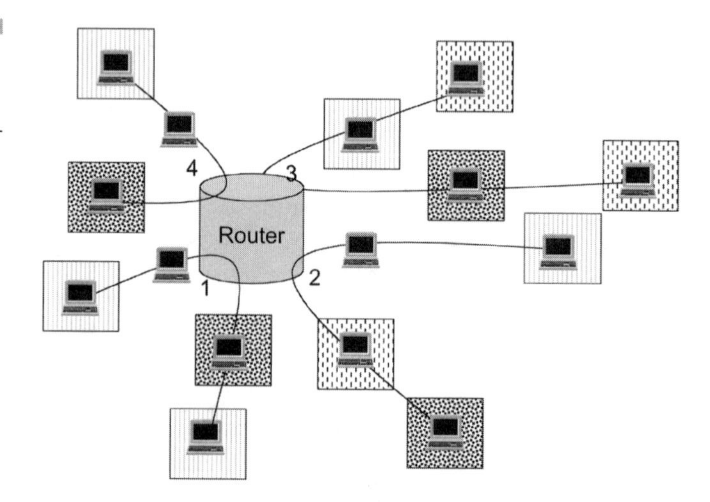

Figure 4-12 VLAN segmentation

Because switches operate at layer two of the OSI Protocol Model, they have less overhead than layer three routers and therefore do not suffer from the same latency issues that plague routers. Switched VLANs divide the network into separate broadcast domains, but without the latency issues found in routed environments. Figure 4-13 is an example of a switched VLAN.

What is important to note here is that although the physical LAN has changed in a minor way through the addition of LAN switches and the deployment of a single router, the logical topology first seen in Figure 4-8 has not. Note also that the LAN tags appear on the router interface, which means that the router becomes a member of all of the VLANs. A router is still required to move traffic between broadcast domains, but note the simplicity of this model. The physical layout and the logical definition of the LAN have been largely preserved.

So why would a company go through the process of establishing a VLAN environment? Consider the network shown in Figure 4-13. Note that all of the devices on VLAN 1 are used to access the computer, typically a highly intensive, traffic-heavy activity. With VLAN capability we can group these traffic-intensive devices into a single broadcast domain, thus allowing us to confine intensive broadcast traffic to a logical LAN segment comprising only those devices that need to see it—even though the devices are physically on separate physical segments. Furthermore, because we are switching traffic rather than routing it, we reduce the latency inherent in routed network environments. Furthermore, security

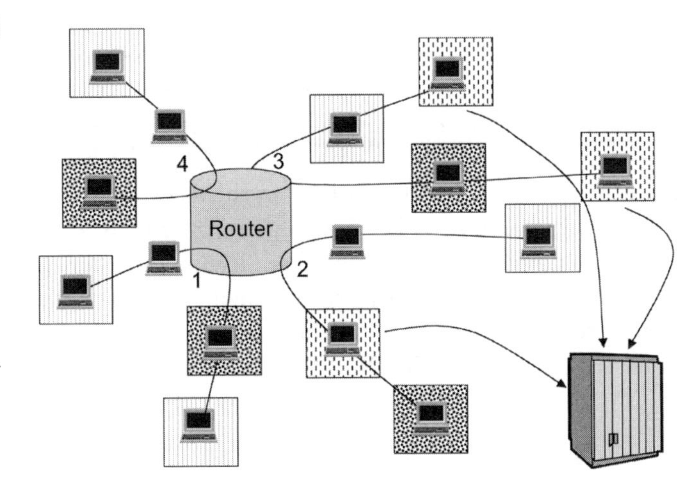

Figure 4-13 VLAN access. Note that all stations on VLAN segment 1 require access to computer resources and can thus be segmented.

is enhanced because we have the ability to control access on a segment-by-segment basis.

By extension, then, we can create logical subnetworks that are independent of the actual physical location of the users. In the case of Champlain College, for example, we could create department-specific VLAN segments for faculty, major-specific VLAN segments for students with similar interests, a VLAN segment for international students, dorm-specific VLAN groupings, and function-specific segments for such entities as finance, student life, and admissions. In other words, we can easily and seamlessly create groupings by function, traffic pattern, and work responsibilities.

Benefits of a VLAN Installation

The benefits of a VLAN implementation include better performance, a network that is easier to manage by IT staff, topological independence, and improved network security.

Because of their minimal overhead requirements, switched networks always increase performance over shared medium environments simply because they reduce collisions. By grouping network users into logical clusters, network administrators can improve performance by limiting the amount of domain-specific broadcast traffic to only those users that need to see it. And because traffic volumes decline, less traffic is routed, further reducing router-induced latency.

VLANs improve network management efficiency because they allow administration personnel to perform centralized configurations on diversely located devices. And because VLANs are logical constructs rather than physical ones, management of the network becomes totally independent of physical device location. Additional ports can be physically added and configured wherever it physically makes sense to do so, and they can be logically allocated in the same way.

Security is enhanced in VLAN environments for one very simple reason. In a shared, contention-based environment, *every station hears every transmission*. Because they are switched, VLANs deliver frames only to their intended recipients and deliver broadcast frames only to other members of the same VLAN.

Downsides of VLANs

As with any technology, VLANs do have a number of limitations, including broadcast limitations, device limitations, and port limits. Because many large VLAN environments rely on ATM transport at the core of the network, and because of the unique way ATM addresses traffic in the network, a special server may be required as part of the ATM infrastructure. The server limits the number of broadcasts that can be forwarded, which can be problematic in broadcast-intensive environments such as AppleTalk and internetwork packet exchange (IPX). The result may be the need to pay particularly close attention to the number of devices on a segment, as well as VLAN "membership."

Ethernet edge devices typically have the ability to support a maximum of 500 unique addresses, which clearly places an upward limit on the size of VLAN "segments." A similar limit exists at the port level; if a hub or switch is connected to a VLAN port, then every port on that hub or switch must belong to the same VLAN. These limitations, while not show-stoppers, do have an impact on the overall design and management of the network and must be taken into account by IT personnel.

The Ethernet Frame, Continued

Before we took our VLAN side trip, we were discussing the makeup of the Ethernet frame. If the four-byte VLAN tag mentioned earlier is present in an Ethernet frame, it is found between the source MAC address and the length/type field. The first two bytes of the tag comprise the 802.1Q tag type, always set to a value of 0×8100. This value indicates the presence of the VLAN tag and indicates to a receiving station that the normal length/type field will be found four bytes later in the frame.

The final two bytes of the VLAN tag are used for a variety of purposes. The first three bits represent a user priority field used to assign a priority level to the information contained in the frame. The next bit is called a canonical format indicator (CFI) and is used in Ethernet to indicate the presence of a routing information field (RIF). The final 12 bits are the VLAN identifier (VID) that identifies the VLAN that the frame belongs

Preamble (7 bytes)	Start frame delimiter (1 byte)	Dest. MAC address (6 bytes)	Source MAC address (6 bytes)	Length/ type = 802.1Q tag type (2 byte)	Tag control information (2 bytes)	Length/ type (2 bytes)	MAC client data (0-n bytes)	Pad (0-p bytes)	Frame check sequence (4 bytes)

Figure 4-14
Ethernet frame with VLAN identifier

to. The format of an Ethernet frame with the VLAN identifier is shown in Figure 4-14.

The *interframe gap* defines a minimum idle period between the transmission of sequential frames. Sometimes called an *interpacket gap,* it provides a recovery period between frames so that devices can prepare to receive the next frame. The minimum interframe gap is 9.6 microseconds for 10 Mbps Ethernet, 960 nanoseconds for 100 Mbps Ethernet, and 96 nanoseconds for 1 Gbps Ethernet.

In 1998, the IEEE approved 802.3ac, the standard that defines frame formats that support VLAN tagging on Ethernet networks.

The Role of Gigabit Ethernet

With the arrival in 1998 of the 802.3z standard for Gigabit Ethernet, a field was added to the end of the frame to ensure that the frame would be long enough to allow collisions to propagate to all stations in the network at high transmission speeds. The field is added as needed to bring the minimum length of the frame to 512 bytes. The Gigabit Ethernet extension field is only required in half-duplex mode, because the collision protocol is not used in full-duplex mode.

Gigabit Ethernet Frame Bursting

When the 802.3z standard was released in 1998, the IEEE also added a *burst mode* option that allows a station to transmit a series of frames without relinquishing control of the transmission medium. It is only used with Gigabit and higher Ethernet speed and is only applicable in half-duplex mode. When a station is transmitting a series of short frames, this mode of operation can dramatically improve the overall performance of Gigabit Ethernet LANs.

The details of burst mode operation are interesting. Once a station has successfully transmitted a single frame, it has the option to continue to transmit additional frames until it reaches a burst limit of 8,192 byte times. An interframe gap, described earlier, is inserted between each frame. In this case, however, the transmitting station populates the interframe gap with extension bits, which are nondata symbols that serve as a "keepalive message" and serve to identify the actual data bits by the receiving station.

A Point of Contention: Jumbo Frames

In mid-1998, Alteon Networks put forward an initiative to increase the maximum size of the data field from 1,500 octets to 9,000 octets. The initiative has been endorsed by a number of companies in the belief that larger frames would result in a more efficient network by reducing the number of frames that have to be processed. Furthermore, the 32-bit CRC used in traditional Ethernet loses its overall effectiveness in frames larger than 12,000 octets.

Alteon's proposal restricted the use of jumbo frames to full-duplex Ethernet links only and described a link negotiation protocol that allowed a station to determine if the station on the other end of the segment was capable of supporting jumbo frames before it chose to transmit them.

Jumbo frames have been implemented in many systems but are not yet widespread. As demand grows and as IP version 6 (IPv6) becomes more common, the technique will become more widely used.

Media Access Control (MAC) in Ethernet

Both half-duplex (one way at a time transmission) and full-duplex (two-way simultaneous) transmission schemes can be implemented in Ethernet networks. Both are discussed in the paragraphs that follow.

Half-Duplex Ethernet Using CSMA/CD

In traditional CSMA/CD-based Ethernet, half-duplex transmission is most commonly used. On LANs that rely on CSMA/CD, two or more stations share a common transmission medium. When a station wishes to transmit a frame, it must wait for an idle period on the medium when no other station is transmitting and transmit the frame over the medium so that all other stations on the network hear the transmission. If another station tries to send data at the same time, a collision occurs. The transmitting station then sends a jamming signal to ensure that all stations are aware that the transmission failed due to a collision. The station then waits a random period of time before attempting to transmit again. This process is repeated until the frame is transmitted successfully.

The rules for transmitting a frame under command of CSMA/CD are simple and straightforward. The station monitors the network for the presence of an idle carrier or the presence of a transmitting station. This is the "carrier sense" part of CSMA/CD. If a transmission is detected, the transmitting station waits but continues to monitor for an idle period. If an active carrier is not immediately detected, and the period during which no carrier is equal to or greater than the interframe gap, the station begins frame transmission.

During the transmission, the frame monitors the medium for collisions. If it detects one, the station immediately stops transmitting and sends a 32-bit jamming sequence. If the collision is detected early in the frame transmission, the station will complete the sending of the frame preamble before beginning the jamming sequence, which is transmitted to ensure that the length of the collision is sufficient to be noticed by the other transmitting stations.

After transmitting the jamming sequence, the station waits a random period of time before attempting to transmit again, a process known as *backoff*. By waiting a random period before transmitting, secondary collisions are reduced.

If multiple collisions occur, the transmission is still repeated, but the random delay time is progressively increased with each attempt. This serves to further reduce the probability of another collision. This process continues until one of the stations successfully transmits a frame without experiencing a collision. Once this has occurred, the station clears its collision counter.

The Backoff Process

As we discussed a moment ago, the backoff process is the process with which a transmitting station determines how long to wait after a collision before attempting to retransmit. Obviously, if all stations waited the same amount of time, secondary collisions would inevitably occur. Stations avoid this eventuality by generating a random number that determines the length of time they must wait before restarting the CSMA/CD process. This period is called the *backoff delay*.

The algorithm used to manage this process in Ethernet is officially known as *truncated binary exponential backoff*. I strongly urge you to memorize that name so that you can casually throw it out during cocktail parties—it's very impressive. When a collision occurs, each station generates a random number that always falls within a specific range of values. The random number determines the number of slot times it must wait before attempting to retransmit. The values increase by an exponential factor following each failed retransmission. In the first attempt, the range is between zero and one. On the second attempt, it is between zero and three. For the third, it falls between zero and seven. The process continues, expanding each time to a larger range of wait times. If multiple collisions occur, the range expands until it reaches ten attempts, at which point the range is between zero and 1,023, an enormous spread. From that point on, the range of values remains fixed between zero and 1,023. Practical limits exist, however: If a station fails to transmit following 16 attempts, the MAC scheme issues an *excessive collision error*. The frame being transmitted is dropped, requiring that the application software reinitiate the transmission process.

Truncated binary exponential backoff is effective as a delay-minimization mechanism when LAN traffic is relatively light. When the traffic is heavy, it is less effective. When traffic is excessively heavy, collisions cause excessive collision errors to be issued, an indication that traffic has increased to the point that a single Ethernet network can no longer effectively manage it. At that point, network administrators should step in and segment the network as part of a traffic management process.

Slot Time

Slot times are critical in half-duplex Ethernet networks. The time varies depending on the bandwidth of the LAN (i.e., the version of Ethernet deployed), but it is defined to be 512 bit times for networks operating at both 10 and 100 Mbps and 4,096 bit times for Gigabit Ethernet. To guarantee the possibility of detecting collisions, the minimum transmission time for a complete frame must be at least the duration of one slot time, whereas the time required for collisions to propagate to all stations on the network must be less than a single slot period. Because of this design, a station cannot complete the transmission of a frame without detecting that a collision has occurred.

When stations transmit frames of data across the LAN, the transmission inevitably suffers unpredictable delays because of transit delays across variable-length segments, delays resulting from the electronics through which they must pass, processing delays, and hub complexities. The length of time it takes for a signal to propagate between the two most separated stations on the LAN is called the maximum propagation delay. This time period is important because the knowledge of it helps network designers ensure that all stations can hear collisions when they occur and respond to them. In order for a station on the LAN to detect that a collision has occurred between the frame it is transmitting and that of another station, the signal must have the opportunity to propagate across the network so that another station can detect the collision and issue a jamming signal. That jamming signal must then make its way back across the network before being detected by the transmitting station. The sum of the maximum round-trip propagation delay and the time required to transmit the jamming signal defines the length of the Ethernet slot time.

Slot times help to set limits on network size in terms of the maximum length of cable segments and the number of repeaters that can exist in a given path. If the network becomes too large, *late collisions* can occur, a phenomenon that occurs when the evidence of a collision arrives too late to be useful to the MAC function. The transmitted frame is dropped and application software must detect its loss and initiate the retransmission sequence. As a result, if a collision occurs, it will be detected within the first 512 bits (or 4,096 in the case of Gigabit Ethernet) of the transmission. This makes it easier for the Ethernet hardware to manage frame retransmissions following a collision. This is important because Ethernet hardware is relatively inexpensive, and because complexity adds cost, simple processes help to keep it inexpensive.

In Gigabit Ethernet environments, signals propagate a very short (and in fact unmanageable) distance within the traditional 512-bit timeframe. At gigabit transmission speeds, a 512-bit timeslot would support a maximum network diameter of about 20 meters. As a result, a carrier extension has been introduced to increase the timeslot to 4,096 bits. This allows Gigabit Ethernet to support networks as broad as 200 meters—plenty for certain applications in many metro environments.

Full-Duplex Ethernet

With the release of 802.3x, full-duplex Ethernet arrived on the scene, a protocol that operates effectively without the benefit of CSMA/CD. A full-duplex transmission allows stations to simultaneously transmit and receive data over a single point-to-point link. As a result, the overall throughput of the link is doubled.

For full-duplex to operate on an Ethernet LAN, facilities must meet the following stringent requirements:

- The medium itself must support simultaneous two-way transmission without interference. Media that meet this requirement include 10BaseT, 10BaseFL, 100BaseTX, 100BaseFX, 100BaseT2, 1000BaseCX, 1000BaseSX, 1000BaseLS, and 1000BaseT. 10Base5, 10Base2, 10BaseFP, 10BaseFB, and 100BaseT4 are incapable of supporting full-duplex transmission.
- Full-duplex operation is restricted to point-to-point connections between two stations. Because no contention exists (there are only two stations after all), collisions cannot occur and CSMA/CD is not required. Frames can be transmitted at any time, limited only by the minimum interframe gap.
- Both stations must be configured for full-duplex transmission.
- Needless to say, full-duplex operation results in a far more efficient network because no collisions occur and therefore no bandwidth is wasted.

Pause Frames in Full-Duplex Transmission

Full-duplex operation includes an optional flow-control mechanism called a *pause frame*. These frames permit one station to temporarily flow-control traffic coming from the other station, except MAC control frames used to manage transmission across the network. If one station transmits a high volume of frames, resulting in serious congestion at the other end, the receiving station can ask the transmitter to temporarily cease transmission, giving it the opportunity to recover from the onslaught. Once the time period specified by the overloaded station expires, the transmitter returns to normal operation. Pause frames are bidirectional, meaning that either of the peer stations on the link can issue them.

Pause frames conform to the standard Ethernet frame format but also include a unique type field, as well as a number of other variations on the traditional theme. For example, the destination address of the frame can be set to either the unique destination address of the station that the Pause command is directed toward or to a globally assigned multicast address, which has been reserved by the IEEE for use in pause frames. For those readers involved with bridging protocols, it is also reserved in the IEEE 802.1D bridging standard as an address type that will not be forwarded by bridges that it passes through, guaranteeing that the frame will not travel beyond its home segment.

The *Type field* is set to 88-08 in hex to indicate to a receiving device that the frame is a MAC control frame.

The *MAC Control Opcode field* is set to 00-01 to indicate that the frame being used is a pause frame. Incidentally, the pause frame is the only type of MAC control frame that is currently defined.

The *MAC Control Parameters field* contains a 16-bit value that specifies the length of the requested pause in units of 512 bit times. Values can range from 00-00 to FF-FF in hexadecimal. If a second pause frame arrives before the current pause time has expired, its parameter will replace the current pause time.

Finally, a 42-byte *reserved field* (transmitted as all zeros) pads the pause frame to the minimum Ethernet frame size. See Figure 4-15.

Preamble (7 bytes)	Start frame delimiter (1 byte)	Dest. MAC address (6 bytes) = (01-80-C2-00-00-01) or unique DA	Source MAC address (6 bytes)	Length/type (2 bytes) = 802.3 MAC control (88-08)	MAC control opcode (2 bytes) = PAUSE (00-01)	MAC control parameters (2 bytes) = (00-00 to FF-FF)	Reserved (42 bytes) = all zeros	Frame check sequence (4 bytes)

Figure 4-15
Full-duplex transmission with pause frame

Link Aggregation

Link aggregation, sometimes referred to as *trunking*, is an Ethernet feature used only in full-duplex mode. Essentially inverse multiplexing, link aggregation improves link performance and bandwidth availability between a pair of stations by allowing multiple physical facilities to serve as a single logical link. The link aggregation standard was created by the IEEE 802.3ad Working Group and entered into the formal standards in 2000.

Prior to the introduction of link aggregation, it was difficult to deploy multiple links between a pair of Ethernet stations because the spanning tree protocol commonly deployed in bridged Ethernet networks is designed to disable secondary paths between two points on the network to prevent loops from occurring on the network. The link aggregation protocol eliminates this problem by allowing multiple links between two devices. They can be between two switches, between a switch and a server, or between a switch and an end-user station, all offering the following advantages:

- Bandwidth can be increased incrementally. Prior to the introduction of link aggregation, bandwidth could only be increased by a factor of 10 by upgrading the 100 Mbps facility with a 1 Gbps facility.
- Link aggregation facilitates load balancing by distributing traffic across multiple facilities. Traffic can either be shared equally across the collection of aggregated facilities or it can be segregated according to priority and transported accordingly.
- Multiple links imply redundant transport capability. Under the link aggregation protocol, if a link in the collection fails, all traffic is simply redirected to another facility in the group, thus ensuring survivability.
- The link aggregation process is completely transparent to higher-layer protocols. It operates by inserting a "shim" protocol between the MAC protocol and the higher-layer protocols that reside above them. Each device in an aggregation of facilities transmits and receives using its own unique MAC address; as the frames reach the link aggregation protocol, the unique MAC addresses are "shielded" so that collectively they look to the higher layers like a single logical port.

Link aggregation is limited to point-to-point links and only works on links that operate in full-duplex mode. Furthermore, all the links in an aggregation group must operate at the same bandwidth level.

The Ethernet Physical Layer

Over the years since Ethernet's arrival on the LAN scene, it has gone through a series of metamorphoses. Today it still operates at a range of bandwidth levels and is found in a variety of applications ranging from the local area to limited long-haul deployments. In the sections that follow, we discuss the various flavors of Ethernet, focusing on the Physical Layer. We begin with the venerable Base5.

10Base5

10Base5 was the first 10 Mbps Ethernet protocol. It operated over 10 mm coaxial cable and was commonly referred to as "Thicknet" because of the diameter of the cable. A word about nomenclature: 10Base5 represents a naming convention crafted during the creation of the 802 Committee standards. The 10 refers to 10 Mbps, the transmission speed; Base means that the transmission is digital baseband; and 5 refers to the 500-meter maximum segment length. Other names in the series are similar.

The original coax used in Thicknet installations was marked every 2.5 meters to indicate the point at which 10Base65 transceivers could be connected to the shared medium. They did not have to be connected at every 2.5-meter interval, but they *did* have to be placed at distances that were multiples of 2.5 meters. This separation served to minimize signal reflections that could degrade the quality of signals transmitted over the segment. Ten mm Thicknet cable was typically bright yellow, with black indicator bands every 2.5 meters.

A Touch of Transylvania

In 10Base5 networks, transceivers were attached using a sort of clamp that wrapped around the cable and used needle-like attachments to pierce the insulation and make contact with the conductors within. These devices (see Figure 4-16) were often called *vampire taps* because of

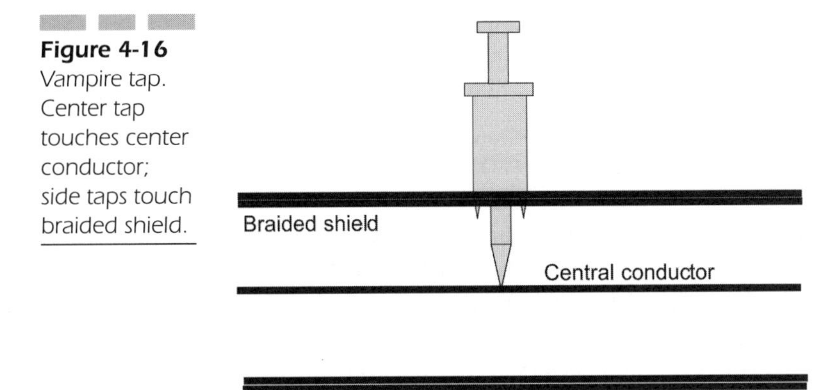

Figure 4-16 Vampire tap. Center tap touches center conductor; side taps touch braided shield.

the way they punctured the shared cable to get at the "lifeblood" within. They were often referred to as nonintrusive taps because there was no requirement to interrupt network operations to install them.

In these early versions of Ethernet, end-user stations were attached to the transceiver via a cable assembly known as an attachment unit interface (AUI). This AUI included the cable as well as a network interface card (NIC) that connected the cable to the NIC using a specialized 15-pin AUI connector. According to IEEE standards, this cable could be up to 50 meters in length, which means that stations could be that distance from the cable segment.

A 10Base5 coaxial cable segment could be up to 500 meters in length, and as many as 100 transceivers could be attached to a single segment, as long as they were separated by multiples of 2.5 meters. Furthermore, a 10Base5 segment could comprise a single piece of cable or could comprise multiple cable sections attached end to end, as long as impedance mismatches were managed properly to prevent signal reflections within the network.

Furthermore, multiple 10Base5 segments could be connected through segment repeaters, forming a large (but single and contiguous) collision domain. The repeaters regenerated the signal by transmitting it from one segment to another, amplifying the signal to maintain its signal strength throughout all the segments. The original standard permitted as many as five segments with four repeaters between any two stations. Three of the segments could be 1,500-meter coaxial segments, whereas the other two had to be point-to-point, 1,000-meter interrepeater links. Each end of the three segments could have a 50-meter AUI cable, resulting in a maximum network length of 2,800 meters.

10Base2

The 10Base2 protocol supports 10 Mbps transmission over 5 mm coaxial cable, a configuration sometimes called Thin-Net. It is also called cheapnet and was the first standard introduced after traditional 10Base5 that relied on a different physical medium.

In many ways, 10Base2 resembles 10Base5. Both use 50-ohm coax, and both use a bus topology. At the Physical Layer, the two standards share common signal transmission standards, collision detection techniques, and signal-encoding parameters. The primary advantage of 10Base2's thinner cable is that it is easier to install, manage, and maintain because it is lighter, more flexible, and less expensive than the thicker medium used by its predecessor. On the downside, the thinner cable offers greater transmission resistance and is therefore less desirable electrically. It supports a maximum segment length of 185 meters, compared to 10Base5's 500 meters, and supports no more than 30 stations per segment compared to 100 stations on 10Base5.

Although the spacing between stations on 10Base5 networks is critical for optimum performance, it is less so with 10Base2. The only limitation is that stations must be separated by no less than half a meter to minimize signal reflection.

Alas, vampires became extinct with the arrival of 10Base2, which uses a British Naval Connector (BNC) T-connector instead of a vampire tap to attach transceivers to the cable segment. As Figure 4-17 shows, the vertical part of the T is a plug connector that attaches directly to the internal or external NIC, whereas the horizontal arm of the T comprises two sockets that can be attached to BNC connectors on each end of the cable segments that are to be connected. When a station is removed from the LAN, the T connector is replaced with a barrel connector that provides a straightthrough connection.

To further reduce signal reflection in the network and the resulting errors that can arise, each end of a 10Base2 segment must be terminated with a 50-ohm BNC terminator, which is electrically grounded for safety.

Figure 4-17 A standard BNC connector

Figure 4-18
BNC connectors used to daisy-chain computers together in a LAN

10Base2 Topologies

Two wiring schemes are supported by 10Base2. The T connectors described earlier support a daisy-chain topology in which the cable segment loops from one computer to the next. Electrical terminators (50-ohm resistors) are installed on the unused connector at each end of the chained segment, as shown in Figure 4-18.

The alternative wiring scheme is a point-to-point model in which the segment connects a single station to a 10Base-2 repeater. This wiring scheme is used in environments where the structure of the building makes it impractical or impossible to daisy-chain multiple computers together. The segment is terminated at the computer station on one end and connected to a repeater on the other end.

10Base2 Summary

10Base2 operates in half-duplex mode at 10 Mbps over a single 5 mm coaxial cable segment. Because the cable is thinner, it is less expensive and easier to install, but because it is a narrower cable than 10Base5, transmission distances over it are somewhat limited compared to its predecessor. Further, when the protocol is deployed over a daisy-chained topology, it can be somewhat difficult to administer and troubleshoot when problems occur. The maximum segment length in 10Base2 is 185 meters, and the maximum 30 stations per segment must be separated by at least one-half meter.

10BaseT

The introduction of 10BaseT marked the arrival of "modern Ethernet" for business applications. It was the first version of the 802.3 standard to leave coax behind in favor of alternative media. 10BaseT provides a 10 Mbps transmission over two pair of Category Three twisted pair, sometimes called voicegrade twisted pair. Because it is designed to use twisted pair as its primary transmission medium, 10BaseT rapidly became the most widely deployed version of Ethernet on the planet.

The 10BaseT protocol uses one pair to transmit data and the other pair to receive. Both pairs are bundled into a single cable that often includes two additional pairs, which are not used in 10BaseT. This so-called multifunction cable may also include a bundle of fibers and one or more coaxial cables so that a single, large cable can transport voice, video, and data. Each end of the cable pairs used for 10BaseT is terminated with an eight-position RJ-45 connector.

Traditional 10BaseT connections are point to point. A 10BaseT cable therefore can have no more than two Ethernet transceivers, one at each end of the cable. In a typical installation, one end of the cable is usually connected to a 10BaseT hub, while the other end is to a computer's NIC or to an external 10BaseT transceiver.

In some cases, a pair of 10BaseT NICs are directly attached to each other using a special crossover cable that attaches the transmit pair of one station to the receive pair of the other, and vice versa. When attaching an NIC to a repeating hub, however, a normal cable is used and the crossover process is conducted inside the hub.

10BaseT Topology and Physical Layer Concerns

When Cat 3 twisted pair is used, the maximum length per segment is 100 meters. Under some circumstances, longer segments can be used as long as they meet signal-quality transport standards. Cat 5 cable, which is higher quality than Cat 3 but is also more expensive, can extend the maximum segment length to 150 meters and still maintain quality stringency, as long as installation requirements are met to ensure engineering compliance.

Although the original Ethernet products used a shared bus topology, modern Ethernet installations rely on a collection of point-to-point links that together make up a star-wired configuration, as shown in Figure 4-19. In this scheme, the point-to-point links connect to a shared central hub that is typically mounted in a central wiring closet to facilitate maintenance, troubleshooting, and provisioning.

Because 10BaseT uses separate transmit and receive pairs, it easily supports full-duplex transmission as an option. Needless to say, all components of the circuit—the media, the NIC, and the hub—must support full-duplex transport.

10BaseT Summary

10BaseT rapidly became the most widely deployed version of Ethernet LAN in the world because of its line speed and the fact that it operates over standard twisted-pair facilities. Also, because it supports a hubbed wiring scheme, it is easy to install, maintain, and troubleshoot. The only disadvantage it has is the fact that it supports shorter segment lengths than its predecessors, but its advantages far outweigh its disadvantages. In the functional domain of business, 10BaseT unquestionably dominates all other contenders.

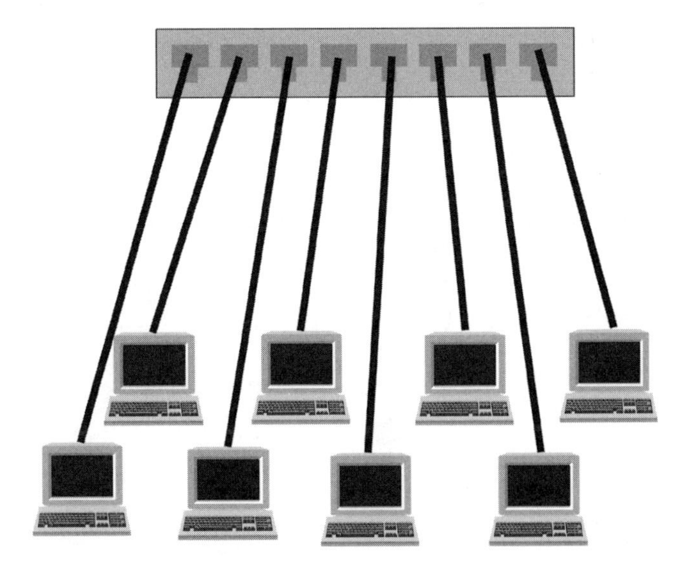

Figure 4-19 A star-wired LAN configuration. Each station is connected to a hub (top) that is typically (but not always) in a wiring closet.

Nomenclature

With all this talk about wiring schemes, topologies, facilities, and logical and physical layouts, I thought it might be useful to take a brief side trip to explain two terms that are often misused and almost always misunderstood. These terms are *topology* and *wiring scheme*.

For the purposes of our discussion, topology refers to the *logical layout of the devices on the shared medium.* On the other hand, wiring scheme refers to the *physical layout of the network and the devices on it.* A simple example will clarify the difference.

Figure 4-20 shows a ring network—perhaps a token ring LAN, for example. Topologically, this network is a ring. However, to guard against ring failure caused by a physical breach, rings are often wired as shown in Figure 4-21. If Station A should become physically disconnected, for example, a relay in the hub closes and seals the breach in the ring. The wiring scheme of this LAN is a star—the form that best describes its physical layout. In fact, this layout is known as a *star-wired ring*. Similarly, the LAN shown in Figure 4-22 relies on a shared bus *topology*. However, because the entire shared medium (the bus) is inside a physi-

Figure 4-20
A ring network

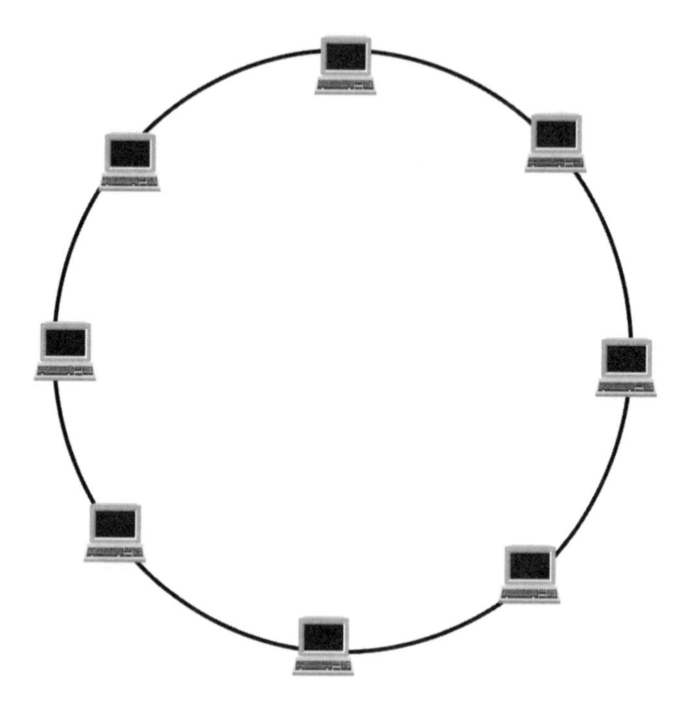

Figure 4-21
A bus-based LAN. All stations share access to the bus and therefore contend for its use.

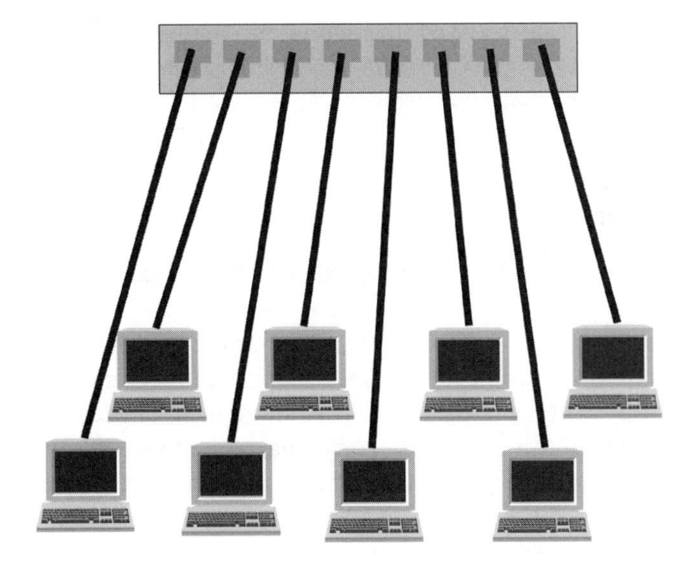

Figure 4-22
A shared bus Ethernet LAN. Note that the shared bus is inside the hub device.

cal hub, the single station LAN segments all emanate outward in a *star wiring scheme*.

10Broad36

Okay, if Ethernet works so well for LANs, perhaps it will work well for other applications, such as broadband video? 10Broad36 was designed to support the transmission of LAN traffic in addition to broadcast-quality video over the coax used in terrestrial Community Antenna Television (CATV) systems. It provides 10 Mbps of bandwidth over a broadband cable system. The 36 in 10Broad36 refers to the maximum span length supported by the standard between two stations—3,600 meters.

CATV, which, contrary to popular belief, does *not* stand for Cable Television, is an analog transmission scheme in which channels are created by dividing the available transmission spectrum into 6 MHz frequency bands and assigning each band to a broadcast channel. By managing the available bandwidth properly, broadband cable systems can support multiple services, including both data and video.

Broadband cable supports significantly longer segments than those supported by the original baseband 10Base5 and 10Base2. 10Broad36 segments can be as long as 1,800 meters, and all segments terminate at a head end (signal origination point). As a result, a span of 3,600 meters can be accommodated (far end of one segment through the head end to the far end of another segment). 10Broad36 transceivers attach directly to the cable both physically and electrically. The attachment cable can be as long as 50 meters, which adds another 100 meters to the overall end-to-end span of the system.

Fiber Optic Interrepeater Link (FOIRL)

As businesses went through their inevitable morphological shifts from monolithic designs to a more distributed functional model in response to the growing demands from customers in the metro marketplace, one of the demands that became somewhat urgent was the need for greater coverage distance without sacrificing performance or bandwidth availability. One of the first major evolutionary stages that emerged was the FOIRL, a two-fiber design that delivered 10 Mbps of bandwidth. It was designed to provide point-to-point connections with greater distances between repeaters. The standard was originally released in 1987, considered an early standard by those who study modern metro networks, but it was soon updated in concert with changes in technological demands placed on it. In 1993, 10BaseFL (fiber link) was released by the IEEE, which further refined and solidified the FOIRL concept.

FOIRL allows spans to be configured as long as 1,000 meters, significantly farther than any previous LAN standard. It should be noted that the newer 10BaseFL standard is backwards compatible with the older FOIRL technology. For example, a 10BaseFL transceiver can be deployed on one end of a facility, while an older FOIRL transceiver is used at the other. In this case, however, the maximum segment length is limited to the length specified by FOIRL (1,000 meters), not the 2,000-meter length supported by 10BaseFL.

10BaseF

The next member of the Ethernet pantheon is 10BaseF. Crafted as an enhancement to the earlier FOIRL standard, 10BaseF defines 10 Mbps operation over fiber. 10BaseF actually defines three segment types: 10BaseFL, 10BaseFB, and 10BaseFP. For reasons that will become clear in the paragraphs that follow, the three are not compatible with each other at the optical interface level.

10BaseFL

10BaseFL offers 10 Mbps transmission over a pair of fibers. The standard augments the FOIRL standard and supports a maximum segment length of 2,000 meters compared with FOIRL's 1,000 meters.

10BaseFL has a variety of applications. It can be used to connect computers, repeaters, or a computer to a repeater. All segments are point to point in 10BaseFL, with a single transceiver at each end of the segment. Typically, a computer attaches to the shared infrastructure via an external 10BaseFL transceiver. The NIC in the computer attaches to the external transceiver through a standard AUI cable. The transceiver, on the other hand, attaches to the two fibers using standard ST connectors. One fiber is used to transmit data, while the other is used to receive.

10BaseFL (fiber link, sometimes fiber loop) was originally designed to work with multimode fiber, because at the time of its creation multimode fiber was affordable and readily available, whereas single mode was only used by large service providers in long-haul installations due to cost.

The independent send and receive paths defined in 10BaseFL allow full-duplex mode to be supported as an optional configuration. In full-duplex mode, 10BaseFL supports 2,000-meter segment lengths. Furthermore, segment lengths are not restricted by the round-trip timing requirements of the CSMA/CD collision domain. In fact, segment lengths as great as 5 kilometers (3 miles) can be supported, and much longer segments can be installed if SMF is deployed.

High-quality multimode fiber and transceivers can support segment lengths of 5 kilometers. Even longer distances can be supported with the more expensive SMF.

10BaseFL is ideal for connecting between buildings. In addition to supporting longer segment lengths, fiber optic cables are immune to electrical hazards such as lightning strikes and ground currents that can occur when connecting separate buildings. Fiber is also immune to electrical noise that can be generated by motors or other electrical equipment.

10BaseFB

10BaseFB (fiber backbone) is primarily used to interconnect repeaters at 10 Mbps, using a unique synchronous signaling protocol. The signaling protocol allows the number of repeaters in a 10 Mbps Ethernet network to be extended. Like 10BaseFL, 10BaseFB segments can be as long as 2,000 meters.

Two factors limit the number of repeaters that can be deployed between two stations. First, each repeater adds a signal delay that can cause the time it takes for collisions to propagate throughout the network to exceed the maximum 512-bit time limit. Second, repeaters introduce random bit loss in the preamble field that can result in an overall reduction of the interframe gap below the required 9.6 microseconds (at 10 Mbps). 10BaseFB increases the number of allowable repeaters by reducing the amount of interframe gap loss. It does this by synchronizing the transmission interplay between the two repeaters. The interframe gap at the output of a normal repeater can be reduced by as many as eight bits. With 10BaseFB repeaters, the loss is reduced to two bits. This simplifies network design dramatically; suddenly, the only limiting factor on the size of the network is the time required for collisions to propagate.

10BaseFB can only be used on point-to-point links between repeaters, and both repeaters must support 10BaseFB. It *cannot* be used to connect a computer directly to a repeater. Furthermore, it does not support full-duplex mode.

10BaseFP

10BaseFP (fiber passive) relies on what is known as a *fiber optic passive star* topology. Based on passive optics, the passive star requires no power, can connect as many as 33 devices, and supports 500-meter segments. The star comprises a bundle of fibers that are fused so that a signal from the ingress fiber can be transmitted to a bundle of egress fibers at 10 Mbps. Although not widely used, this model is gaining acceptance in cable television infrastructures because it requires no local power for signal propagation. In effect, the star serves as a passive hub that receives signals from 10BaseFP transceivers and distributes them to all the other 10BaseFP transceivers. It supports half-duplex transmission only.

Fast Ethernet (100BaseT)

100BaseT refers to the specifications and media standards for 100 Mbps Ethernet, often called *Fast Ethernet*. Four standards have been defined: 100BaseTX, 100BaseFX, 100BaseT4, and 100BaseT2. Each will be discussed in the following sections.

All 100BaseT standards share a common MAC scheme, but each has its own unique Physical Layer specification, sometimes referred to as a PHY, and usually defining the characteristics of the transceiver used in the system. The PHY can be integrated directly into a repeater or NIC, or attached externally.

100BaseTX 100BaseTX is a 100 Mbps Ethernet transmission scheme using two pairs of twisted pair, one for transmit, one for receive. A typical 100BaseTX wiring scheme uses a cable that often houses four pairs. However, because of the technology's high sensitivity to crosstalk, it is strongly recommended that the two remaining pairs be left unused. The standard supports 100-meter transmission over 100-ohm Cat 5 UTP cable. Category 5 is a "higher-quality" conductor than the Cat 3 cabling typically used with 10BaseT and is rated for frequencies up to 100 MHz, whereas Cat 3 supports transmission up to 16 MHz.

All 100BaseTX segments are point to point with a single transceiver at each end of the cable. Most connections are designed to connect a computer to a hub; these hubs typically have an integrated transceiver in 100BaseTX installations.

100BaseFX 100BaseFX is a two-fiber architecture that also operates at 100 Mbps. The segment maximum length is 412 meters for half-duplex links and 2,000 meters (or longer) for full-duplex designs. 100BaseFX is basically an optical equivalent of the 100BaseTX standard described earlier; the cabling and connectors found in 100BaseTX installations are simply replaced with fiber and optical connectors. Both standards use the same 4B/5B signal encoding scheme.

100BaseFX segments are point-to-point facilities with a single transceiver at each end of the link, and the transceivers are unique to the standard. 100BaseFX allows full-duplex mode, which allows the longer 2,000-meter segment design to be implemented. As before, longer distances can be accommodated with SMF.

100BaseT4 100BaseT4 supports 100 Mbps transmission over four pairs of Cat 3 (or higher) twisted pair, a strong advantage given the low cost of Cat 3. In metro operations, where cost is a factor, such advantages tend to be noticed.

In 100BaseT4, one of the four pairs serves as a transmit pair, one as a receive pair, and two bidirectional pairs used to transmit or receive data. This model seems odd but actually makes a lot of sense. It ensures that one pair is always available for collision detection, while the three remaining pairs are available for data transport. 100BaseT4 does not support the full-duplex mode of operation because it cannot support simultaneous transmit and receive at 100 Mbps.

100BaseT2 100BaseT2 is the only standard that supports 100 Mbps transmission over two pairs of Cat 3 twisted pair. And unlike 100BaseTX, if additional pairs are available in the cable sheath, 100BaseT2 permits them to be used to transport voice or other LAN traffic. Even though 100BaseT2 is an innovative and powerful standard, it is not widely deployed today.

Gigabit Ethernet (1000BaseX)

1000BaseX, otherwise known as Gigabit Ethernet, is a relatively new arrival on the scene that has turned the LAN world on its ear and had an extraordinary impact on the metro world. Suddenly, corporate LANs are no longer hobbled by their glacially slow 10 Mbps networks or even their all too slow 100 Mbps LANs. As with the Fast Ethernet series, Gigabit Ethernet standards actually define a family of standards, described in this section. Interestingly, these standards are based on Physical Layer specifications adapted from the American National Standards Institute (ANSI) X3.230-1994 standard for Fibre Channel, a high-speed interface specification that has long been used for storage applications. By adopting the existing Fibre Channel specifications, 1000BaseX products found the market significantly easier to penetrate. It is important to note, however, that 1000BaseT does not use the Fibre Channel Physical Layer specifications and is not part of the 1000BaseX family of standards. 1000BaseT is described later.

1000BaseLX 1000BaseLX is a fascinating standard, and the technologies that underlie it are equally interesting. The L in LX stands for long, because it relies on long-wavelength lasers to transmit significant amounts of data over long-haul fiber. The lasers specified by the standard operate in the 1,270- to 1,355-nanometer range, and both SMF and multimode fiber are supported. Although long-wavelength lasers are more expensive than short-wavelength devices, they can transmit reliably over longer distances.

1000BaseSX The antithesis of 1000BaseLX, 1000BaseSX ("short") relies on short-wavelength lasers operating in the 770- to 860-nanometer range to transmit data over multimode fiber. These lasers are less expensive than their long-wavelength cousins and are ideal for short distances, such as many of the fiber runs required in metro environments. The maximum segment length for these systems is between 275 and 550 meters, depending on the transmission mode (half or full-duplex) and the type of fiber deployed.

1000BaseCX The more things change, the more they stay the same. The C in 1000BaseCX stands for copper because it uses specially designed, shielded, balanced copper cables, sometimes called twinax. Segment distances are limited to 25 meters, but even with such a short run there is a clear application for high-speed copper for the interconnection of network components within a building.

1000Base-T In June 1999, the IEEE released 802.3ab, the formal definition for 1000BaseT. The standard defines Gigabit Ethernet with transmission distances up to 100 meters over Cat-5-balanced copper facilities.

The 1000BaseT PHY relies on full-duplex baseband transmissions over four pairs of Cat 5 cable. The aggregate data rate of 1,000 Mbps is achieved by transmitting 250 Mbps over each pair. Hybrids and echo cancellation devices support full-duplex transmissions by allowing data to be transmitted and received on the same pairs simultaneously.

10 Gigabit Ethernet

To confirm the high-speed future of Ethernet, the IEEE 802.3ae Working Group has defined two Physical Layer standards for 10 Gbps optical Ethernet. These include a LAN PHY and a WAN PHY. The LAN PHY does not differ from the existing LAN standard, whereas the WAN PHY is new. It specifies asynchronous 10 Gbps transmissions using the traditional 802.3 Ethernet frame format, stipulates the minimum and maximum frame sizes, and recommends a MAC scheme at the client interface. At 1,550 nanometers, it will initially reach 40 Km, expanded to 80 Km in coming months.

Currently, 10 Gbps Ethernet is standardized for transport over fiber only; no copper twisted-pair option is available as there is in Gigabit Ethernet. When the standard was first conceived, little thought was given to a need for that kind of bandwidth to the desktop. However, the need is beginning to emerge and some chip manufacturers are now planning for 10 Gbps Ethernet over copper, using a design that will transmit 100 meters without a repeater. Working closely with the standards bodies, this capability will no doubt be reached soon.

Ethernet Summary

So why has Ethernet suddenly become such a powerful contender in the metro space? There are numerous reasons. Ethernet has been around for a long time and is understood, easy to use, and trusted. It is a low-cost solution that offers scalability, rapid and easy provisioning, granular bandwidth management, and simplified network management due to its single-protocol nature. As its popularity has increased and it has edged into the WAN environment, it now supports such applications as large-scale backup, streaming media server access, Web hosting, ASP connectivity, SANs, video, and disaster recovery.

VLANs are critically important in environments that require logical network segmentation, and at Champlain College this has proven to be a major issue—and an advantage. Often characterized as a VPN made up of a securely interconnected collection of LANs, a number of vendors now offer VLANs as a service. TLAN or target LAN server (TLS) services have gained the attention of the industry; they allow multiple LANs to be interconnected in a point-to-point fashion. An example is shown in Figure 4-23 where we see a company that has multiple locations inter-

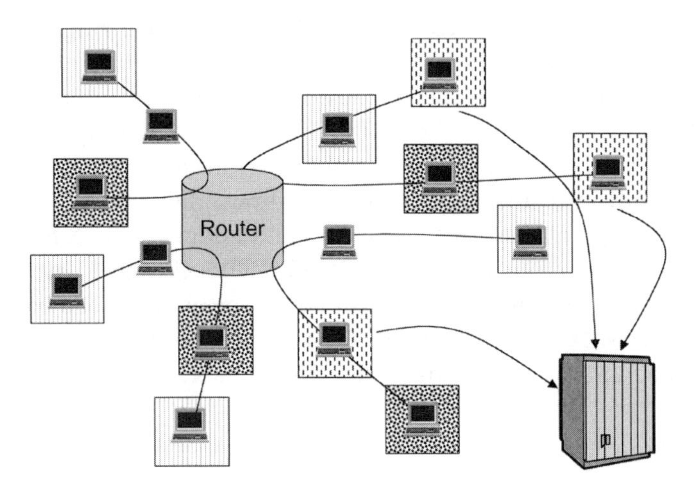

Figure 4-23 A typical VLAN. The PCs with different "background colors" belong to different VLANs.

connected via a high-speed facility to create a VLAN. Users in both locations can interconnect with each other, and the high speed of the network connection gives the appearance of collocation.

Many vendors are now selling Ethernet Private Line (EPL), an emerging replacement for legacy dedicated private lines. Implemented over SONET/SDH, the service combines SONET/SDH's reliability, security, and wide presence with Ethernet's rapid, flexible provisioning ability. These companies' products support the transport of SNA, video, SAN, and VoIP application traffic.

Ethernet Disadvantages (Yes—There Are Some!)

Of course, Ethernet presents some disadvantages. These include the potential for packet loss and the resultant inferior QoS. However, two new standards have emerged that will go a long way toward eliminating this problem. The IEEE 802.17 Resilient Packet Ring (RPR) standard is designed to give SONET/SDH-like protection and reliability to Ethernet using a Data Link Layer protocol that is ideal for packet services across not only LANs, but MANs and WANs as well. Primarily targeted at voice, video, and IP traffic, RPR will restore a breached ring within 50 ms. The IEEE began the 802.17 RPR standards project in December 2000 with the intention of creating a new MAC layer. Fiber rings are widely used in both MANs and WANs; these topologies unfortunately are dependent on protocols that are not scalable for the demands of packet-switched networks such as IP. The 802.17 RPR Working Group promotes RPR as a technology for device connectivity. Issues currently being addressed are bandwidth allocation and throughput, deployment speed, equipment, and operational costs. The second standard, ITU X.86, maps Ethernet (OSI layer two) frames into SONET/SDH Physical Layer (OSI layer one) frames for high-speed transmission across the broad base of legacy networks. Table 4-3 compares the relative characteristics of Gigabit Ethernet and the soon-to-be-released 10 Gigabit Ethernet.

Back to VLANs: 802.1Q

The 802.1Q standard defines the interoperability requirements for vendors of LAN equipment wishing to offer VLAN capabilities. The standard was crafted to simplify the automation, configuration, and management of VLANs, regardless of the switch or end-station vendor.

Table 4-3 Ethernet Media Options

Characteristic	Gigabit Ethernet	10-Gigabit Ethernet
Physical media	SM and MM fiber, WideWave Division Multiplexing (WWDM, 802.3z)	SM, MM fiber
Distance	3.11 miles	40 Km
MAC	Full, half-duplex	Full duplex
Coding	8B/10B	64B/66B, SONET/SDH
Physical media dependencies	850 nm, 1300 nm, 1550 nm WWDM	850 nm, 1300 nm, 1550 nm
Supports 10 Gbps TDM and DWDM?	No	Yes
Packet size	64–1,514 octets	64–1,514 octets

Source: 10 Gbps Ethernet Alliance

Like multiprotocol label switching (MPLS), 802.1Q relies on the use of priority tags that indicate service classes within the LAN. These tags form part of the frame header and use three bits to uniquely identify eight service classes. These classes, as proposed by the IEEE, are shown in Table 4-4.

802.1p

Closely associated with 802.1Q is 802.1p,[3] which enables the three QoS bits in the 802.1Q VLAN header to specify QoS requirements. It is primarily used by layer 2 bridges to filter and prioritize multicast traffic. The 802.1p QoS bits can be set by intelligence in the client machine, as dictated by network policy established by the network management organization. In practical application, 802.1p can be converted to DiffServ for QoS integration across the wide area. After all, 802.1p is really a QoS specification for LAN environments, most typically Ethernet. Therefore, the DiffServ byte in the IP header can be encoded at the edge of the network by the ingress router, based on information contained in the 802.1p field in the Ethernet frame header. At the egress router, the opposite occurs, guaranteeing end-to-end QoS across the wide area.

Table 4-4
IEEE LAN Service Classes

Priority	Binary Value	Traffic Type
7	111	Network control
6	110	Interactive voice
5	101	Interactive multimedia
4	100	Streaming multimedia
3	011	Excellent effort
2	010	Spare
1	001	Background

[3]The fact that the Q is uppercase and p lowercase is not an accident. IEEE 802.1p is an adjunct standard and does not stand on its own. IEEE 802.1Q, on the other hand, is an independent standard.

Of course, these standards only address the requirements of LANs, which will inevitably interconnect with WAN protocols such as ATM or IP. Consequently, the Internet Engineering Task Force (IETF) DiffServ committee has developed standards for interoperability between 802.1Q and wide area protocols such as IP's DiffServ, whereas the ATM Forum has a similar effort underway to map 802.1Q priority levels to ATM service classes.

Differentiated Services (DiffServ)

DiffServ has the ability to prioritize packets through the use of bits in the IP header known as the Differential Services Code Point (DSCP), formerly part of the Type of Service (TOS) field. It relies on per hop behaviors (PHBs), which define the traffic characteristics that must be accommodated. The best known of these is the Expedited Forwarding PHB, designed to be used for services that require minimum delay and jitter such as voice and video. DiffServ then is a technique for classifying packets according to QoS requirements. Because the classification process occurs at the edge of the network, it scales well as the network grows.

DiffServ breaks the responsibilities of traffic management into four key areas, based on the overall architecture of the network, as illustrated in Figure 4-24. At the customer's access router, traffic is managed according to flow requirements and is clustered for delivery to the service provider's network. The traffic is then handed to the service provider's

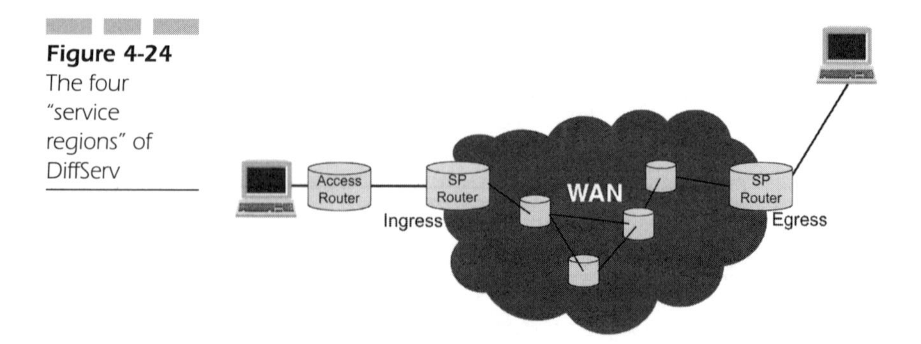

Figure 4-24 The four "service regions" of DiffServ

ingress router, which sits at the edge of the network and is responsible for implementing the Service Level Agreement (SLA) between the customer and the service provider.

Once the edge routers have classified the traffic, the core routers can handle it according to the DSCP "markers" they assign to each packet. Within the network, then, core transit routers simply route the traffic as required. By the time the traffic arrives at the core, the edge devices have already classified it, and the core router simply handles the interior routing function. Ultimately, the traffic reaches the service provider's egress router, another edge device, which performs additional traffic-shaping functions to ensure compliance with the SLA. Thus, the core can be extremely fast because the classification process has already been done at the point of ingress.

DiffServ therefore is *not* an end-to-end protocol. Traffic, perhaps from a LAN, arrives at the edge of the network, where DiffServ's domain begins. The ingress and egress routers manage and shape traffic flows, with the freedom to use packet discard if appropriate.

Riding above IP is the layer where IT considerations are brought to the fore. These include media-related services such as voice and video, and operational services such as call setup, user privacy, authentication, and various forms of location management. Voice digitization is the most important application to be managed at this level because it is among the most sensitive to delay of all the IP-related applications. For traditional uncompressed voice, the signal should be digitized using G.711 and transmitted at 64 Kbps. However, compressed voice is a different beast. G.729 is the most common compression algorithm, although its broadband cousin, G.722, should also be available. It should be noted that not all vendors implement both 722 and 729; compatibility issues can arise, so ask early and often.

For videoconferencing, which requires significant levels of bandwidth for proper QoS support, compression will also be called for. ISDN is the most common technique used today for videoconferencing transport, relying on H.261 and H.263. H.264 has recently been introduced by a few vendors but is not as commonly used. However, its enhanced capabilities will undoubtedly make it more common, and it must be watched closely.

The operational services we mentioned earlier are largely signaling related. Call setup and related services are typically managed by the signaling protocols described earlier that perform much the same function as the legacy Signaling System 7 (SS7). These protocols include the IETF's SIP, the ITU's H.323, the Media Gateway Control Protocol (MGCP), also from the IETF, and MeGaCo/H.248, jointly deployed by the ITU and the IETF. H.323 was the first to be deployed, but as we observed earlier SIP has crept up on it and is now the preferred signaling protocol for converged network operations. MGCP and H.248, while interesting, seem to be losing favor and will probably give up in favor of SIP. Other functions performed at this level include user privacy, location-based services, which include presence management, and user authentication.

Service-Affecting Issues

Because of the architecture of the PSTN, the number of service-affecting issues is relatively limited. Because of its reliance on dedicated (or seemingly dedicated) resources, the list is restricted to physical network failures and the occasional security breach for the most part. IP telephony, however, has a long and noble list of potential "gotchas" that must be addressed by network managers if VoIP is to offer a carrier-grade lookalike. They include latency, jitter, packet loss, bandwidth availability, infrastructure reliability, and network security.

Latency

Latency is a measure of the time it takes a packet to make its way across the network from the source to the destination. For the most part, latency does not present an overwhelming problem other than an annoying discontinuity between the two communicating parties. Excessive latency can result in a "synch problem" between the two communicating endpoints, similar to what is occasionally experienced with mobile telephony when one speaker "talks over" the other because of unequal delay in each direction. The standards call for an end-to-end latency of less than 150 ms if a call is to emulate carrier-grade PSTN quality. To achieve this, network designers must take into account a number of factors that can contribute to latency. These include the following:

- The process of actually creating the packet on the sending side and populating it with data to be transmitted, a function performed by the coder/decoder (CODEC), and the process on the receiving side, during which the receiving CODEC must "depacketize" the data for reassembly. As long as the packets are small (a desirable characteristic in VoIP networks), they are relatively easy to process and should not create delays in excess of 30 ms.

- The delay inherent in dealing with the physical transmission facility. As long as the transmission facility is of an adequately high bandwidth level, the delay that is incurred in the process of serialization will not be objectionable.
- The interval required for the transmitted signal to traverse the network, typically referred to as propagation delay. For most enterprise installations, propagation delay is not noticeable unless (1) facility bandwidth is low and (2) the distance between the two communicating entities is extreme.
- The period of time during which a packet waits for transmission from the source, typically called *queuing delay*. Queuing delay is the result of either excess traffic impinging on an overloaded (and therefore improperly engineered) server or a configuration parameter that has been improperly set. Naturally, a facility that is improperly configured or engineered ultimately results in latency that can grow objectionable if left untended.
- The interval required for an intermediate device to receive a packet, open and examine it, and make a forwarding decision based on the destination address contained in the packet. This particular issue can be complicated by such factors as security limitations related to forwarding, access control lists that require additional processing, and so on.

Jitter

Jitter is an interesting measure in networks. It is fundamentally a comparison of the expected and actual packet arrival times and the variability between the two. Put another way, jitter is a measure of the variability in delay from one packet to the next, a characteristic that can result in unacceptable voice quality.

The relationship between delay and jitter is interesting. As long as the majority of the packets are delayed by the same amount, QoS experiences no perceptible impact. If, however, the delay is variable, then problems arise. That's jitter.

Jitter occurs for a variety of reasons, the most common of which is variations in queuing delay caused by unexpected shifts in network traffic. A secondary—but equally common—cause is routing variability. If one packet takes a two-hop route to the destination and the next takes a seven-hop route, jitter will result because of the variable delay that will occur from packet to packet.

Media gateways typically have buffers (see Figure 4-25) that collect a stream of packets in a logical "bucket" and then meter them out to the end-user device at a measured rate to ensure proper levels of performance. This process does not eliminate jitter, but it does reduce its impact, sometimes dramatically. However, if the jitter becomes extreme, "the bucket can go dry," resulting in an inadequate packet supply and substandard voice quality.

Packet Loss

We are talking about a connectionless network, after all, and packets do occasionally get lost. Packet loss has rational causes such as buffer overflows, during which routers discard packets as a survival mechanism. This can be catastrophic for real-time applications such as video and voice, but less so for non-real-time applications such as Web access and

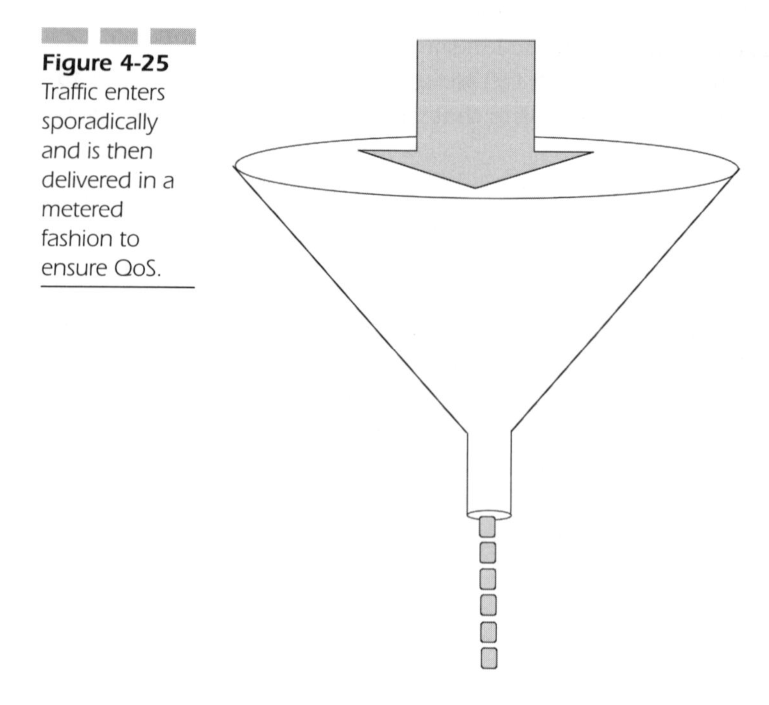

Figure 4-25 Traffic enters sporadically and is then delivered in a metered fashion to ensure QoS.

e-mail. Of course, voice packets must have (to a certain extent) their own rules. Signaling and voice packets clearly should not be lost or discarded because of the impact of degraded QoS. Packet prioritization and class of service (CoS) control are therefore powerful tools that the network manager can use to his or her advantage. This is not to say that packet loss cannot be tolerated, but limits do exist. Roughly speaking, as long as the rate of loss is kept below 5 to 10 percent and is spread across a reasonably large population of users, it will be undetectable for the most part.

Bandwidth

Bandwidth considerations become less than straightforward in converged networks because of the sometimes unpredictable, bursty behavior of packets. If we consider the most basic *characteristics* of voice and data, we see why this can be a difficult task. Voice is inordinately forgiving of packet loss but highly intolerant of delay. Data, on the other hand, has the opposite requirements: It easily tolerates delay but is not the least bit forgiving of packet loss, because each packet contains data that is necessary to reconstruct the original message. If the available bandwidth is not properly allocated and the voice and data packets are not prioritized as they should be, QoS becomes an issue.

That being said, a converged network based on IP is the right way to go for several good reasons. If the VoIP network were to use the same CODEC encoding parameters as those used in the PSTN to ensure the delivery of toll-quality voice, the bandwidth required for the transport of the voice and signaling content would be higher than that required in the PSTN. This would be because of the excessive protocol overhead that IP layers onto its transported payload. However, because VoIP networks rely on such capabilities as silence suppression and digital compression to reduce their total bandwidth requirements, and because they are not built on a fixed bandwidth model like the PSTN, applications running on converged networks actually require less bandwidth than their circuit-switched counterparts.

It is critical therefore that network administrators not only allocate bandwidth based on peak calling times as they do when engineering the PSTN, but they must also set aside bandwidth for signaling traffic. Signaling activities generate traffic on the network and the traffic volume naturally increases as a function of the calling volume at any point in time. Generally speaking, signaling should be engineered into the system at approximately three to five percent of the voice and data traffic that the network is expected to carry.

Infrastructure

"Our greatest concern as we designed the network was to guarantee that we didn't have single points of failure," said Aaron Videtto. He pointed to a drawing of the campus network, pointing out the various areas where they focused a great deal of attention during the design phase. "I mentioned earlier that we went to great pains to locate the redundant call servers on opposite ends of the campus to protect ourselves from a physical failure that could take down both devices. But interestingly enough, the architecture of the converged network itself turned out to be the greatest protection we could have. Because it's a distributed model, there is no single point that could fail and take down large parts of the campus. Naturally, a device could fail in a building, which would isolate the people working there, but we have battery backup everywhere and the backbone is carefully designed to avoid areas where disruption could occur. All in all, we feel very good about the way the network behaves. And even when we have had problems, they were minor and were rapidly resolved."

For larger installations or environments where disruption of the network can result in the loss of hundreds of thousands of dollars per minute, such as in a brokerage house or major order-taking center, failover strategies are a good idea as an additional redundancy measure. For example, redundant facilities are often installed between critical areas in the network, and backup devices such as power supplies and processor cards are often installed in failover mode such that the failure of a device will cause an automatic "rollover" to the backup device, thus reducing the impact of the failure on the user. Naturally, converged networks rely heavily on gateways and gateway controllers, and network designers should therefore take into account the level of redundancy they wish to have in the network for those devices.

By using the rich capabilities of the IP routing protocols that establish the path through the network for routed traffic, gateway devices can monitor and detect the status of the next hop (most likely the default gateway) before transmission, thus incorporating predictive behavior into their failover strategy. Similarly, the hop count can be substantially reduced in some installations by simply collocating the devices, thus eliminating the possibility of a link failure, which isolates large swaths of network users.

Security

As voice becomes just another component of the data stream, it also becomes vulnerable to the same kinds of intrusive attacks that plague data. At the most basic level, the enterprise can protect itself through the deployment of Secure Shell (SSH) protocols and Remote Authentication Dial-In User Service (RADIUS), an authentication and accounting system that requires users to enter their username and password, which are then handed off to a RADIUS server. The RADIUS server verifies the correctness of the information before providing authorized access. SSH and RADIUS, however, cannot provide 100 percent security against such threats as Denial-of-Service (DoS) attacks. Other techniques are required. One is to use private addressing to securely identify the media gateways and call-processing servers. Private addressing is a form of proxy service; the private address is not advertised to the public and devices identified in this manner are therefore not identifiable (or addressable!) by the outside world.

Firewalls should also be put into place as an added layer of security. All devices—call-processing servers, gateways, and messaging systems—should be "firewalled" to protect them from DoS attacks. The firewall must be able to dynamically open VoIP ports when they are required, but they must also be configured to immediately close them when the call is complete to protect against unauthorized access to the network behind the firewall. Firewall policies should be designed to control VoIP activities, restricting its use based on communication with authorized devices or vetted IP addresses. For tracking purposes, firewalls should generate activity logs that can be used for attack analysis and to provide forensic backup should it be required.

Managing the IT Integration

With all the attention being paid in the technology industry today to VoIP and convergence, it's sometimes easy to forget that the overall equation has another side. Convergence isn't just a telecom phenomenon; it has staggering impacts on the IT environment as well, and because voice becomes an IT application during the convergence process, it is prudent to stop for a moment and think about the IT implications. For example, what about the in-place IT infrastructure? Can the installed base of servers, management systems, PCs, application software, and operating systems work as is, or will they require (at least in part) an upgrade to a more enhanced or capable version? And assuming that the application software is fully functional (or can be easily upgraded), how many of the current applications will be ported over to the new system because some of them will be integrated? Is the network infrastructure (LAN, MAN, WAN) robust enough in terms of redundancy, survivability, and robustness to support the addition of voice and potentially video to the traffic mix that they will be required to transport? Similarly, how capable are the existing interfaces to the PSTN, the user's applications (where appropriate), and other carriers?

In the final analysis, the key consideration is the return on investment (ROI). ROI calculations are tricky under the best of circumstances, but they are important and do provide insight into the economic gains that can be expected from such a significant capital and expense outlay. And although the ROI is really a measure of future performance, a current indicator is simply the offset to the total cost of ownership, measured in reduced capital expenditure (CAPEX) and operating expenditures (OPEX), as well as more efficient and effective business practices that result from the migration. This is why the benchmark measures described earlier are so important; they provide a starting point from which advancement is measured.

A Plethora of Considerations

A conversion of this magnitude requires answers to a number of complex questions before a final decision can be made. In addition to the vendor-related issues discussed previously, it is also important to ask questions that are more business oriented. For example, are the requirements that will be placed on the new infrastructure appreciably different from those that were placed on the earlier system? In most cases, that will not be the case, but what *will* be the case is that the users of the system will expect the same level of performance that they are accustomed to with the prior system. One question, then, revolves around power: Are there situations within the enterprise that require carrier-grade telephone service but that will not have access to PoE? If so, then alternative backup power will have to be installed at a perhaps significant expense. What about the nature of the access devices that the firm's workers use? Are they exclusively Windows or are there also Mac and Linux machines that will require access to the system? If so, what will be the nature of the softphone client that these devices require?

Similarly, infrastructure requirements related to the dispersed nature of the business deserve consideration. Do you intend to converge all servers into a common location or will they be distributed as a way to ensure survivability? Is the application set so critical that a backup site will be required, or are network and device diversity with appropriate levels of power and data backup adequate? And finally, how diversely does the business operate in terms of geography? Will telephony access be required in multiple countries or is the operation primarily domestic in nature?

Applications also yield their own set of questions. Is one intent of the converged infrastructure that of centralizing voice mail, eliminating the cost of commercial audio conferencing, introducing company-wide unified messaging, and integrating the call center functions with Web access? If so, then it deserves additional attention.

Economic Considerations

Economic considerations come into play as well. What is the value of the investment that your firm already has in digital devices that can be redeployed across the new system? Although not an option in Champlain's case, it is important to consider whether the company's existing PBX can be retrofitted for IP operation. If so, what is the cost to do so versus engaging in a full-scale replacement? What about the relative state of the voice and data networks? Data equipment tends to age and become obsolete faster than voice equipment; does this conversion call into question any issues related to the two being out of economic lockstep? Will the company incur additional expense if the data side of the house needs upgrading before the voice side?

Also, personnel and HR-related considerations cannot be underestimated. For example, to what degree is the workforce geographically distributed? If the answer is "more than substantially," it may make sense to converge your current applications set, including telephony, secure instant messaging, application sharing, and video as a way to reduce cost and ensure the delivery of functionality to remote workers. It is highly likely that the converged infrastructure will yield greater advantages for some workers than for others, and that's okay. Remember that the needs of the business must be met first, and if enhancing the workplace for remote workers yields substantial returns in terms of customer service, then the perceived imbalance of benefit may be perfectly justified. Ultimately, the move comes down to a measurable value: If migrating to a converged model of operation is good for the business, and the cost can be justified with measurable future value for the enterprise, then it deserves serious consideration.

Functional and Organizational Impacts

As the process of designing and "bounding" the project continues, a number of steps must be taken in sequence to ensure that all the appropriate considerations have been taken into account. The business reasons, discussed in the previous section, must be considered first; they are driven by the "big four:"

- Will this conversion increase our revenues?
- Will this conversion reduce our CAPEX and OPEX spending?
- Will this conversion stabilize our competitive position?
- Will this conversion somehow mitigate whatever downside risk we face from operating in an increasingly competitive market?

At this stage, attention must be paid to the overall scope of the project, taking into account such issues as geographic scope, the nature of the work force, and the evolving manner in which employees and customers rely on the network. For example, will the conversion affect a single site or will it affect multiple locations? Are they geographically contiguous or are they spread across a broader area that could have toll implications? Will the conversion affect only the corporate LAN, including wireless and desktop users, or will it also have an impact on the wide area because of remote or small-office, home-office (SOHO) workers?

Other considerations exist as well. What percentage of the workforce is mobile? How many users will be regularly and routinely changing their work location or roaming, thereby requiring secure remote access to the network? This becomes important, simply because an argument can be made that mobile or remote workers receive greater benefits from a converged network infrastructure than office workers who rarely change location. Finally, if a call center is involved in the enterprise, is it physical or virtual? Are there multiple locations, or is there a single site?

"We also wanted to take into account the impact that this conversion could have on the other organizations that we partner with," said Dusini. "For example, we have a number of cooperative arrangements with other schools in the area that are both academic and commercial. For example, we share in the Cat Scratch program." Such a program offers a student charge card that is accepted by area businesses, allowing students to buy products and services without having to carry cash or a credit card. "And it occurred to us that we could do all kinds of things for that program with the new system. We could give parents the ability to recharge their kid's card over the Web using a secure link and a credit card, we could allow area vendors to advertise specials for students, and so on. We began to realize that this thing's impact had the potential to go way beyond Champlain College and to have a positive impact on the entire Burlington area."

Of course, other considerations have a potential cost associated with them. This is not a bad thing, but it requires careful planning to avoid unanticipated surprises in the middle of the conversion process. Most of these impacts are people related. For example, what impact will the conversion have on IT operations personnel who have to monitor and maintain the network? How will human resources have to adapt to ensure we have the right people who are properly trained and in the right places to take advantage of the system's expanded capabilities? What unexpected benefits—or, possibly, liabilities—will emerge and confront organizations throughout the campus? And what about security considerations?

"That's a whole new set of problems," conceded Dusini. "Because this is an IP network, and because voice packets are like any other data packet, we realized that if we did a bad job with the up-front design work as it relates to security, it would become relatively easy to capture someone's phone conversation, or worse. Luckily our campus security organization is appropriately paranoid, and when we began to design the VLAN architecture they came to us and told us that they were (justifiably) worried about students having the ability to record their roommates' conversations. The way the network operates in the dorms, the PC is typically plugged into the phone, so it uses one of the available switch ports. On a topologically flat network you'd have issues, but because we segment traffic using the VLAN, we have eliminated that concern for the most part. But we do have other concerns and we consider them our top operational priorities. Keep in mind that we have call manager and E911 servers on the network as well. So what happens when somebody breaks into the network and floods the Cisco CallManager with a Denial-of-Service attack? That could bring the network to its knees. So we spent a great deal of time working out management scenarios for every possible development. It's not pleasant thinking about such things, but we have no choice—it's part of the job."

Taking Care of FCAPS

Network management falls into five distinct areas of responsibility: fault management, configuration management, accounting management, performance management, and security management. These five responsibilities (FCAPS) are equally important if not more so in converged networks where both voice and data reside. As rich applications evolve and the enterprise becomes more dependent on them, network and IT managers must put into place a management program that ensures the availability of network-based voice and data services and must treat them as the highest-priority activity for which they have responsibility. As part of the up-front planning process, management must discuss the strategic implications of convergence and how the environment will be managed postconversion. Questions should be asked in each of the five areas:

- **Fault management:** What procedures will be put into place to ensure that faults are detected, isolated, reported, and corrected in the shortest interval possible? What is the ideal reporting mechanism for fault resolution? What kind of database must be created to track troubles and generate trouble history reports?
- **Configuration management:** What documentation procedures should be put into place to ensure that configuration changes are properly tracked? How is the configuration database logically linked to the trouble-reporting database should problems arise from configuration errors or discrepancies?
- **Accounting management:** How are delivered services billed? Are they billed on a flat-rate basis or is the billing algorithm usage based? Is there a chargeback mechanism in place or are telecom and IT services centrally funded? Are there procedures in place to handle exceptions, such as inordinately high costs resulting from a work group with higher than normal MAC requirements?
- **Performance management:** What are the performance benchmarks for each delivered application? Are they reasonable? How often are they tested against reality? How often are they tested to ensure that they comply with the needs of the business? What

reporting procedures are in place to ensure that users have a feedback mechanism to report performance problems? Is there a maintenance schedule in place and published that sets aside time specifically for maintenance related to performance?

- **Security management:** What specific security requirements have been identified and accepted as critical components of secure service delivery? What secure mechanisms (IP Security [IPSec], secure VPN tunneling) are used for remote access? How are remote and branch offices connected securely to centralized resources? What audit procedures are in place to ensure password expiration, secure wireless access, and VLAN management? How discretely are firewalls managed? Are there any specific, unique security requirements that must be dealt with on an individual basis?

Application Support

Why is it so important to manage the forces listed in the prior section? Because most well-run organizations are bound to have a collection of applications running in their functional pantheon that provides day-to-day functionality for the business and that represents the company's window on the market—and their customers. They may include the following:

- A PC-based IP telephony application that supports mobile users and functions under Windows, Mac OS, and Linux.
- A collection of call-control services that facilitate outbound dialing, enterprise application-driven screen pops to the user device, call screening, and the ability to generate analytical management reports.
- A messaging application that offers full-featured, centrally managed voicemail that provides such functions as message forwarding, e-mail integration (unified messaging), voice mail, and fax support, as well as the ability to integrate applications and create hybrid applications—the ability to reply to an e-mail message with a voice response, for example, or to speak to a machine and have it intuitively interpret what is said. Aaron Videtto explained: "We use a Cisco speech-to-text program called Nuance in our call director, for example. When a user speaks the name of the department they wish to connect to, we have programmed the system to verify the

department name first. For example, as soon as a caller speaks the word 'Admissions,' the system looks up the number that we have programmed for that department and then double-checks with the user to ensure that it has correctly understood their request by reading back the name in the database. This makes the customer happy, but it also makes *me* happy. It prevents me from having to record the names of all of the departments on campus."

- Follow-me services with which system users can be located based on user-defined caller types. Here are a few examples:
  - "I am not in the office. Press one to have my agent locate me."
  - Follow a sequential "path" to reach the user in at least three locations: first a cell phone, then a home phone, and finally an alternate number such as a Skype or Vonage identity.
  - Determine call handling based on caller ID information.
- IP-based interapplication conferencing that combines the capabilities of instant messaging, audio conferencing, videoconferencing, and file and document sharing. Visual collaborations in the form of such well-known applications as screen-sharing and shared whiteboarding have become commonly used collaboration tools.
- IP contact-center software to control customer interactions by leveraging IP telephony systems with industry-leading call routing, management, and monitoring capabilities from vendors such as Aspect, Genesis, and Siebel.

The typical corporation, by now well into its own convergence of internal telecommunications and IT organizations, will also need a collection of support functions. These include the following:

- Centralized LAN, WAN, and IP telephony management.
- A common configuration and administration function for all applications.
- Well-tested survivability strategies for engaging distributed call control in the event of call controller, IP network, or PSTN failure.
- Generation and collection of call detail records for accounting and inventory purposes, as well as for compliance with Sarbanes-Oxley and other regulatory mandates.

True network convergence, which integrates the functions of IP telephony and the data network, requires

- That the corporate or organizational LAN be able to automatically detect IP phones when they go live on the network, and invoke traffic prioritization within the switching infrastructure.
- That LAN and VoIP applications interoperate to enable E911. At Champlain College (and many other venues), this is done by fundamentally linking the MAC address of a VoIP phone to its IP address and then linking it to a campus information database for instantaneous location processing.

Chapter Summary

The conversion from a PSTN-based calling environment to a converged, all-in-one IP-based architecture is not for the faint of heart. In a well-thought-out plan, its results will be nothing less than stellar, but it should not be viewed as a magic bullet. Remember the telco mantra: If it ain't broke, don't fix it. VoIP is a good idea, but only if it's a good idea!

In our next chapter, we examine the operational considerations of a VoIP environment, taking into account such issues as regulation and management.

CHAPTER 5

Cable Plant

Aaron Videtto shakes his head and laughs when he describes the process of assessing the cable plant and then deciding how to bring it up to speed for the voice over IP (VoIP) conversion that the school decided to pursue. "The cable plant here at the school was designed for delivering telephone service to the buildings on campus—and not much else," he explains. "Basically, we have an 1,800-pair cable that snakes its way through the campus, providing connectivity to all of the buildings. Every time we pass a building with the cable there's a splice, and at the splice we drop off 50 pairs." He gestures out the window at the collection of classroom and administrative buildings, Lake Champlain shimmering under a blanket of ice in the distance. "This campus has 42 buildings. That's a lot of splices. And keep in mind that the cable is 14 years old. Also, don't forget that Vermont has five seasons: Spring, Summer, Fall, Winter, and Mud Season. During mud season everything melts and the splices fill with water. In the winter they freeze, and expansion creates all kinds of strange phenomena, including arcing and other electrical problems. It was a real nightmare because calls were dropped and there was unpredictable but often serious noise on the line. Because the plant was so old a significant percentage of the wire pairs were marginal or bad, so whenever I added or moved a phone I'd have to look for a good pair. It was like having a poltergeist in the network: I'd have a dead signal, or the lights on the phone would randomly flash on and off. Then, to really spice things up you'd often hear echo and crosstalk.

"It was not a pretty picture, and we had to do something. We needed a catalyst, a triggering event, and that came when we decided to build the three new buildings."

The construction of the three new buildings—the SD Ireland Center for Global Business and Technology, the Main Street Suites and Conference Center, and the new Student Life Complex—provided the impetus to reexamine at the overall design philosophy of the campuswide network. The cable plant that interconnects the various functional areas of the enterprise (or in this case, of the college) is clearly one of the most critical components of the entire installation—and among the most vulnerable. It must be configured and sized to assume growth and changes in the applications it will transport, something that most enterprises know far more about today than they did a decade or more ago.

"When we decided to look at our options for moving forward, we found that we had a couple," says Videtto. "All the buildings were wired in 1981 with cabling, which at the time was state of the art—but unfortunately far below par for our purposes. In fact, it was pre-CAT3. So that presented us with a serious challenge. Cisco, one of the vendors we worked closely with on the project, came in, took a look at the plant, and told us that we may have a problem.

"Let me describe what we had before we went through the conversion. In each of the rooms we had one jack plate. We had one four-pair wire from the basement to each jack. Each jack plate had two RJ-11s, with two pair wired to one RJ-11, and two on the other. In the basement we had a punch-down block. From there, two pairs cross-connected to the PBX [to the 1,800-pair cable described earlier].

"This is where the whole thing gets a little hinky. We took a CAT5 cable, cut off one end, plugged the other end into the switch, frayed the cut end, and plugged it into the punch-down block. That's how the network cables get to each room. The phones plugged into the RJ-11 jack without a problem. But now we have a challenge: We have four students in each room, sharing one connection. So to counter the inevitable contention, we have a little hub hanging off the cable that provides four additional connections. So we brought in Cisco, and they took one look at it and laughed. They told us that it probably wouldn't work, but they agreed to work with us to at least see if we could get it to perform. So what we ended up doing was actually pretty clever. We cut the cable, terminated it with RJ-45 connections, and put a powered hub in the basement to feed them. This allows us to deliver power via Ethernet to the rooms, so we can ensure service continuity in the event of a commercial power failure. So this was the final design, and it actually works quite well: We go from the phone to the wall, then from the wall to the hub, and then the students plug into the hub. And because we had to do this to a large number of locations, we went to Panduit, a local cable vendor, and had them build us a special cable that consists of two CAT5 cables bonded together. That way the students have only one wire crossing the floor. Panduit made the cables 265 feet long, so they can be snaked across the floor of the dorms to just about anywhere they need to be. Students plug their computers into the hub, and everybody's happy."

Of course, the story doesn't end there. "Meanwhile, Cisco took a piece of the wire back to their labs because they had to have the business unit approve our rather unusual wiring scheme before they would sell their solution to us. After a short while they called and told us that it would work. In fact, they sounded as surprised as we were! So we signed the contract, and at that point we were off and running."

All of the campus buildings are connected by a fiber backbone, and the various powered hubs and switches are connected to the backbone, providing high-speed, high-volume (and growth-resistant) connectivity throughout the campus. Interconnecting the various segments of the backbone are three Cisco fiber switches. Furthermore, every basement has anywhere from one to six switches, depending on the population of the building, and all are daisy-chained together with a single fiber that homes back to the building where our central management capability is located."

Videtto and the installation team had to be somewhat judicious in their overall design because of Ethernet cabling restrictions. For example, the maximum length of twisted-pair cable that interconnects Ethernet switches to other devices such as Internet Protocol (IP) phones and computers is about 100 yards. Optical connections are also limited, although they can run significantly farther—up to about 500 yards.

I asked Videtto about the level of uninterruptible power that the campus requires. "Because we have to offer 911 service, we must have uninterruptible power. Originally we had a turbine; now, we have decentralized power plant that is in keeping with our decentralized service delivery model. In every basement we have a 2-hour battery backup device on every powered switch. We went with 2 hours of backup power because we discussed the situation with the housing people. We asked them one question: How long will they wait to move people to alternative housing in the event of a long-term power failure? Their response was 2 hours."

Private Branch Exchange (PBX) Cabling

In today's enterprise network, CAT5 twisted-pair cable is the standard for most installed infrastructures. It has the ability to transport both voice and data and is relatively cost effective for hybrid voice-data installations.

The interface used in these installations does not have to resemble that described for Champlain College. Desktop devices require different interfaces; telephones, for example, require an RJ-11 jack, whereas Ethernet devices (PCs) require a larger RJ-45 connector.

Moves, adds, and changes in the PBX often require changes to the wiring and *always* require database updates. Most corporate telephony administrators claim that about 15 percent of their users move each year, and the typical cost per move ranges from $150 to $200. As the corporation grows and the number of moves, adds, and changes grows with them, this cost can grow to a rather significant level (not including the cost of the disruption caused by all that movement). And although many

PBX installations are wired in such a way that all connections are "home-run" back to a central location and terminated on an Ethernet switch to reduce management complexity, the cost is still high. As we will see, VoIP environments reduce this cost significantly—by effectively eliminating it.

In Support of Call Centers—and Customers

Perhaps the single most compelling reason to migrate to a converged network infrastructure is the enhanced ability to support call-center operations and, by extension, customers.

The advantages of an IP-based voice communication system over a traditional PBX system for call center or *customer relations management* (CRM) applications include greater flexibility in distributing calls and easier integration of voice and data. As described previously, in an IP-based system, voice goes where the IP network goes. Therefore, in call-center environments calls may be distributed to users anywhere on the IP network. Although distributing calls between local and remote offices is also possible with traditional PBX systems, the ability to easily move users from one location to another is a strong advantage of IP-based voice systems. This allows, for example, a customer support expert to move from a central site to a home office, and easily remain connected to the call distribution group. This flexibility is becoming extremely important as businesses define their CRM strategies. The Internet is making this requirement even more urgent. With less business being done in person and more attention being paid to customer satisfaction, it is critical that your voice communications system work with key CRM applications to provide the highest level of customer service possible. This is key to building customer loyalty. Solutions for managing customer contact and ongoing relationships range from informal contact-center capabilities to a highly sophisticated call-center solution built around an *automatic call distribution* (ACD) module. The ACD provides the function of queuing and managing the distribution of calls to the appropriate agents. Calls entering the contact center can initially be handled by the *interactive voice response* (IVR) module, which helps to determine how to best service the customer. Often, the customer is handed off to an available agent with the appropriate information. At that point, a screen pop-up delivers customer database information to the agent's PC.

Today, the call center is transitioning and taking on a much broader role in the enterprise. This new expanded role is being described within the umbrella concept of CRM, which enhances call-center technology with applications tailored to specific business functions, such as sales force automation and customer support services. The expanded role of CRM is driving improvements in the way call centers (or, more appropriately, call noncenters) need to work: CRM must expand to include distributed employees: With the growing number of remote workers, mobile workers, and teleworkers, it is important that the CRM system be able to leverage agent skills, regardless of where they sit in the organization. Distributed, location-independent CRM delivered wherever there is IP access enables more employees to take responsibility for managing customer relationships. Systems must be able to handle any media type: With the rise of the Internet it is no longer possible to assume that voice (telephony) will be the only way of communicating with businesses. E-mail, web forms, instant messaging, and chat rooms are all widely used, and all these tools provide an opportunity to exceed customer expectations and build loyalty. Queues must link to agent skills. Customers with a particular interest or requirement do not want to be routed from one agent to another; they want to deal with one person who can provide the assistance they need. Similarly businesses must use the most appropriate person for a call, based on their business objectives (cost, workload, and results). Getting the right person to handle the call ultimately costs less and enhances the customer experience.

System Survivability

We have already discussed the importance of uninterruptible power in converged systems. There are, however, other factors that must be considered to ensure that the system is truly carrier-grade and can provide the intended applications in the face of environmental problems. These factors fall into five categories: backup, system redundancy, physical access, security, and network access.

Backup

As IT and telecom converge, the need to perform scheduled backups of critical databases becomes even more compelling. Daily incremental backups should be performed on all systems, and full backups should be performed following any major change to the system. Complete archival (image) backups should be performed weekly. Backup media should be archived at an offsite facility.

System Redundancy

Critical hardware devices that are central to the routine functionality of the network should be configured with redundant components such as power supplies, cooling fans, and backplanes. A sufficient spares inventory should be kept on hand at all times. If possible, servers should support hot-swap capability, allowing maintenance personnel to replace failed or marginal components without bringing down the entire system. Network interfaces should be redundant as well, facilitating automatic failover in the event of a network outage on one or more links. Network management procedures should be established so that notifications and alerts of failures or degraded service situations occur in real time, and failover occurs at a subsecond interval.

Closely related is the issue of personnel training. All management personnel should be properly trained on procedures related to troubleshooting, problem resolution, escalation, notification, documentation, and other critical areas.

Physical Access

Physical access could be lumped under the section on security, but there is more to it than that. All server rooms should be secured against unauthorized entry and should have 24-hour surveillance to protect against the possibility of a door being inadvertently left open.

Security

Security covers a sweeping range of responsibilities, including both physical access, described previously, and logical access. The unauthorized access to a corporate network can be catastrophic, particularly when that network hosts both voice and data applications. Most corporations have taken steps to physically and logically secure their enterprise networks, and although their efforts are effective they are also costly. "Complete security" is not really possible (after all, the only system that is truly secure is one that has no connection to the outside world, rendering it relatively useless to anyone who isn't sitting at its keyboard). However, reasonable trade-offs exist between *securing* the network and the *cost* of securing the network. Carnegie Mellon's *Computer Emergency Response Team* (CERT) serves as a nexus for Internet security problem reports. The organization provides technical support and coordinates responses to broad-spectrum security threats such as worms, viruses, and Trojan horses; identifies and tracks trends in intruder behavior; works closely with other organizations to resolve security problems; and publishes volumes of information on network and computer security. It's good that they do what they do because their single biggest adversary is the Internet itself. Designed as a peer-to-peer network, the Internet essentially allows any device on the network to see and interact with any other device on the network. This characteristic is commonly exploited by hackers and other ne'er-do-wells who use a variety of attack techniques to identify and exploit system vulnerabilities. These techniques include IP spoofing, in which packets are "disguised" to look like they were sent from a trusted source; denial-of-service attacks, under which a network is inundated with a flood of packets serving to disable certain network components and making it easier for attackers to gain access to the network; and synchronization (SYN) floods, through which a target device (router) is flooded with Transmission Control Protocol (TCP) connection requests, resulting in a slowdown or failure of the target machine.

Paul Dusini and Aaron Videtto were very concerned about the potential for system penetration when they worked on their own network design for Champlain College. They knew that IP telephony introduces its own set of critical vulnerabilities, which meant that it was critical that they did the best job possible in the early stages of network design to address potential security flaws. Because their network calls for the PC to be plugged into the IP phone, they knew they had to use a virtual local area network (VLAN) design to eliminate the potential for one person capturing someone else's data. And because they knew they would have call managers and Enhanced 911 (E911) servers, both crucial components of a carrier-grade network, they had to build in protection and redundancy against such things as flood or denial-of-service attacks.

Network managers can take a number of steps to secure their networks against outside attack without creating an environment that is onerous for the users. For example, network administrators can establish rule sets that allow end-user devices to access outside Web sites, but only if they do so via a web proxy server so that source addresses are hidden. Similarly, e-mail messages can be required to pass through a proxy device for the same reason.

Naturally, more restrictive rules can be applied, but they come with a cost. E-mail messages can be opened automatically upon arrival so that they can be scanned for viruses or other destructive attachments, but this introduces delay, cost, and questions about user privacy. Network managers must therefore strike a balance between security and paranoia.

Network Access

The enterprise network must have access to the public switched telephone network (PSTN) to facilitate on-net to off-net communication. Network access must therefore be managed and monitored. For example, network managers may want to install multiple PSTN connections (and in some cases wide area network [WAN] connections) to protect against network failures. Many network managers put into place a process for automatic cutover and notification in the event of a failure.

An Important Aside: Billing as a Critical Service

One area that is often overlooked when companies look to improve the quality of the services they provide to their customers is billing. And although it is not typically viewed as a strategic competitive advantage, studies have shown that customers view it as one of the top considerations when assessing capability in a service provider.

Billing offers the potential to strengthen customer relationships, improve long-term business health, cement customer loyalty, and generally make businesses more competitive. However, for billing to achieve its maximum benefit and strategic value, it must be fully integrated with a company's other operation support systems, including network- and service-provisioning systems, installation support, repair, network management, and sales and marketing. If done properly the billing system becomes an integral component of a service suite that allows the service provider to quickly and efficiently introduce new and improved services in logical bundles; improve business indicators such as service timeliness, billing accuracy, and cost; offer custom service programs to individual customers based on individual service profiles; and transparently migrate from legacy service platforms to the so-called next generation network.

In order for billing as a strategic service to work successfully, service providers must build a business plan and migration strategy that takes into account integration with existing operations support systems, business process interaction, the role of IT personnel and processes, and postimplementation testing to ensure compliance with strategic goals stipulated at the beginning of the project.

Billing and Usage Capture

Telephony and data services delivered through an enterprise PBX are necessary to the proper care and feeding of the enterprise and its employees. However, it does represent a significant cost, and that cost must be managed. The good news is that PBXs, like all phone switches, generate call detail records, controlled by settings input at the management console, that can be studied and analyzed to determine usage patterns. Usage data can be generated regarding a user, department, time of day, and called number, and custom reports can be generated as well.

Another Aside: The Odd World of Regulation and VoIP

VoIP regulation is a highly misunderstood topic that is critically important to telecom managers and administrators, regardless of the size of their network or the perceived importance of the services it delivers. The regulation of IP networks continues to be a topic of serious discussion among regulatory bodies, content providers, service providers, and various consumer protection agencies.

Before we dive into the specifics of regulation as it relates to IP networks, let's take a few minutes to talk about regulation in general. Misconceptions about the role of regulation must be clarified if regulation of VoIP networks is to be effective. We begin with an overview of recent regulatory activities.

Recent Regulatory Actions

In the early months of 2003, the FCC dramatically rewrote the basics of telecom regulation. Following the 2003 Triennial Review that concluded in February of that year, the FCC released a series of high-impact decisions that pleased everyone—and no one.

In an effort to resolve the question of which *unbundled network elements* (UNEs) that the incumbent local exchange carriers (ILECs) had to make available to their competitors, the FCC essentially turned the issue over to the states for local resolution. They also ruled that ILECs no longer had to provide switching as an unbundled element to competitive local exchange carriers (CLECs) for business customers unless state regulatory agencies could prove within 90 days of the order that this resulted in an overly onerous impairment to their ability to do business.

In the residence market, the FCC outlined specific criteria that each state had to use to determine whether CLECs were impaired without unbundled switching. They also concluded that ILECs did not have to provide competitors with SONET-level transport services (OC-N), but shifted responsibility to the states to determine whether ILECs should be required to unbundle dark fiber and DS3 transport on a route-by-route basis.

For broadband, the FCC's decision was clearer. They ruled that all new broadband buildouts, including both fiber builds and hybrid loops, were exempt from unbundling requirements as well. They also ruled that line sharing would no longer be classified as an unbundled network element, a decision that put considerable pressure on competitive carriers, such as Covad Communications, that rely on line sharing.

Apparently, the decisions reached by the commission were far from universal. Commissioner Kevin Martin was the only member who agreed with the entire order. Chairman Powell (who has since left the agency) and Commissioner Kathleen Abernathy disagreed with the decisions about line sharing and unbundled switching, while Jonathan Adelstein and Michael Copps disagreed with the broadband decision.

The feedback from various industry segments about the decisions was, as would be expected, varied. CLECs grudgingly agreed for the most part that it was about as good as it could have been, and it could have been far worse. ILECs, on the other hand, saw the decision to turn much of the decision making over to the states as a step in the wrong direction, concluding that it would extend the period of time over which definitive decisions would be made.

What we do know is this. While the decisions made by the fragmented FCC were not ideal for all players (how could they possibly be), they did move the industry forward and shook up the players in a positive way. ILECs now had a renewed incentive to invest in broadband infrastructure, while competitive providers had incentives to invest in alternative technologies such as cable and wireless (both fixed and mobile) in their competitive efforts. Digital subscriber line (DSL) rollouts were accelerated, and as penetration climbed alternative solutions were invoked. So while the results were not as comfortable as they could be for the industry players, they led to the appropriate marketplace behavior—and that's a good thing.

Consider the following scenario. Regulators, after a great deal of wrangling, remove unbundling requirements, providing the appropriate degree of incentive to incumbent providers to accelerate broadband deployment. ILECs publicly commit to universal broadband (DSL) deployment throughout their operating areas. In response, cable and wireless players accelerate their own rollouts, preparing for the price wars that will inevitably come.

The broadband deployment effort involves infrastructure. Shortly after the service announcement, ILECs issue RFPs and RFQs for DSL access multiplexer (DSLAMs), optical extension hardware, and DSL modems, and they begin to jockey for content alliances since DSL provides the necessary bandwidth for television-signal delivery in addition to voice and high-speed data. Customers, meanwhile, excited by the prospect of higher-speed access and all that it will bring them in the way of enhanced capability, buy upgraded PCs and broadband service packages. Hardware manufacturers and service providers applaud the evolution because they know that the typical broadband user generates 13 times the traffic volume that a dial-up user generates. And while service providers rarely charge by the transported megabyte, the increased traffic volume requires capability upgrades to the network—capital expenditure (CAPEX), in other words.

Meanwhile, software manufacturers, mobile appliance makers, and content owners scramble to develop products for wireline and wireless broadband delivery. Upgrades occur, innovation happens, prices come down. Cable and wireless players march along with their own parallel efforts, and soon this great, dynamic money engine known as the telecom industry starts to turn again, slowly at first but building rapidly as it feeds on its own self-generated fuel. And that is the ultimate end state—a self-perpetuating industry that evolves and changes in concert with market demand, sustained by a forward-thinking, reasonable regulatory environment.

What this boils down to is Michael Powell's ultimate goal: to force the industry at large to become facilities based. The fact is that the local loop is a natural monopoly—there's no getting around that. The UNE rules are basically unnatural and simply don't work as well as what was envisioned. Want to compete with the incumbent service providers? Use a different technology—like cable or broadband wireless. Clearly this strategy is working in the cable industry's favor.

Other Regulatory Activity

On May 12, 2003, the FCC commissioners were presented with a proposal to modify the existing rules that govern media ownership. On the surface, this does not appear to affect the technology sector per se, but in fact it potentially has a significant impact because of a blurring of the lines between the sectors. Under the terms of the proposal, ownership of broadcast rights could change dramatically. Among the changes are the following.

A proposal to allow a single company to own television stations that reach as many as 45 percent of U.S. households, today capped at 35 percent. Needless to say, the major networks are in favor of eliminating the existing 35 percent cap. This decision would favor companies like News Corporation (the owner of Fox) and Viacom (the owner of CBS and UPN), which are already in violation of the 35 percent limit because of M&A activity that put them (slightly) over the top. Some inviolable restrictions were part of the decision: The four major networks (CBS, NBC, ABC, and Fox) are prohibited from merging, and ownership of more than eight broadcasting stations in a single market remains prohibited.

Two cross-ownership rules are under scrutiny and proposals to modify them are included in the proposal. One prevents a company from owning a newspaper *and* a radio or television station in the same city; the other limits ownership of both radio and television stations in the same market. Under the terms of the proposal, these two rules would be combined into a single rule that eliminates most of the existing restrictions. Cross-ownership would be allowed in large and medium-size markets but would be restricted or perhaps even banned in smaller markets.

Under the terms of federal communications law, the FCC is required to consider reasonable changes to the rules it oversees that affect the communications marketplace and the public it serves. Since some of these laws were first written into law over 50 years ago, they are in dire need of revision. In many cases the changes are driven by growing influence of the cable and Internet-dominated sectors.

More Recent Events

In March 2004, an appeals court struck down FCC rules for how regional telephone companies must open their networks to competitors. Federal law originally required regional phone companies to lease parts of their networks (the UNE-P mandates) to competitors at reasonable rates, set by the states. The ILECs have long contended that they have been forced to give competitors rates that are below their actual cost.

In its decision, the appeals court found that the FCC wrongly gave power to state regulators to decide which parts of the telephone network had to be unbundled. The court also upheld an earlier FCC decision that the ILECs are not required to lease their high-speed facilities to competitors at discount rates the way they do their standard phone lines.

In April 2004, regulators rejected AT&T's petition to eliminate the requirement that they pay long-distance fees on calls transported partially over the Internet.

The decision means that AT&T could be required to pay hundreds of millions of dollars in unpaid retroactive fees to the ILECs.

Earlier in 2004 the FCC ruled that calls originating and terminating on the Internet, such as those made using Skype, Vonage, and other voice-over-Internet service providers, are free from the fees and taxes that traditional phone companies are required to pay (for example, as support for E911 and Universal Service Fund). The FCC said that because calls that travel over the Internet don't provide anything in the way of enhanced features, standard rules apply. Furthermore, in a recent decision, the FCC consolidated all regulatory power over VoIP service at the federal level, wresting it from the states.

The key here is that regulatory agencies are trying to balance the need for a regulated telecommunications marketplace with the need for an unfettered development environment that can technologically innovate without fear of onerous fees and taxes. The FCC wants to engender a spirit of facilities-based competition, rather than the UNE-based environment that has not worked as well as hoped.

Perhaps the reason for this is a realization on the part of the regulators of exactly what their role should be. For the longest time it appeared that regulators had come to believe that their primary responsibility was to create competition. In fact, the primary goals of regulation are twofold: first and foremost, to ensure that users of the regulated service (whether it be telecommunications, power, or airlines) have the best possible experience while using the service and, second, to "fill in the gaps" where the market fails to create the appropriate environment. Now regulators can use competition as a way to achieve these goals, but competition is a means to an end, not the end itself.

VoIP Implications

Regulators should not be in the business of regulating technology per se, but rather the application of technologies according to the mandates listed previously. Today, significant regulatory attention is being paid to VoIP. But consider this: VoIP is a technology, not a service. What comes out of the phone is still the same voice that has always come out of the phone. And although VoIP today enjoys a certain amount of immunity from regulatory forbearance, this is a temporary situation. The time will come, and not all that long from now, when VoIP providers will reach "critical mass" and will find themselves under the same degree of regulatory scrutiny as other more traditional providers find themselves under.

Ongoing discussions about how VoIP should be treated continue to rage at both the federal and state levels. A number of factors are fueling the fire, including the fact that large, incumbent, and important companies depend for their very survival on the system of taxation and regulatory continuity that has guided the telecom industry for many years. And it isn't just the telephone companies: government agencies, interested in preserving their ability to levy taxes, install legal wiretaps under the auspices of the Community Assistance for Law Enforcement Act (CALEA) and guarantee the availability of such mandates as universal service and E911.

Another key issue revolves around general misunderstandings of just what VoIP and its technological entourage are capable of doing for society at large. Regulators, legislators, and other policymakers need to understand that voice is just one more form of content and that it is becoming part of a services "borg" that will yield untold value in the form of converged and unified application sets.

Of course, there is the fear factor. As focus shifts from carriage to content, so does the realization that the money lies not so much in the network but rather in what the network transports. The conduit is another commodity, and the differentiation that all players strive for lies in the content. As VoIP (via Vonage and Skype) destroys the telco voice services fiefdom, IP-based videoconferencing and messaging erode the data side of the business—leaving them with transport as their primary business service—unless they change the very definition of their role as a *service provider*. Clearly the incumbent players will jump on the IP bandwagon and will become major VoIP providers, but that represents one service with shrinking revenues. The real threat comes from convergence and from the growing number of subscribers, both residences and enterprises, that rely on their PCs and laptops, their gateways and gatekeepers, and their Session Initiation Protocol (SIP) servers to deliver customer satisfaction without involving traditional telephony infrastructures. It is critical, therefore, that policymakers take on the responsibility of educating themselves about the world in which they make decisions. This is not an option, and these new technologies will *not* go away. As I sit here writing this, I can see a press release on my desk from Vonage that came out today. It announces that they have reached 500,000 subscribers and are signing up 15,000 new subscribers every week. Now *that* is a force to contend with.

Current Challenges

A collection of challenges face the industry today. They include CALEA, E911, Universal Service Funds, and other long-standing decisions that don't work well within a VoIP construct.

Enhanced 911 (E911)

E911 requirements mandate that local telcos provide emergency location services so that EMS and other first responders can determine the address of the calling party in the case of an emergency. Unfortunately, as we now know from Champlain College's experiences, an IP address is not tied to a physical address in the same way a telephone number is associated with a physical twisted pair. Consider the case of Vonage that I described earlier: I selected a Vermont area code for my telephone number because I live in Vermont; I could have just as easily chosen a code from Alabama, or Virginia, or Idaho—in spite of the fact that I live in Vermont. When I travel to Africa and plug my laptop into the broadband connection in Durban, I guarantee that if I dial 911 the ambulance won't find me—in spite of the fact that my number shows a Vermont area code. In some cases, state public utility commissions (PUCs) have asked VoIP service providers to publish the home addresses of their subscribers, but many people use a VoIP provider precisely because they travel, in which case the local address would provide little value. In other cases, manufacturers have proposed the installation of GPS chips into laptops and other mobile devices as a way to track the physical location of a calling party, but privacy concerns have stalled that effort.

CALEA

The Communications Assistance for Law Enforcement Act of 1994, commonly known as CALEA, guarantees the right of law enforcement to wiretap a telephone line with the appropriate judicial forbearance. The process of installing a wiretap on a traditional circuit-switched line is trivially easy. In a VoIP environment, however, all bets are off. SIP-based VoIP calls, for example, using DSL or broadband cable, comprise a stream of Real-Time Transport Protocol (RTP) packets that flow directly between the endpoints and bypass the IP telephony provider's SoftSwitch. As a result the SoftSwitch can monitor or record only SIP signaling data, *not* the voice conversations that they control. In the connectionless IP domain, where packets route according to whim and whimsy, it is virtually impossible to wiretap a call.

This has not yet become a major issue because the numbers of subscribers we're talking about remain small relative to those served by the PSTN. However, as VoIP becomes a more popular delivery modality, the focus will shift.

For the time being, the FCC allows Internet telephony service providers, sometimes called ITSPs, to be exempt from CALEA when wiretap requirements are technologically impossible to satisfy. The concerns continue, however, and debate over homeland security and antiterror will add fuel to the fire.

Taxation and Revenue

Telephony taxes are collected at both the state and federal levels. They include Universal Service Fund fees, federal excise taxes, disabled access surcharges, and miscellaneous state levies. As VoIP edges out services delivered via the PSTN, and the customer base for circuit switching diminishes, a different revenue model will have to be created to preserve universal service, network maintenance, community access requirements, and other valuable services made possible by these levies.

If taxes are collected on VoIP, government agencies will be faced with a rather interesting problem: IP-based service providers will be easy taxation targets, but what about the users that place peer-to-peer calls from their PCs and laptops and who cannot be seen? After all, a packet is a packet is a packet—there is no way to tell that the data stream emanating from a user contains voice packets without a great deal of work. Consider this: Based on the scenario just described, there is no reason why a VoIP service provider wouldn't move its network control offshore to avoid paying VoIP taxes! So, friend regulator—how do we deal with *that* problem?

Local service quality is managed by state public utilities commissions or public service boards. They serve as the negotiation body between customer issues and local service providers, providing telco oversight and a sounding board for customers who have a complaint against their service provider. Some states have concluded that they should locally regulate VoIP providers to ensure the delivery of highest-quality customer service, but of course this isn't possible because of the way VoIP works! In fact, Michael Powell, in a sweeping and gutsy move, recently consolidated all VoIP regulation at the federal level, wresting it from the states and handing it to the FCC.

Chapter Summary

The FCC seems to get it: They understand what their charter has become and are making strides to realize the goals of their mandate. They are trying to walk a fine line between mandate and hands-off regulation, have been appropriately delicate with the information services versus telecom provider battle, and have steadfastly refused to impose Internet taxation in favor of more of a laissez-faire approach to the Internet. These are all good things. At the time of this writing Michael Powell has left his position as the chair of the FCC, and speculation exists as to what will follow. Most believe that the road to further liberalization will continue to be the FCC's preferred path; we'll see.

Some believe that recent decisions by the Senate (but not the full Congress) to allow state commissions to levy universal service and access charges on any applications that traverse the Internet, including instant messaging (IM), are a bad idea. If accepted, the law would subject IP telephone calls that never traverse the circuit-switched PSTN to taxes that serve as PSTN subsidies. The bill stalled in 2004, but count on a replay in 2005 as legislators gear up for a very big and important game.

Stand by—this is going to be one to watch.

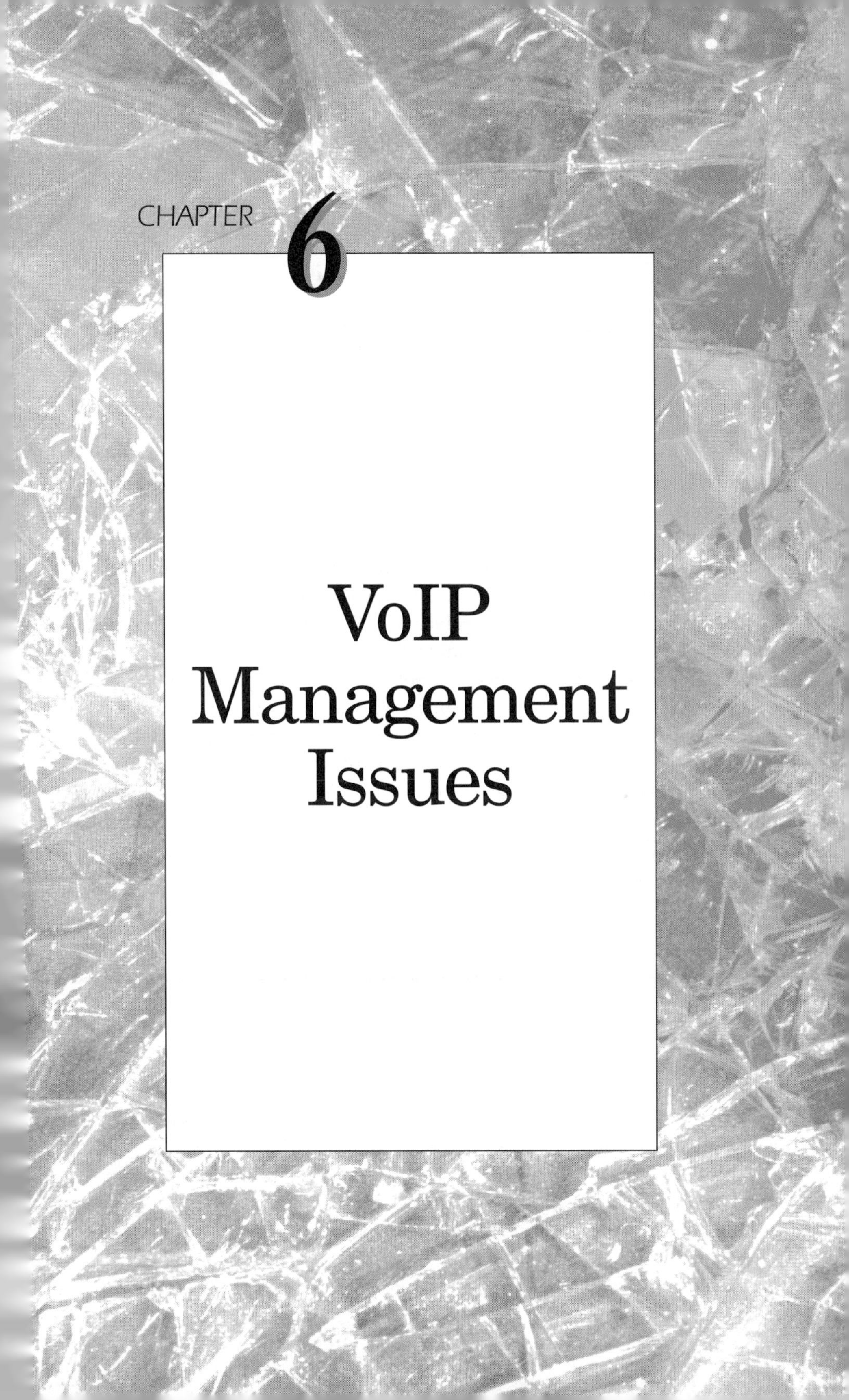

CHAPTER 6

VoIP Management Issues

In this final chapter we examine the managerial and human capital issues that must be dealt with if the cutover is to be successful. To successfully manage voice over IP (VoIP), the enterprise must do whatever it can to ensure that both short- and long-term success factors are met.

Viewing the Field

Before making the decision to implement a VoIP solution, it is important to perform a functional audit of the preexisting network to assess its capabilities and deficiencies. For example, voice packets, because of their sensitivity to delay, must receive higher-priority treatment than data traffic such as Web access and e-mail. Network conditions that cause data packets to be delayed or resent result in a minor annoyance on the part of the user and *no* impact on the application, whereas the same conditions can prove catastrophic for voice. Because most data applications that traverse the Internet today rely on the Transmission Control Protocol (TCP) for overall end-to-end quality control, they benefit from the protocol's embedded error recovery and retransmission capabilities. Those capabilities, however, come with a price: The protocol overhead they create is costly (the famous packet tax) and can result in throughput delays. Voice applications, therefore, rely on an alternative control protocol called the Real-Time Transport Protocol (RTP), which is based on the User Datagram Protocol (UDP), a much-reduced connectionless counterpart to TCP. UDP does not retransmit packets in the event of error, thus eliminating the delays that result from recovery processing.

Quality of Service (QoS) Preparations

The network that is to become a transport mechanism for VoIP should be capable of supporting a variety of QoS protocols that are designed for real-time applications. At layer three (IP Layer), *Differentiated Services* (DiffServ) and the *Resource Reservation Protocol* (RSVP) are most commonly used for the prioritization of voice traffic in a mixed voice-data IP network. The two protocols are different in their approach: DiffServ can grant prioritization of voice traffic over data traffic, whereas RSVP provides end-to-end reservation of bandwidth for the voice traffic. At layer two (the Data Link Layer), the 802.1p and 802.1Q protocols, designed as traffic classification schemes, ensure classification and prioritization of Ethernet frames on the local area network (LAN) that carry VoIP packets. Of course having the protocols and reacting to them are two entirely different things. The routers and switches in the network must also support these protocols and be able to perform prioritization and assign appropriate levels of bandwidth if they are to ensure end-to-end QoS.

QoS Control and Assurance

In the same way that the traditional private branch exchange (PBX) collects usage data and provides fodder for management reports, VoIP environments must be able to provide a similar level of usage oversight. In some cases this is provided by hardware-based collection devices; in other cases it is software based. The nature of the approach is less important than what comes out of the effort; either way the information emerging from the system must be useful and timely. Some systems, for example, provide real-time usage assessment and programmed notification of usage violations or excessive activity; others offer analytical content and detailed management reports. Naturally, the more current (real-time) the data, the better, although a great deal of value can be gleaned from the time taken to carefully analyze the data.

A clearly stated and managed QoS policy is critical to ensure optimal operation of the voice-enabled enterprise network. Unless proper safeguards are put into place, unexpected traffic bursts can bring the production network to its knees in the same way that storage applications often affected production environments prior to the implementation of *storage area networks* (SANs). And because voice is necessarily a real-time application, its coexistence on the network with its more forgiving data-application counterparts requires constant, active monitoring of network and application performance.

The View from the Field

There is a good news, bad news aspect to the user's view from the field. On the one hand, users have become accustomed to the QoS delivered by the iron-clad public switched telephone network (PSTN) and, as a result, expect the same level of service quality from an alternative enterprise infrastructure like VoIP. On the other hand, they have become accustomed to the inferior call quality delivered by their mobile phones, which means that to a certain extent they have become inured to a slightly different definition of acceptable. The fact is, in the enterprise world, the single most important—and vexing—issue is maintenance of consistent call quality. Some IT managers have attempted to hide inferior voice performance behind a smoke screen of value-added bells and whistles, and their efforts have resulted in spectacular and very visible failures. All the added capabilities in the world will not overcome poor call quality. Ultimately, users expect a level of quality that at least emulates that which they have become accustomed to from the public circuit-switched network. They expect a phone that is always on, immediate dial tone, rapid call cut-through, and a handful of dialing options that have always been available to them. If these few things fail to materialize as part of the overall VoIP conversion, it will be deemed a failure regardless of how many features have been implemented. Part of the management reporting capability discussed earlier must include the five basic tenets of network management: *fault detection, configuration management, accounting, network performance, and security* (our FCAPS acronym described earlier in the book). As long as these five functions are addressed, even at a minimal but visible level, the user community will feel somewhat satisfied with the system and its delivered services.

Hardware Management

Part of the configuration management activity just mentioned includes inventory control and management for the network's pool of active and spare devices. Depending on the nature of the underlying network architecture a variety of devices may need to be accounted for and monitored, and it is imperative that IT staff put into place a set of procedural guidelines for doing so on an ongoing, scheduled basis. Some systems are analog and use analog customer premises equipment (CPE) and PBX, while others are digital and rely on a distributed array of servers. For the most part, however, the enterprise will have both.

As voice becomes another IT application riding on the corporate network, the VoIP infrastructure must be treated by IT staff as a managed infrastructure that is just as important, if not more so, as the computer infrastructures they are responsible for. Minor failures in the network can rapidly escalate into major impacts if not detected and corrected quickly, so it is incumbent on IT staff to treat their VoIP assets with the appropriate level of urgency.

Software Management

Equal in importance to device management is software management at both operating-system level and application level. And because the typical VoIP network infrastructure is often highly distributed and virtual, many believe that software currency and maintenance is actually more important than hardware management. Most VoIP implementations are server based and require a server-specific operating system such as Linux or Windows for proper operation. Applications work only if the underlying operating system is running at peak performance, and some evidence exists to suggest that the bulk of software-related trouble reports in VoIP systems are operating system related rather than being the result of some flaw in the application software. In fact, there are now open-source (as in Freeware) soft PBXs that can be downloaded and implemented with great success.

The software management process must provide for *centralized control* of a *distributed resource*. Traffic analysis and impact, version control, debug, and other functions should be part of a routine quality-management program.

Security

Some would argue that security falls within the bailiwick of hardware and software management, but because it is so important—and becoming even more so—we treat it here as a stand-alone consideration.

Security can be implemented at many levels of the network and can range from transparent to downright onerous. Some systems, for example, require users to enter a password or some other unique identifier prior to authentication. Other more secure (paranoid?) environments encrypt all calls to avoid the risk of interception or eavesdropping.

Keep in mind that (1) a VoIP application resides on an Internet Protocol (IP) network, and (2) VoIP is just one more IP-based data application as far as the network is concerned. It can therefore be treated as a data application *as long as service quality is not adversely affected.* IP networks invariably rely on firewalls for protection against unauthorized intrusion, so VoIP applications must be able to operate on a trans-firewall basis while at the same time preserving the integrity provided by the firewall. The VoIP system must also be able to detect abnormal events and pass them along through alarms, traps, and notifications to a collection agent, typically through interaction with a management protocol such as the Simple Network Management Protocol (SNMP).

Because the firewall is such a crucial element of the overall network, and because the addition of VoIP to the network can have so much impact, the sizing of the server must become part of the initial assessment of network capacity and capability. Servers should also be set up in such a way that they have the ability to respond to an abnormal situation and send an alarm to the appropriate response device or enterprise agency.

Event Management

It is an undeniable fact of IT life that even the most well managed, best designed, most proactively maintained networks will suffer unpredictable failures and events—what some data center managers refer to as “unscheduled outages.” (Is there such a thing as a scheduled outage?) I remember a catastrophe from my data center days that occurred when a squirrel chewed through a primary power feed at a southern California computer facility and dropped the entire data center—hard. That is to say, we think it was a squirrel—whatever it was pretty much vaporized. The point is that some failures simply cannot be anticipated. Management systems, therefore, must be as proactive as they can possibly be to collect, monitor, assess, and react as quickly and effectively as possible in the event of a system failure. Network managers should have both audible and visual alarms available to them and should be ale to configure them in the most timely and effective fashion, including using automated escalation, automatic corrective action when possible, automated diagnostic reporting, and the ability to access both logical and physical maps of the overall network.

Moves, Adds, and Changes (MACs): The Bane of the Network Manager

"This job would be so much easier if it weren't for all those pesky customers." How many times have we all said that or something like it? Networks would be so much easier to manage if people would just quit moving around so much! Unfortunately, that just isn't going to happen. Networks, like the people they serve, are dynamic creatures and constantly subject to change. As work groups evolve, people come and go, and offices are added to, closed, or moved, the network must adapt. Furthermore, the arrival of VoIP adds a whole new set of complex issues that network managers must learn to work around. IP, Session Initiation Protocol (SIP), and H.323 (along with a host of others) are open standards, which invites and even encourages the existence of a multivendor network.

Applications will also evolve in this environment, and they will become more complex rather than easier to manage. Converged or unified applications such as voicemail, fax, and e-mail will become de rigeur on the PBX, as will click-to-call, click-to-conference, and Web chat, all of which will become integrated into the overall delivered application complement.

At the same time, while open standards and converging applications become a reality, the user community will come to recognize that it has untold choices in terms of terminal type, application set, network interface (customized Web access, speech-to-text conversion, and so on). IT will be required to respond.

Training Issues

When staffing considerations became the topic of conversation at Champlain College, a number of factors came to the fore. "One area that concerned us," says Paul Dusini, "was that the PBX was such old technology that it pretty much just ran. In the traditional telephony world, if a card burned, the vendor would just send out a technician who would replace the card. This service model is no longer as effective as it used to be."

Aaron Videtto agrees. "Unfortunately, the issue here is that the networking staff has more to deal with because they now have VLANs and voice traffic to manage. And because VoIP is somewhat immature, the vendors don't support it the way they support the old PBX. It isn't that they don't want to, it's just that the service model is different. Today, we dial a toll-free number and within seconds somebody in Ireland or India or South Africa is talking us through the problem.

"It is important that your staff be trained—and trained well—on the new technology," says Dusini. "We have one person doing that [Aaron]. We now have multiple servers that have to be configured, and they come from different vendors, which means that we have to deal with each a little differently. As we add additional software to support different applications, much of it open, life gets even more complicated. After all, it's open software, so we can get in there 'under the hood.'

"Applications development is important too. We have a Web developer on staff that we can really rely on. There will be some canned applications that find their way into the system, but we want to be able to build hooks between our systems and the phone systems. After all, some of that you just want to do yourself." He is quick to caution, however, that a fair balance must be struck between what the customer does and what the vendor provides in the way of support.

Dusini also identifies a number of risk factors that must be taken into account. "You really have to think about all the different kinds of support that you'll need during the conversion process. This includes such things as commitment from the technical staff as well as implementation support that you'll get from vendors. When we started, the engineers were *consumed* with trying to eliminate near-end echo and didn't really care if we were able to forward a phone to Savannah. We had lots of multiline phones, so we had some migration issues. But most of the things we found we had to watch out for were support and implementation factors.

"Training became a big issue for us as well [and] affected a number of departments," explains Videtto. "We added a lot of new features with the system. For example, we added queuing in our highest call volume departments (Financial Aid, Student Accounts, and Advising and Registration), and we had to teach the people in those organizations how to operate it. They had a lot of questions: What does it mean whether you're logged in or not? The bills go out in June, and only two people work in student accounts. If they were both on the phone, all their incoming calls would go to voicemail, and as soon as they hung up from a 20-minute call they would have 15 voicemail messages waiting for them. That was awful customer service! So we decided to add queuing. Unfortunately, the department director didn't want it because of fears about poor customer service. But we convinced them to use it, and at the same time gave callers the ability to leave a voicemail message if they preferred. So we worked with them and 'proved in' the system, and in 18 months only six or seven people have left voicemail messages—almost all callers prefer to wait in queue."

The Pain of Conversion

Aaron Videtto grimaces when I ask him about the actual conversion. "Well, I'd be lying if I told you it went perfectly, because it didn't. We had some big failures. For example, on the first day of cutover, the main college number didn't work! I don't know about you, but I'd call that a big problem!" He shakes his head and laughs. "It was all about timing. We just didn't do enough training on queuing and other factors to do it successfully on opening day, which was in July, a very, very busy time for us. The queues didn't work properly, and some of the user-specific stuff didn't work as advertised.

"We also had to do a lot of customer training. The concept of unified messaging was accepted without a problem, but there's a difference between accepting the idea and being able to use it immediately. For example, because your voicemailbox and e-mail mailbox are logically the same place, our users had some problems. If they hit mailbox quota on voicemail and were looking for e-mail, there wouldn't be any—because their quote had been exceeded. So we had to do some fast training to make sure that the students, faculty, and staff understood how the system actually worked."

I asked about the cutover interval. "We did this in four months," says Videtto. "That turned out to be not enough time because our implementer wasn't as effective as they could have been. This had nothing to do with their ability—they were very good—but in a sense we were their guinea pigs, so we learned together. We were promised an implementation team that was based out of one [nearby] city, and when it came time to do the work, they sent a team from a different city, and we felt that that was wrong, considering the importance of the project and the fact that we had been promised a more local cutover team.

"We also had some vendor training issues. For example, the person who came in to train us on how to use the telephones had never actually seen the phones—but she was a wizard with the manuals! We assumed that everyone we would work with would be fully apprised and trained on the system, but that turned out to be a slightly inaccurate assumption. It's not that it was bad; it was just not what we expected and we didn't have enough experience to ask the right questions. But in the end it worked out alright because we learned to question everything.

"When we got to the days leading up to the conversion, things got frantic around here. The actual conversion was done between the beginning of April and the beginning of July, with a cutover date of July 17. That day was the beginning of the longest weekend I have ever worked. We worked straight through the weekend without sleep, but ultimately everything came up and we were online. We did run parallel systems for a period of time to ensure that everything would continue to work. But other than a few minor failures—like the loss of the main campus number on cutover day—we were okay.

"I suppose that the most important message I can leave your readers with is this," says Aaron Videtto. "We went into this conversion process because we had a solid business reason. We had a critical piece of hardware that was about to fail, and we had a choice to make. We could upgrade it and get it to limp along for a while longer, or we could take a bit of a risk and go with the new technology. We opted for the latter because we realized just how much it would affect our ability to serve the customer. And in the end, it turned out to be the best decision we could have made. However, be warned. The conversion was a long, complicated process, and it went as well as it did for us because we did our homework. We're not the largest installation out there, but we're not insignificant, either. We took our time, looked at the business and economic drivers, studied the layout of the campus, and finally made a decision. I'm glad we did."

Chapter Summary

In the end, Champlain College's decision to evolve from a traditional PBX to an IP-based converged network was a good one. Their rationale was solid and driven by the appropriate business and service drivers that should be part of such a major decision. However, their business case is not unique: The economic, competitive, and marketplace forces that drove the college to go the VoIP route are identical to those that would drive any other medium to large business with plans for growth, evolution, and enhanced customer service. The key to a successful launch is planning, consideration from multiple angles, and, as we said earlier in the book, an emotionless set of decisions based on well-grounded questions.

There is no question that VoIP is the ultimate end state of today's network, and the fundamental question is "when." However, it is early enough to recognize that "whether" is still a perfectly valid question. Until the business case clearly points to a converged solution as a compelling, business-affecting reason to convert from one platform to the next, go slowly. "If it ain't broke, don't fix it" makes an awful lot of sense.

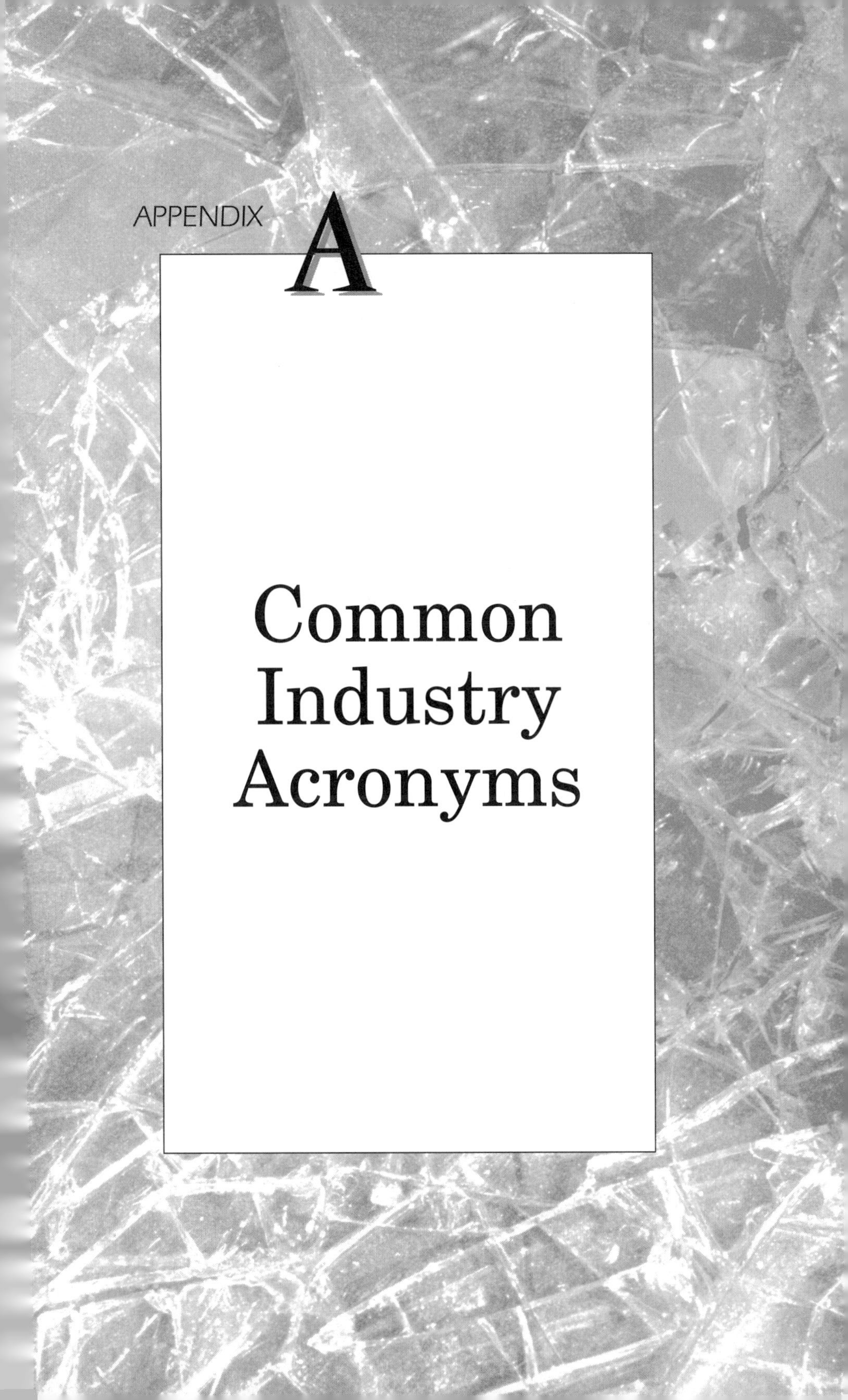

APPENDIX A

Common Industry Acronyms

AAL ATM Adaptation Layer
AARP AppleTalk Address Resolution Protocol
ABM asynchronous balanced mode
ABR available bit rate
AC alternating current
ACD automatic call distribution
ACELP algebraic code-excited linear prediction
ACF advanced communication function
ACK acknowledgment
ACM address complete message
ACSE association control service element
ACTLU activate logical unit
ACTPU activate physical unit
ADCCP advanced data communications control procedures
ADM add-drop multiplexer
ADPCM adaptive differential pulse code modulation
ADSL asymmetric digital subscriber line
AFI application family identifier (RFID)
AFI authority and format identifier
AI application identifier
AIN advanced intelligent network
AIS alarm indication signal
ALU arithmetic logic unit
AM administrative module (Lucent 5ESS)
AM amplitude modulation
AMI alternate mark inversion
AMP administrative module processor
AMPS advanced mobile phone system
ANI automatic number identification (SS7)
ANSI American National Standards Institute
APD avalanche photodiode
API application programming interface
APPC advanced program-to-program communication
APPN advanced peer-to-peer networking
APS automatic protection switching
ARE all routes explorer (source route bridging)
ARM asynchronous response mode
ARP Address Resolution Protocol (IETF)
ARPA Advanced Research Projects Agency
ARPANET Advanced Research Projects Agency Network
ARPU average revenue per user

ARQ automatic repeat request
ASCII American Standard Code for Information Interchange
ASI alternate space inversion
ASIC application-specific integrated circuit
ASK amplitude shift keying
ASN abstract syntax notation
ASP application service provider
AT&T American Telephone and Telegraph
ATDM asynchronous time division multiplexing
ATM asynchronous transfer mode
ATMF ATM Forum
ATQA answer to request A (RFID)
ATQB answer to request B (RFID)
ATS answer to select (RFID)
ATTRIB attribute (RFID)
AU administrative unit (SDH)
AUG administrative unit group (SDH)
AWG American wire gauge
B2B business-to-business
B2C business-to-consumer
B8ZS binary 8 zero substitution
BANCS Bell Administrative Network Communications System
BBN Bolt, Beranak, and Newman
BBS bulletin board service
Bc committed burst size
BCC blocked calls cleared
BCC block check character
BCD blocked calls delayed
BCDIC Binary Coded Decimal Interchange Code
Be excess burst size
BECN backward explicit congestion notification
BER bit error rate
BERT bit error rate test
BGP Border Gateway Protocol (IETF)
BIB backward indicator bit (SS7)
B-ICI broadband intercarrier interface
BIOS basic input/output system
BIP bit interleaved parity
B-ISDN broadband integrated services digital network
BISYNC Binary Synchronous Communications Protocol
BITNET Because It's Time Network

BITS building integrated timing supply
BLSR bidirectional line switched ring
BOC Bell Operating Company
BPRZ bipolar return to zero
bps bits per second
BRI basic rate interface
BRITE basic rate interface transmission equipment
BSC binary synchronous communications
BSN backward sequence number (SS7)
BSRF Bell System Reference Frequency
BTAM basic telecommunications access method
BUS broadcast unknown server
C/R command/response
CAD computer-aided design
CAE computer-aided engineering
CAGR compound annual growth rate
CAM computer-aided manufacturing
CAP carrierless amplitude/phase modulation
CAP competitive access provider
CAPEX capital expense
CAPEX capital expenditure
CARICOM Caribbean Community and Common Market
CASE common application service element
CASE computer-aided software engineering
CASPIAN Consumers Against Privacy Invasion and Numbering (RFID)
CAT computer-aided tomography
CATIA computer-assisted three-dimensional interactive application
CATV community antenna television
CBEMA Computer and Business Equipment Manufacturers Association
CBR constant bit rate
CBT computer-based training
CC cluster controller
CCIR International Radio Consultative Committee
CCIS common channel interoffice signaling
CCITT International Telegraph and Telephone Consultative Committee
CCS common channel signaling
CCS hundred call seconds per hour
CD collision detection

CD compact disc
CDC Control Data Corporation
CDMA code division multiple access
CDPD cellular digital packet data
CD-ROM compact disc read-only memory
CDVT cell delay variation tolerance
CEI comparably efficient interconnection
CEPT Conference of European Postal and Telecommunications Administrations
CERN European Council for Nuclear Research
CERT computer emergency response team
CES circuit emulation service
CEV controlled environmental vault
CGI common gateway interface (Internet)
CHAP Challenge Handshake Authentication Protocol
CHL chain home low RADAR
CICS customer information control system
CICS/VS customer information control system/virtual storage
CID card identifier (RFID)
CIDR classless interdomain routing (IETF)
CIF cells in frames
CIR committed information rate
CISC complex instruction set computer
CIX commercial Internet exchange
CLASS custom local area signaling services (Bellcore)
CLEC competitive local exchange carrier
CLLM consolidated link layer management
CLNP Connectionless Network Protocol
CLNS connectionless network service
CLP cell loss priority
CM communications module (Lucent 5ESS)
CMIP Common Management Information Protocol
CMISE common management information service element
CMOL CMIP over LLC
CMOS complementary metal oxide semiconductor
CMOT CMIP over TCP/IP
CMP communications module processor
CNE certified NetWare engineer
CNM customer network management
CNR carrier-to-noise ratio
CO central office

CoCOM Coordinating Committee on Export Controls
CODEC coder/decoder
COMC communications controller
CONS connection-oriented network service
CORBA common object request brokered architecture
COS class of service (APPN)
COS Corporation for Open Systems
CPE customer premises equipment
CPU central processing unit
CRC cyclic redundancy check
CRM customer relationship management
CRT cathode ray tube
CRV call reference value
CS convergence sublayer
CSA carrier serving area
CSMA carrier sense multiple access
CSMA/CA carrier sense multiple access with collision avoidance
CSMA/CD carrier sense multiple access with collision detection
CSU channel service unit
CTI computer telephony integration
CTIA Cellular Telecommunications Industry Association
CTS clear to send
CU control unit
CVSD continuously variable slope delta modulation
CWDM coarse wavelength division multiplexing
D/A digital-to-analog
DA destination address
DAC dual attachment concentrator (FDDI)
DACS digital access and cross-connect system
DARPA Defense Advanced Research Projects Agency
DAS dual attachment station (FDDI)
DAS direct attached storage
DASD direct access storage device
DB decibel
DBS direct broadcast satellite
DC direct current
DCC data communications channel (SONET)
DCE data circuit-terminating equipment
DCN data communications network
DCS digital cross-connect system
DCT discrete cosine transform

DDCMP Digital Data Communications Management Protocol (DNA)
DDD direct distance dialing
DDP Datagram Delivery Protocol
DDS DATAPHONE digital service (sometimes digital data service)
DDS digital data service
DE discard eligibility (LAPF)
DECT Digital European Cordless Telephone
DES Data Encryption Standard (NIST)
DHCP Dynamic Host Configuration Protocol
DID direct inward dialing
DIP dual inline package
DLC digital loop carrier
DLCI data link connection identifier
DLE data link escape
DLSw data link switching
DM delta modulation
DM disconnected mode
DM data mining
DMA direct memory access (computers)
DMAC direct memory access control
DME distributed management environment
DMS digital multiplex switch
DMT discrete multitone
DNA digital network architecture
DNIC Data Network Identification Code (X.121)
DNIS dialed number identification service
DNS domain name service
DNS domain name system (IETF)
DOD direct outward dialing
DOD Department of Defense
DOJ Department of Justice
DOV data over voice
DPSK differential phase shift keying
DQDB distributed queue dual bus
DR data rate send (RFID)
DRAM dynamic random access memory
DS data rate send (RFID)
DSAP destination service access point
DSF dispersion-shifted fiber
DSI digital speech interpolation

DSL digital subscriber line
DSLAM digital subscriber line access multiplexer
DSP digital signal processing
DSR data set ready
DSS digital satellite system
DSS digital subscriber signaling system
DSSS direct sequence spread spectrum
DSU data service unit
DTE data terminal equipment
DTMF dual tone multifrequency
DTR data terminal ready
DVRN dense virtual routed networking (Crescent)
DWDM dense wavelength division multiplexing
DXI data exchange interface
E/O electrical-to-optical
EAN European Article Numbering System
EBCDIC Extended Binary Coded Decimal Interchange Code
EBITDA earnings before interest, tax, depreciation, and amortization
ECMA European Computer Manufacturer Association
ECN explicit congestion notification
ECSA Exchange Carriers Standards Association
EDFA erbium-doped fiber amplifier
EDI electronic data interchange
EDIBANX EDI Bank Alliance Network Exchange
EDIFACT Electronic Data Interchange for Administration, Commerce, and Trade (ANSI)
EFCI explicit forward congestion indicator
EFTA European Free Trade Association
EGP Exterior Gateway Protocol (IETF)
EIA Electronics Industry Association
EIGRP Enhanced Interior Gateway Routing Protocol
EIR excess information rate
EMBARC electronic mail broadcast to a roaming computer
EMI electromagnetic interference
EMS element management system
EN end node
ENIAC electronic numerical integrator and computer
EO end office
EOC embedded operations channel (SONET)
EOT end of transmission (BISYNC)

EPC Electronic Product Code
EPROM erasable programmable read-only memory
EPS earnings per share
ERP enterprise resource planning
ESCON enterprise system connection (IBM)
ESF extended superframe format
ESOP employee stock ownership plan
ESP enhanced service provider
ESS electronic switching system
ETSI European Telecommunications Standards Institute
ETX end of text (BISYNC)
EVA economic value added
EWOS European Workshop for Open Systems
FACTR Fujitsu Access and Transport System
FAQ frequently asked questions
FASB Financial Accounting Standards Board
FAT file allocation table
FCF free cash flow
FCS frame check sequence
FDA Food and Drug Administration
FDD frequency division duplex
FDDI fiber distributed data interface
FDM frequency division multiplexing
FDMA frequency division multiple access
FDX full-duplex
FEBE far-end block error (SONET)
FEC forward error correction
FEC forward equivalence class
FECN forward explicit congestion notification
FEP front-end processor
Farfel far-end receive failure (SONET)
FET field effect transistor
FHSS frequency hopping spread spectrum
FIB forward indicator bit (SS7)
FIFO first in, first out
FITL fiber in the loop
FLAG fiber link across the globe
FM frequency modulation
FOIRL fiber optic interrepeater link
FPGA Field Programmable Gate Array
FR frame relay

FRAD frame-relay access device
FRBS frame-relay bearer service
FSDI frame size device integer (RFID)
FSK frequency shift keying
FSN forward sequence number (SS7)
FTAM file transfer, access, and management
FTP File Transfer Protocol (IETF)
FTTC fiber to the curb
FTTH fiber to the home
FUNI frame user-to-network interface
FWI frame waiting integer (RFID)
FWM four-wave mixing
GAAP generally accepted accounting principles
GATT general agreement on tariffs and trade
GbE Gigabit Ethernet
Gbps gigabits per second (billion bits per second)
GDMO Guidelines for the Development of Managed Objects
GDP gross domestic product
GEOS geosynchronous Earth orbit satellites
GFC generic flow control (ATM)
GFI general format identifier (X.25)
GFP generic framing procedure
GFP-F generic framing procedure—frame-based
GFP-X generic framing procedure—transparent
GMPLS generalized MPLS
GOSIP government open systems interconnection profile
GPS global positioning system
GRIN graded index (fiber)
GSM Global System for Mobile communications
GTIN global trade item number
GUI graphical user interface
HDB3 high-density, bipolar 3 (E-Carrier)
HDLC high-level data link control
HDSL high-bit-rate digital subscriber line
HDTV high-definition television
HDX half-duplex
HEC header error control (ATM)
HFC hybrid fiber/coax
HFS hierarchical file storage
HIPAA Health Insurance Portability and Accountability Act
HLR home location register

HPPI high-performance parallel interface
HSSI high-speed serial interface (ANSI)
HTML Hypertext Markup Language
HTTP Hypertext Transfer Protocol (IETF)
HTU HDSL transmission unit
I intrapictures
IAB Internet Architecture Board (formerly Internet Activities Board)
IACS integrated access and cross-connect system
IAD integrated access device
IAM initial address message (SS7)
IANA Internet Address Naming Authority
ICMP Internet Control Message Protocol (IETF)
IDP Internet Datagram Protocol
IEC interexchange carrier (also IXC)
IEC International Electrotechnical Commission
IEEE Institute of Electrical and Electronics Engineers
IETF Internet Engineering Task Force
IFRB International Frequency Registration Board
IGP Interior Gateway Protocol (IETF)
IGRP Interior Gateway Routing Protocol
ILEC incumbent local exchange carrier
IM instant messenger (AOL)
IML initial microcode load
IMP interface message processor (ARPANET)
IMS information management system
InARP Inverse Address Resolution Protocol (IETF)
InATMARP inverse ATMARP
INMARSAT International Maritime Satellite Organization
INP internet nodal processor
InterNIC Internet Network Information Center
IP intellectual property
IP Internet Protocol (IETF)
IPO initial product offer
IP PBX IP-based private branch exchange
IPX internetwork packet exchange (NetWare)
IRU indefeasible rights of use
IS information systems
ISDN integrated services digital network
ISO International Organization for Standardization
ISO Information Systems Organization

ISOC Internet Society
ISP Internet service provider
ISUP ISDN user part (SS7)
IT information technology
ITU International Telecommunication Union
ITU-R International Telecommunication Union—Radio Communication Sector
IVD inside vapor deposition
IVR interactive voice response
IXC interexchange carrier
JAN Japanese Article Numbering System
JEPI joint electronic paynets initiative
JES job entry system
JIT just in time
JPEG Joint Photographic Experts Group
JTC Joint Technical Committee
KB kilobytes
Kbps kilobits per second (thousand bits per second)
KLTN potassium lithium tantalate niobate
KM knowledge management
LAN local area network
LANE LAN emulation
LAP link access procedure (X.25)
LAPB link access procedure balanced (X.25)
LAPD link access procedure for the D-channel
LAPF link access procedure to frame mode bearer services
LAPF-Core core aspects of the link access procedure to frame mode bearer services
LAPM link access procedure for modems
LAPX link access procedure half-duplex
LASER light amplification by the stimulated emission of radiation
LATA local access and transport area
LCD liquid crystal display
LCGN logical channel group number
LCM line concentrator module
LCN local communications network
LD laser diode
LDAP Lightweight Directory Access Protocol (X.500)
LEAF® Large Effective Area Fiber® (Corning product)
LEC local exchange carrier
LED light-emitting diode

LENS lightwave efficient network solution (Centerpoint)
LEOS low Earth orbit satellites
LER label edge router
LI length indicator
LIDB line information database
LIFO last in, first out
LIS logical IP subnet
LLC logical link control
LMDS local multipoint distribution system
LMI local management interface
LMOS loop maintenance operations system
LORAN long-range radio navigation
LPC linear predictive coding
LPP Lightweight Presentation Protocol
LRC longitudinal redundancy check (BISYNC)
LS link state
LSI large scale integration
LSP label switched path
LSR label switched router
LU line unit
LU logical unit (SNA)
MAC media access control
MAN metropolitan area network
MAP Manufacturing Automation Protocol
MAU medium attachment unit (Ethernet)
MAU multistation access unit (Token Ring)
MB megabytes
MBA™ metro business access™ (Ocular)
Mbps megabits per second (million bits per second)
MD message digest (MD2, MD4, MD5) (IETF)
MDF main distribution frame
MDU multidwelling unit
MeGaCo media gateway control
MEMS micro electrical mechanical system
MF multifrequency
MFJ modified final judgment
MGCP Media Gateway Control Protocol
MHS message-handling system (X.400)
MIB management information base
MIC medium interface connector (FDDI)
MIME multipurpose Internet mail extensions (IETF)

MIPS millions of instructions per second
MIS management information systems
MITI Ministry of International Trade and Industry (Japan)
MITS Micro Instrumentation and Telemetry Systems
ML-PPP Multilink Point-to-Point Protocol
MMDS multichannel, multipoint distribution system
MMF multimode fiber
MNP Microcom Networking Protocol
MON metropolitan optical network
MoU memorandum of understanding
MP Multilink PPP
MPEG Motion Picture Experts Group
MPLS multiprotocol label switching
MPOA multiprotocol over ATM
MPλS multiprotocol lambda switching
MRI magnetic resonance imaging
MSB most significant bit
MSC mobile switching center
MSO mobile switching office
MSPP multiservice provisioning platform
MSVC meta-signaling virtual channel
MTA major trading area
MTBF mean time between failure
MTP message transfer part (SS7)
MTSO mobile telephone switching office
MTTR mean time to repair
MTU maximum transmission unit
MTU multitenant unit
MVNO mobile virtual network operator
MVS multiple virtual storage
NAD node address (RFID)
NAFTA North American Free Trade Agreement
NAK negative acknowledgment (BISYNC, DDCMP)
NAP network access point (Internet)
NARUC National Association of Regulatory Utility Commissioners
NAS network attached storage
NASA National Aeronautics and Space Administration
NASDAQ National Association of Securities Dealers Automated Quotations
NAT network address translation
NATA North American Telecommunications Association

NATO North Atlantic Treaty Organization
NAU network accessible unit
NCP network control program
NCSA National Center for Supercomputer Applications
NCTA National Cable Television Association
NDIS network driver interface specifications
NDSF non-dispersion-shifted fiber
NetBEUI NetBIOS extended user interface
NetBIOS network basic input/output system
NFS network file system (Sun)
NIC network interface card
NII national information infrastructure
NIST National Institute of Standards and Technology (formerly NBS)
NIU network interface unit
NLPID network layer protocol identifier
NLSP NetWare Link Services Protocol
NM network module
Nm nanometer
NMC network management center
NMS network management system
NMT Nordic Mobile Telephone
NMVT Network Management Vector Transport Protocol
NNI network node interface
NNI network-to-network interface
NOC network operations center
NOCC network operations control center
NOPAT net operating profit after tax
NOS network operating system
NPA numbering plan area
NPN negative-positive-negative
NREN National Research and Education Network
NRZ nonreturn to zero
NRZI nonreturn to zero inverted
NSA National Security Agency
NSAP network service access point
NSAPA network service access point address
NSF National Science Foundation
NTSC National Television Systems Committee
NTT Nippon Telephone and Telegraph
NVB number of valid bits (RFID)

NVOD near video on demand
NZDSF nonzero dispersion-shifted fiber
OADM optical add-drop multiplexer
OAM operations, administration, and maintenance
OAM&P operations, administration, maintenance, and provisioning
OAN optical area network
OBS optical burst switching
OC optical carrier
OEM original equipment manufacturer
O-E-O optical-electrical-optical
OLS optical line system (Lucent)
OMAP operations, maintenance, and administration part (SS7)
ONA open network architecture
ONS object name service
ONU optical network unit
OOF out of frame
OPEX operating expense
OS operating system
OSF Open Software Foundation
OSI open systems interconnection (ISO, ITU-T)
OSI RM open systems interconnection reference model
OSPF Open Shortest Path First Protocol (IETF)
OSS operation support systems
OTDM optical time division multiplexing
OTDR optical time-domain reflectometer
OUI organizationally unique identifier (SNAP)
OVD outside vapor deposition
OXC optical cross-connect
P/F poll/final (HDLC)
PAD packet assembler/disassembler (X.25)
PAL phase alternate line
PAM pulse amplitude modulation
PANS pretty amazing new stuff
PBX private branch exchange
PCB protocol control byte (RFID)
PCI pulse code modulation
PCI peripheral component interface
PCMCIA Personal Computer Memory Card International Association
PCN personal communications network
PCS personal communications services

PDA personal digital assistant
PDH Plesiochronous Digital Hierarchy
PDU protocol data unit
PIN positive-intrinsic-negative
PING packet Internet groper (TCP/IP)
PKC public key cryptography
PLCP Physical Layer Convergence Protocol
PLP Packet Layer Protocol (X.25)
PM phase modulation
PMD physical medium dependent (FDDI)
PML Physical Markup Language
PNNI private network node interface (ATM)
PON passive optical networking
POP point of presence
POSIT profiles for open systems interworking technologies
POSIX portable operating system interface for UNIX
POTS plain old telephone service
PPM pulse position modulation
PPP Point-to-Point Protocol (IETF)
PPS protocol parameter selection (RFID)
PRC primary reference clock
PRI primary rate interface
PROFS professional office system
PROM programmable read-only memory
PSDN packet-switched data network
PSK phase shift keying
PSPDN packet-switched public data network
PSTN public switched telephone network
PTI payload type identifier (ATM)
PTT post, telephone, and telegraph
PU physical unit (SNA)
PUC public utility commission
PUPI pseudo-unique PICC identifier
PVC permanent virtual circuit
QAM quadrature amplitude modulation
Q-bit qualified data bit (X.25)
QLLC qualified logical link control (SNA)
QoS quality of service
QPSK quadrature phase shift keying
QPSX queued packet synchronous exchange
R&D research and development

RADAR radio detection and ranging
RADSL rate adaptive digital subscriber line
RAID redundant array of inexpensive disks
RAM random access memory
RARP Reverse Address Resolution Protocol (IETF)
RAS remote access server
RATS request for answer to select (RFID)
RBOC regional Bell operating company
READ_DATA read data from transponder (RFID)
REQA request A (RFID)
REQB request B (RFID)
REQUEST_SNR request serial number (RFID)
RF radio frequency
RFC request for comments (IETF)
RFH remote frame handler (ISDN)
RFI radio frequency interference
RFID radio frequency identification
RFP request for proposal
RFQ request for quote
RFx request for x, where "x" can be proposal, quote, information, comment, etc.
RIIC regional holding company
RHK Ryan, Hankin, and Kent (Consultancy)
RIP Routing Information Protocol (IETF)
RISC reduced instruction set computer
RJE remote job entry
RNR receive not ready (HDLC)
ROA return on assets
ROE return on equity
ROI return on investment
ROM read-only memory
RO-RO roll-on, roll-off
ROSE remote operation service element
RPC remote procedure call
RPR resilient packet ring
RR receive ready (HDLC)
RSA Rivest, Shamir, and Aleman
RTS request to send (EIA-232-E)
S/DMS SONET/digital multiplex system
S/N signal-to-noise ratio
SAA systems application architecture (IBM)

SAAL signaling ATM adaptation layer (ATM)
SABM set asynchronous balanced mode (HDLC)
SABME set asynchronous balanced mode extended (HDLC)
SAC single attachment concentrator (FDDI)
SAK select acknowledge (RFID)
SAN storage area network
SAP service access point (generic)
SAPI service access point identifier (LAPD)
SAR segmentation and reassembly (ATM)
SAS single attachment station (FDDI)
SASE specific applications service element (subset of CASE, Application Layer)
SATAN system administrator tool for analyzing networks
SBS stimulated Brillouin scattering
SCCP signaling connection control point (SS7)
SCM supply chain management
SCP service control point (SS7)
SCREAM™ scalable control of a rearrangeable extensible array of mirrors (Calient)
SCSI small computer systems interface
SCTE serial clock transmit external (EIA-232-E)
SDH synchronous digital hierarchy (ITU-T)
SDLC synchronous data link control (IBM)
SDS scientific data systems
SEC Securities and Exchange Commission
SECAM sequential color with memory
SELECT select transponder (RFID)
SELECT_ACKNOWLEDGE acknowledge selection (RFID)
SELECT_SNR select serial number (RFID)
SF superframe format (T1)
SFGI startup frame guard integer (RFID)
SGML Standard Generalized Markup Language
SGMP Simple Gateway Management Protocol (IETF)
SHDSL symmetric HDSL
S-HTTP secure HTTP (IETF)
SIF signaling information field
SIG special interest group
SIO service information octet
SIP Serial Interface Protocol
SIP Session Initiation Protocol
SIR sustained information rate (SMDS)

SLA service level agreement
SLIP Serial Line Interface Protocol (IETF)
SM switching module
SMAP system management application part
SMDS switched multimegabit data service
SMF single mode fiber
SMP Simple Management Protocol
SMP switching module processor
SMR specialized mobile radio
SMS standard management system (SS7)
SMTP Simple Mail Transfer Protocol (IETF)
SNA systems network architecture (IBM)
SNAP Subnetwork Access Protocol
SNI subscriber network interface (SMDS)
SNMP Simple Network Management Protocol (IETF)
SNP sequence number protection
SNR serial number (RFID)
SOHO small-office, home-office
SONET synchronous optical network
SPAG Standards Promotion and Application Group
SPARC scalable performance architecture
SPE synchronous payload envelope (SONET)
SPID service profile identifier (ISDN)
SPM self-phase modulation
SPOC single point of contact
SPX sequenced packet exchange (NetWare)
SQL Structured Query Language
SRB source route bridging
SRP Spatial Reuse Protocol
SRS stimulated Raman scattering
SRT source routing transparent
SS7 Signaling System 7
SSCC serial shipping container code
SSL secure socket layer (IETF)
SSP service switching point (SS7)
SSR secondary surveillance RADAR
SST spread-spectrum transmission
STDM statistical time-division multiplexing
STM synchronous transfer mode
STM synchronous transport module (SDH)
STP signal transfer point (SS7)

STP shielded twisted pair
STS synchronous transport signal (SONET)
STX start of text (BISYNC)
SVC signaling virtual channel (ATM)
SVC switched virtual circuit
SXS step-by-step switching
SYN synchronization
SYNTRAN synchronous transmission
TA terminal adapter (ISDN)
TAG technical advisory group
TASI time-assigned speech interpolation
TAXI transparent asynchronous transmitter/receiver interface (Physical Layer)
TCAP transaction capabilities application part (SS7)
TCM time-compression multiplexing
TCM trellis coding modulation
TCP Transmission Control Protocol (IETF)
TDD time division duplexing
TDM time division multiplexing
TDMA time division multiple access
TDR time domain reflectometer
TE1 terminal equipment type 1 (ISDN capable)
TE2 terminal equipment type 2 (non-ISDN capable)
TEI terminal endpoint identifier (LAPD)
TELRIC total element long-run incremental cost
TFTP Trivial File Transfer Protocol
TIA Telecommunications Industry Association
TIRIS TI RF identification systems (Texas Instruments)
TIRKS trunk-integrated record-keeping system
TL1 Transaction Language 1
TLAN transparent LAN
TM terminal multiplexer
TMN telecommunications management network
TMS time-multiplexed switch
TOH transport overhead (SONET)
TOP technical and office protocol
TOS type of service (IP)
TP twisted pair
TR token ring
TRA traffic-routing administration
TSI time slot interchange

TSLRIC total service long-run incremental cost
TSO terminating screening office
TSO time-sharing option (IBM)
TSR terminate and stay resident
TSS telecommunication standardization sector (ITU-T)
TST time-space-time switching
TSTS time-space-time-space switching
TTL time to live
TU tributary unit (SDH)
TUG tributary unit group (SDH)
TUP telephone user part (SS7)
UA unnumbered acknowledgment (HDLC)
UART universal asynchronous receiver transmitter
UBR unspecified bit rate (ATM)
UCC Uniform Code Council
UDI unrestricted digital information (ISDN)
UDP User Datagram Protocol (IETF)
UHF ultra-high frequency
UI unnumbered information (HDLC)
UNI user-to-network interface (ATM, FR)
UNIT™ Unified Network Interface Technology™ (Ocular)
UNMA unified network management architecture
UNSELECT unselect transponder (RFID)
UPC universal product code
UPS uninterruptable power supply
UPSR unidirectional path switched ring
UPT universal personal telecommunications
URL uniform resource locator
USART universal synchronous asynchronous receiver transmitter
USB universal serial bus
UTC coordinated universal time
UTP unshielded twisted pair (Physical Layer)
UUCP UNIX-UNIX copy
VAN value-added network
VAX virtual address extension (DEC)
vBNS very high-speed backbone network service
VBR variable bit rate (ATM)
VBR-NRT variable bit rate non-real-time (ATM)
VBR-RT variable bit rate real-time (ATM)
VC venture capital

VC virtual channel (ATM)
VC virtual circuit (PSN)
VC virtual container (SDH)
VCC virtual channel connection (ATM)
VCI virtual channel identifier (ATM)
VCSEL vertical cavity surface-emitting laser
VDSL very high-speed digital subscriber line
VDSL very high bit rate digital subscriber line
VERONICA very easy rodent-oriented netwide index to computerized archives (Internet)
VGA variable graphics array
VHF very high frequency
VHS video home system
VID VLAN ID
VIN vehicle identification number
VINES virtual networking system (Banyan)
VIP VINES Internet Protocol
VLAN virtual LAN
VLF very low frequency
VLR visitor location register (Wireless)
VLSI very large-scale integration
VM virtual machine (IBM)
VM virtual memory
VMS virtual memory system (DEC)
VOD video-on-demand
VP virtual path
VPC virtual path connection
VPI virtual path identifier
VPN virtual private network
VR virtual reality
VSAT very small aperture terminal
VSB vestigial sideband
VSELP vector-sum excited linear prediction
VT virtual tributary
VTAM virtual telecommunications access method (SNA)
VTOA voice and telephony over ATM
VTP Virtual Terminal Protocol (ISO)
WACK wait acknowledgment (BISYNC)
WACS wireless access communications system
WAIS wide area information server (IETF)

WAN wide area network
WAP Wireless Application Protocol (or wrong approach to portability)
WARC World Administrative Radio Conference
WATS wide area telecommunications service
WDM wavelength division multiplexing
WIN wireless in-building network
WISP wireless ISP
WTO World Trade Organization
WWW World Wide Web (IETF)
WYSIWYG what you see is what you get
xDSL x-type digital subscriber line
XID exchange identification (HDLC)
XML Extensible Markup Language
XNS Xerox Network Systems
XPM cross-phase modulation
ZBTSI zero-byte time slot interchange
ZCS zero code suppression

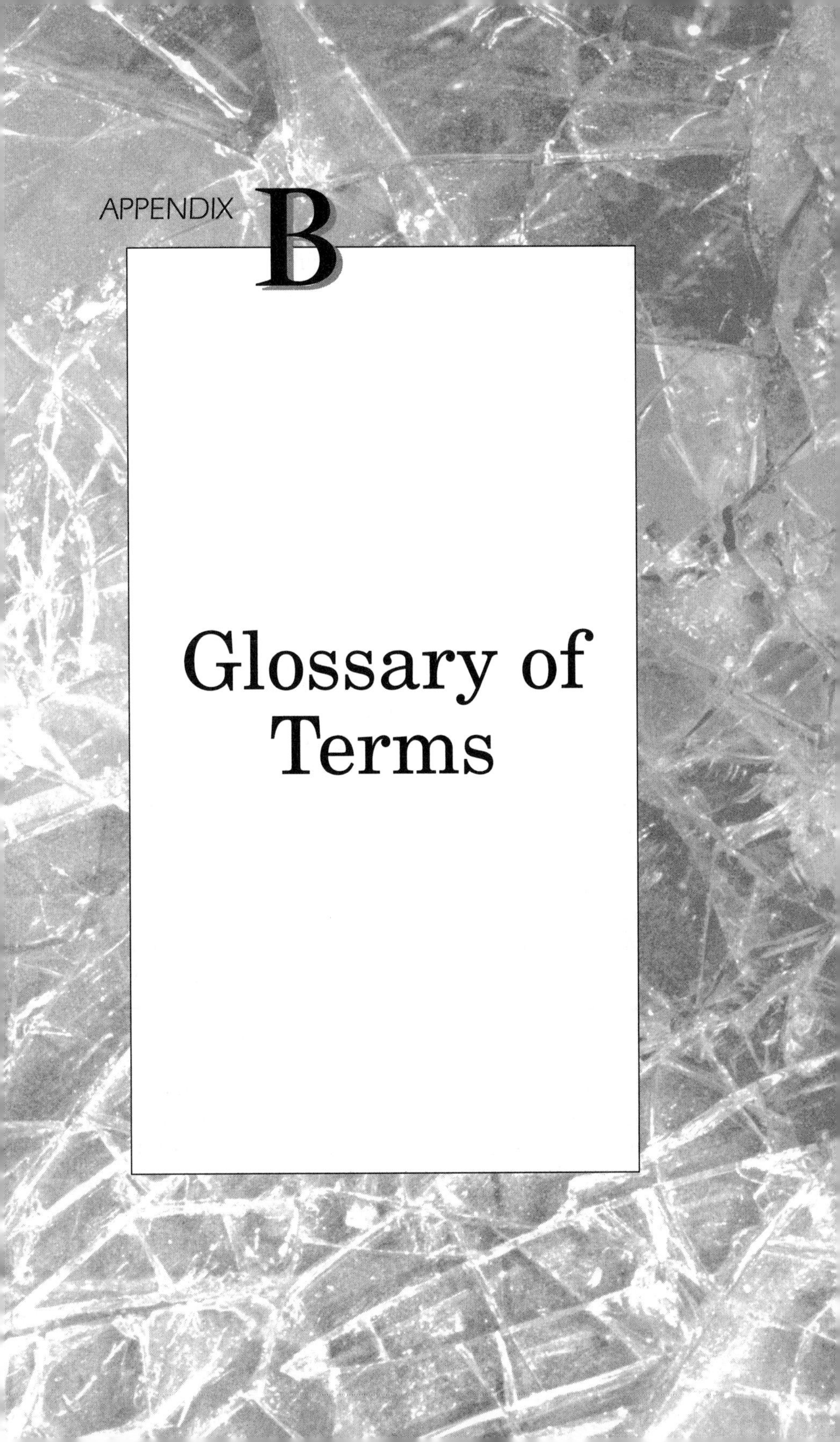

APPENDIX B

Glossary of Terms

Numerical

3G Third-generation (3G) systems will provide access to a wide range of telecommunication services supported by both fixed telecommunication networks and other services specific to mobile users. A range of mobile terminal types will be supported and may be designed for mobile or fixed use. Key features of 3G systems include the compatibility of services, small terminals with worldwide roaming capability, Internet and other multimedia applications, high bandwidth, and a wide range of services and terminals.
4G Fourth-generation (4G) networks extend the capacity of 3G networks by an order of magnitude, rely entirely on a packet infrastructure, use network elements that are 100 percent digital, and offer extremely high bandwidth.

A

abend A contraction of the words "abnormal end" used to describe a computer crash in the mainframe world.
absorption A form of optical attenuation in which optical energy is converted into an alternative form, often heat. Often caused by impurities in the fiber, hydroxyl absorption is the best-known form.
acceptance angle The critical angle within which incident light is totally internally reflected inside the core of an optical fiber.
access The set of technologies used to reach the network by a user.
accounts payable Amounts owed to suppliers and vendors for products and/or services that have been delivered on credit. Most accounts payable agreements call for the credit to be reconciled within 30 to 60 days.
accounts receivable Money that is owed to the corporation.
add-drop multiplexer (ADM) A device used in SONET and SDH systems that has the ability to add and remove signal components without having to demultiplex the entire transmitted stream, a significant advantage over legacy multiplexing systems such as DS3.
aerial plant Transmission equipment (including media, amplifiers, and splice cases) that is suspended in the air between poles.
ALOHA The name given to the first LAN, designed and implemented in Hawaii and used to interconnect the various campuses of the state's university system.

alternate mark inversion The encoding scheme used in T1. Every other "one" is inverted in polarity from the one that preceded or follows it.
ALU Arithmetic logic unit; the "brain" of a CPU chip.
amplifier 1. A device that increases the transmitted power of a signal. Amplifiers are typically spaced at carefully selected intervals along a transmission span. 2. A device used in analog networks to strengthen data signals.
amplitude modulation 1. A signal-encoding technique in which the amplitude of the carrier is modified according to the behavior of the signal that it is transporting. 2. The process of causing an electromagnetic wave to carry information by changing or modulating the amplitude or loudness of the wave.
AMPS Advanced mobile phone service; the modern analog cellular network.
analog 1. A signal that is continuously varying in time. Functionally, the opposite of digital. 2. A word that means "constantly varying in time."
angular misalignment The reason for loss that occurs at the fiber ingress point. If the light source is improperly aligned with the fiber's core, some of the incident light will be lost, leading to reduced signal strength.
APD avalanche photodiode, an optical receptor that can amplify the strength of a weak received signal.
armor The rigid protective coating on some fiber cables that protects them from crushing and from chewing by rodents.
ASCII American Standards Code for Information Interchange. A 7-bit data encoding scheme.
ASIC Application-specific integrated circuit, which is a specially designed IC created for a specific application.
asset What the company owns.
asynchronous Data that is transmitted between two devices that do not share a common clock source.
asynchronous transfer mode (ATM) 1. A standard for switching and multiplexing that relies on the transport of fixed-size data entities called cells, which are 53 octets in length. ATM has enjoyed a great deal of attention lately because its internal workings allow it to provide quality of service (QoS), a much demanded option in modern data networks. 2. One of the family of so-called fast packet technologies characterized by low error rates, high speed, and low cost. ATM is designed to connect seamlessly with SONET and SDH.

ATM adaptation layer (AAL) In ATM, the layer responsible for matching the payload being transported to a requested quality of service (QoS) level by assigning an ALL Type, which the network responds to.
ATM See *asynchronous transfer mode.*
attenuation The reduction in signal strength in optical fiber that results from absorption and scattering effects.
avalanche photodiode (APD) An optical semiconductor receiver that has the ability to amplify weak, received optical signals by multiplying the number of received photons to intensify the strength of the received signal. APDs are used in transmission systems where receiver sensitivity is a critical issue.
average revenue per user (ARPU) The average amount of revenue generated by each customer, calculated by dividing total revenue by the total number of subscribers.
axis The center line of an optical fiber.

B

backscattering The problem that occurs when light is scattered backward into the transmitter of an optical system. This impairment is analogous to echo, which occurs in copper-based systems.
balance sheet The balance sheet provides a view of what a company owns (its assets) and what it owes to creditors (its liabilities). The assets always equal the sum of the liabilities and shareholder equity. Liabilities represent obligations the firm has against its own assets. Accounts payable, for example, represent funds owed to someone or to another company that is outside the corporation, but that are balanced by some service or physical asset that has been provided to the company.
bandwidth 1. A measure of the number of bits per second that can be transmitted down a channel. 2. The range of frequencies within which a transmission system operates.
barcode A machine-scannable product identification label, comprising a pattern of alternating thick and thin lines that uniquely identify the product to which they are affixed.
baseband In signaling, any technique that uses digital signal representation.
baud The *signaling rate* of a transmission system. This is one of the most misunderstood terms in all of telecommunications. Often used

synonymously with bits per second, baud usually has a very different meaning. By using multibit encoding techniques, a single signal can simultaneously represent multiple bits. Thus, the bit rate can be many times the signaling rate.

beam splitter An optical device used to direct a single signal in multiple directions through the use of a partially reflective mirror or some form of an optical filter.

BECN Backward explicit congestion notification; a bit used in frame relay for notifying a device that it is transmitting too much information into the network and is therefore in violation of its service agreement with the switch.

Bell System Reference Frequency (BSRF) In the early days of the Bell System, a single timing source in the Midwest provided a timing signal for all central-office equipment in the country. This signal, delivered from a very expensive cesium clock source, was known as the BSRF. Today, GPS is used as the main reference clock source.

bend radius The maximum degree to which a fiber can be bent before serious signal loss or fiber breakage occurs. Bend radius is one of the functional characteristics of most fiber products.

bending loss Loss that occurs when a fiber is bent far enough that its maximum allowable bend radius is exceeded. In this case, some of the light escapes from the waveguide, resulting in signal degradation.

bidirectional A system that is capable of transmitting simultaneously in both directions.

binary A counting scheme that uses Base 2.

bit rate Bits per second.

Bluetooth An open wireless standard designed to operate at a gross transmission level of 1 Mbps. Bluetooth is being positioned as a connectivity standard for personal area networks.

Bragg grating A device that relies on the formation of interference patterns to filter specific wavelengths of light from a transmitted signal. In optical systems, Bragg gratings are usually created by wrapping a grating of the correct size around a piece of fiber that has been made photosensitive. The fiber is then exposed to strong ultraviolet light, which passes through the grating, forming areas of high and low refractive indices. Bragg gratings (or filters, as they are often called) are used for selecting certain wavelengths of a transmitted signal, and are often used in optical switches, DWDM systems, and tunable lasers.

broadband 1. Historically, broadband meant "any signal that is faster than the ISDN primary rate (T1 or E1)." Today, it means "big pipe," or in other words, a very high transmission speed. 2. In signaling the term means analog; in data transmission it means "big pipe" (high bandwidth).

buffer A coating that surrounds optical fiber in a cable and offers protection from water, abrasion, etc.

building integrated timing supply (BITS) The central office device that receives the clock signal from GPS or another source and feeds it to the devices in the office it controls.

bull's eye code The earliest form of bar code, comprising a series of concentric circles so that the code could be read from any angle.

bundling A product sales strategy in which multiple services (voice, video, entertainment, Internet, wireless, etc.) are sold as a converged package and invoiced with a single, easy-to-understand bill.

bus The parallel cable that interconnects the components of a computer.

butt splice A technique in which two fibers are joined end to end by fusing them with heat or optical cement.

C

cable An assembly made up of multiple optical or electrical conductors as well as other inclusions such as strength members, waterproofing materials, and armor.

cable assembly A complete optical cable that includes the fiber itself and terminators on each end to make it capable of attaching to a transmission or receiving device.

cable plant The entire collection of transmission equipment in a system, including the signal emitters, the transport media, the switching and multiplexing equipment, and the receive devices.

cable vault The subterranean room in a central office where cables enter and leave the building.

call center A room in which operators receive calls from customers.

capacitance An electrical phenomenon by which an electric charge is stored in a circuit.

capacitive coupling The transfer of electromagnetic energy from one circuit to another through mutual capacitance, which is nothing more than the ability of a surface to store an electric charge. Capacitance is simply a measure of the electrical storage capacity between the circuits. Similar to the inductive coupling phenomenon described

earlier, capacitive coupling can be both intentional and unplanned.
capital expenditures (CAPEX) Wealth in the form of money or property, typically accumulated in a business by a person, partnership, or corporation. In most cases capital expenditures can be amortized over a period of several years, most commonly 5 years.
capital intensity A measure that has begun to appear as a valid measure of financial performance for large telecom operators. It is calculated by dividing capital spending (CAPEX) by revenue.
cash burn A term that became a part of the common lexicon during the dot-com years. It refers to the rate at which companies consume their available cash.
cash flow One of the most common measures of valuation for public and private companies. True cash flow is exactly that—a measure of the cash that flows through a company during some defined time period after factoring out all fixed expenses. In many cases cash flow is equated to EBITDA. Usually, cash flow is defined as income after taxes minus preferred dividends plus depreciation and amortization.
CCITT Consultative Committee on International Telegraphy and Telephony. Now defunct and replaced by the ITU-TSS.
CDMA Code division multiple access, one of several digital cellular access schemes. CDMA relies on frequency hopping and noise modulation to encode conversations.
cell The standard protocol data unit in ATM networks. It comprises a five-byte header and a 48-octet payload field.
cell loss priority (CLP) In ATM, a rudimentary single-bit field used to assign priority to transported payloads.
cell relay service (CRS) In ATM, the most primitive service offered by service providers, consisting of nothing more than raw bit transport with no assigned ATM adaption layer types.
cellular telephony The wireless telephony system characterized by low-power cells, frequency reuse, handoff, and central administration.
center wavelength The central operating wavelength of a laser used for data transmission.
central office A building that houses shared telephony equipment such as switches, multiplexers, and cable distribution hardware.
central office terminal (COT) In loop carrier systems, the device located in the central office that provides multiplexing and demultiplexing services. It is connected to the remote terminal.
chained layers The lower three layers of the OSI Model that provide for connectivity.

chirp A problem that occurs in laser diodes when the center wavelength shifts momentarily during the transmission of a single pulse. Chirp is due to instability of the laser itself.

chromatic dispersion Because the wavelength of transmitted light determines its propagation speed in an optical fiber, different wavelengths of light will travel at different speeds during transmission. As a result, the multiwavelength pulse will tend to "spread out" during transmission, causing difficulties for the receive device. Material dispersion, waveguide dispersion, and profile dispersion all contribute to the problem.

CIR Committed information rate; the volume of data that a frame-relay provider absolutely guarantees it will transport for a customer.

circuit emulation service (CES) In ATM, a service that emulates private-line service by modifying (1) the number of cells transmitted per second and (2) the number of bytes of data contained in the payload of each cell.

cladding The fused silica coating that surrounds the core of an optical fiber. It typically has a different index of refraction than the core has, causing light that escapes from the core into the cladding to be refracted back into the core.

CLEC Competitive local exchange carrier; a small telephone company that competes with the incumbent player in its own marketplace.

close-coupling smart card A card that is defined by extremely short read ranges and are in fact similar to contact-based smart cards. These devices are designed to be used with an insertion reader, similar to what is often seen in modern hotel room doors.

CMOS Complimentary metal oxide semiconductor, a form of integrated circuit technology that is typically used in low-speed and low-power applications.

coating The plastic substance that covers the cladding of an optical fiber. It is used to prevent damage to the fiber itself through abrasion.

coherent A form of emitted light in which all the rays of the transmitted light align themselves along the same transmission axis, resulting in a narrow, tightly focused beam. Lasers emit coherent light.

compression The process of reducing the size of a transmitted file without losing the integrity of the content by eliminating redundant information prior to transmitting or storing.

concatenation The technique used in SONET and SDH in which multiple payloads are ganged together to form a super-rate frame capable of transporting payloads greater in size than the basic transmission speed of the system. Thus, an OC-12c provides 622.08 Mbps of total bandwidth, as opposed to an OC-12, which also offers 622.08 Mbps, but in increments of OC-1 (51.84 Mbps).

conditioning The process of "doctoring" a dedicated circuit to eliminate the known and predictable results of distortion.

congestion The condition that results when traffic arrives faster than it can be processed by a server.

connectivity The process of providing an electrical transport of data.

connector A device, usually mechanical, used to connect a fiber to a transmit or receive device or to bond two fibers.

core 1. The central portion of an optical fiber that provides the primary transmission path for an optical signal. It usually has a higher index of refraction than the cladding. 2. The central high-speed transport region of the network.

counter-rotating ring A form of transmission system that comprises two rings operating in opposite directions. Typically, one ring serves as the active path while the other serves as the protect or backup path.

CPU Central processing unit. Literally the chipset in a computer that provides the intelligence.

CRC Cyclic redundancy check. A mathematical technique for checking the integrity of the bits in a transmitted file.

critical angle The angle at which total internal reflection occurs.

cross-phase modulation (XPM) A problem that occurs in optical fiber that results from the nonlinear index of refraction of the silica in the fiber. Because the index of refraction varies according to the strength of the transmitted signal, some signals interact with each other in destructive ways. XPM is considered to be a fiber nonlinearity.

CSMA/CD Carrier sense, multiple access with collision detection; the medium access scheme used in Ethernet LANs and characterized by an "if it feels good, do it" approach.

current assets Those assets on the balance sheet that are typically expected to be converted to cash within a year of the publication date of the balance sheet. Current assets typically include such line items as accounts receivable, cash, inventories and supplies, any marketable securities held by the corporation, prepaid expenses, and a variety of other less critical items that typically fall into the "Other" line item.

current liabilities Obligations that must be repaid within a year.
current ratio Calculated by dividing the current assets for a financial period by the current liabilities for the same period. Be careful: A climbing current ratio might be a good indicator of improving financial performance, but could also indicate that warehoused product volumes are climbing.
customer relationship management (CRM) A technique for managing the relationship between a service provider and a customer through the discrete management of knowledge about the customer.
cutoff wavelength The wavelength below which single-mode fiber ceases to be single mode.
cylinder A stack of tracks to which data can be logically written on a hard drive.

D

dark fiber Optical fiber that is sometimes leased to a client that is not connected to a transmitter or receiver. In a dark fiber installation, it is the customer's responsibility to terminate the fiber.
data Raw, unprocessed zeroes and ones.
data communications The science of moving data between two or more communicating devices.
data mining A technique in which enterprises extract information about customer behavior by analyzing data contained in their stored transaction records.
datagram The service provided by a connectionless network. Often said to be unreliable, this service makes no guarantees with regard to latency or sequentiality.
DCE Data circuit-terminating equipment; a modem or other device that delineates the end of the service provider's circuit.
DE Discard eligibility bit; a primitive single-bit technique for prioritizing traffic that is to be transmitted.
debt to equity ratio Calculated by dividing the total debt for a particular fiscal year by the total shareholder equity for the same financial period.
decibel (dB) A logarithmic measure of the strength of a transmitted signal. Because it is a logarithmic measure, a 20 dB loss would indicate that the received signal is one one-hundredth its original strength.

dense wavelength-division multiplexing (DWDM) A form of frequency-division multiplexing in which multiple wavelengths of light are transmitted across the same optical fiber. These DWDM systems typically operate in the so-called L-Band (1625 nm) and have channels that are spaced between 50 and 100 GHz apart. Newly announced products may dramatically reduce this spacing.
detector An optical receive device that converts an optical signal into an electrical signal so that it can be handed off to a switch, router, multiplexer, or other electrical transmission device. These devices are usually either NPN or APDs.
diameter mismatch loss Loss that occurs when the diameter of a light emitter and the diameter of the ingress fiber's core are dramatically different.
dichroic filter A filter that transmits light in a wavelength-specific fashion, reflecting nonselected wavelengths.
dielectric A substance that is nonconducting.
diffraction grating A grid of closely spaced lines that are used to selectively direct specific wavelengths of light as required.
digital A signal characterized by discrete states. The opposite of analog.
digital hierarchy In North America, the multiplexing hierarchy that allows 64 Kbps DS0 signals to be combined to form DS3 signals for high-bit-rate transport.
digital Literally, discrete.
digital subscriber line access multiplexer (DSLAM) The multiplexer in the central office that receives voice and data signals on separate channels, relaying voice to the local switch and data to a router elsewhere in the office.
diode A semiconductor device that allows current to flow only in a single direction.
direct attached storage (DAS) A storage option in which the storage media (hard drives, CDs, etc.) are either integral to the server (internally mounted) or are directly connected to one of the servers.
dispersion compensating fiber (DCF) A segment of fiber that exhibits the opposite dispersion effect of the fiber to which it is coupled. DCF is used to counteract the dispersion of the other fiber.
dispersion The spreading of a light signal over time that results from modal or chromatic inefficiencies in the fiber.
dispersion-shifted fiber (DSF) A form of optical fiber that is designed to exhibit zero dispersion within the C-Band (1,550 nm).

DSF does not work well for DWDM because of four-wave mixing problems; nonzero dispersion shifted fiber is used instead.

distortion A known and measurable (and therefore correctable) impairment on transmission facilities.

dopant Substances used to lower the refractive index of the silica used in optical fiber.

DS0 Digital signal level 0, a 64 Kbps signal.

DS1 Digital signal level 1, a 1.544 Mbps signal.

DS2 Digital signal level 2, a 6.312 Mbps signal.

DS3 A 44.736 Mbps signal format found in the North American Digital Hierarchy.

DSL Digital subscriber line, a technique for transporting high-speed digital data across the analog local loop while (in some cases) transporting voice simultaneously.

DTE Data terminal equipment; user equipment that is connected to a DCE.

DTMF Dual-tone, multifrequency; the set of tones used in modern phones to signal dialed digits to the switch. Each button triggers a pair of tones.

duopoly The current regulatory model for cellular systems; two providers are assigned to each market. One is the wireline provider (typically the local ILEC), the other an independent provider.

DWDM Dense wavelength-division multiplexing; a form of frequency-division multiplexing that allows multiple optical signals to be transported simultaneously across a single fiber.

E

E1 The 2.048 Mbps transmission standard found in Europe and other parts of the world. It is analogous to the North American T1.

earnings before interest, tax, depreciation, and amortization (EBITDA) EBITDA, sometimes called *operating cash flow,* is used to evaluate a firm's operating profitability before subtracting nonoperating expenses such as interest and other core, nonbusiness expenses and noncash charges. Long ago, cable companies and other highly capital-intensive industries substituted EBITDA for traditional cash flow as a *temporary* measure of financial performance without adding in the cost of building new infrastructure. By excluding all interest due on borrowed capital as well as the inevitable depreciation of assets, EBITDA was seen as a temporary better gauge of potential future performance.

earnings per share (EPS) Calculated by dividing annual earnings by the total number of outstanding shares.
EBCDIC Extended binary coded decimal interchange code; an eight-bit data-encoding scheme.
edge The periphery of the network where aggregation, QoS, and IP implementation take place. This is also where most of the intelligence in the network resides.
EDGE Enhanced data for global evolution; a 384 Kbps enhancement to GSM.
edge-emitting diode A diode that emits light from the edge of the device rather than from the surface, resulting in a more coherent and directed beam of light.
effective area The cross-section of a single-mode fiber that carries the optical signal.
EIR Excess information rate; the amount of data that is being transmitted by a user *above* the CIR in frame relay.
encryption The process of modifying a text or image file to prevent unauthorized users from viewing the content.
end-to-end layers The upper four layers of the OSI Model, which provide interoperability.
enterprise resource planning (ERP) A technique for managing customer interactions through data mining, knowledge management, and customer relationship management.
erbium-doped fiber amplifier (EDFA) A form of optical amplifier that uses the element erbium to bring about the amplification process. Erbium has the enviable quality that when struck by light operating at 980 nm, it emits photons in the 1,550 nm range, thus providing agnostic amplification for signals operating in the same transmission window.
ESF Extended Superframe; the framing technique used in modern T-carrier systems that provides a dedicated data channel for nonintrusive testing of customer facilities.
Ethernet A LAN product developed by Xerox that relies on a CSMA/CD medium-access scheme.
evanescent wave Light that travels down the inner layer of the cladding instead of down the fiber core.
extrinsic loss Loss that occurs at splice points in an optical fiber.
eye pattern A measure of the degree to which bit errors are occurring in optical transmission systems. The width of the "eyes" (eye patterns look like figure eights lying on their sides) indicates the relative bit-error rate.

F

facility A circuit.
facilities-based A regulatory term that refers to the requirement that CLECs own their own physical facilities instead of relying on those of the ILEC for service delivery.
Faraday effect Sometimes called the magneto-optical effect, the Faraday effect describes the degree to which some materials can cause the polarization angle of incident light to change when placed within a magnetic field that is parallel to the propagation direction.
fast Ethernet A version of Ethernet that operates at 100 Mbps.
fast packet Technologies characterized by low error rates, high speed, and low cost.
FDMA Frequency-division multiple access; the access technique used in analog AMPS cellular systems.
FEC Forward error correction; an error-correction technique that sends enough additional overhead information along with the transmitted data that a receiver not only detect an error, but actually fixes it without requesting a resend.
FECN Forward explicit congestion notification; a bit in the header of a frame-relay frame that can be used to notify a distant switch that the frame experienced severe congestion on its way to the destination.
ferrule A rigid or semirigid tube that surrounds optical fibers and protects them.
fiber grating A segment of photosensitive optical fiber that has been treated with ultraviolet light to create a refractive index within the fiber that varies periodically along its length. It operates analogously to a fiber grating and is used to select specific wavelengths of light for transmission.
fiber to the curb (FTTC) A transmission architecture for service delivery in which a fiber is installed in a neighborhood and terminated at a junction box. From there, coaxial cable or twisted pair can be cross-connected from the optical-electical (O-E) converter to the customer premises. If coax is used, the system is called hybrid fiber coax (HFC); twisted-pair-based systems are called switched digital video (SDV).
fiber to the home (FTTH) Similar to FTTC, except that FTTH extends the optical fiber all the way to the customer premises.
fibre channel A set of standards for a serial I/O bus that supports a range of port speeds, including 133 Mbps, 266 Mbps, 530 Mbps, 1

Gbps, and soon, 4 Gbps. The standard supports point-to-point connections, switched topologies, and arbitrated loop architecture.

Financial Accounting Standards Board (FASB) The officially recognized entity that establishes standards for accounting organizations to ensure commonality among countries and international accounting organizations.

four-wave mixing (FWM) The nastiest of the so-called fiber nonlinearities. FWM is commonly seen in DWDM systems and occurs when the closely spaced channels mix and generate the equivalent of optical sidebands. The number of these sidebands can be expressed by the equation $1/2(n^3 - n^2)$, where n is the number of original channels in the system. Thus a 16-channel DWDM system will potentially generate 1,920 interfering sidebands!

frame A variable-size data transport entity.

frame-relay bearer service (FRBS) In ATM, a service that allows a frame-relay frame to be transported across an ATM network.

frame relay One of the family of so-called fast-packet technologies characterized by low error rates, high speed, and low cost.

freespace optics A metro transport technique that uses a narrow unlicensed optical beam to transport high-speed data.

frequency modulation The process of causing an electromagnetic wave to carry information by changing or modulating the frequency of the wave.

frequency-agile The ability of a receiving or transmitting device to change its frequency in order to take advantage of alternate channels.

frequency-division multiplexing The process of assigning specific frequencies to specific users.

Fresnel loss The loss that occurs at the interface between the head of the fiber and the light source to which it is attached. At air-glass interfaces, the loss usually equates to about 4 percent.

full-duplex Two-way simultaneous transmission.

fused fiber A group of fibers that are fused together so that they will remain in alignment. They are often used in one-to-many distribution systems for the propagation of a single signal to multiple destinations. Fused fiber devices play a key role in passive optical networking (PON).

fusion splice A splice made by melting the ends of the fibers together.

G

generally accepted accounting principles (GAAP) Those commonly recognized accounting practices that ensure financial accounting standardization across multiple global entities.
generic flow control (GFC) In ATM, the first field in the cell header. It is largely unused except when it is overwritten in network node interface (NNI) cells, in which case it becomes additional space for virtual path addressing.
GEOS Geosynchronous Earth orbit satellite; a family of satellites that orbit above the equator at an altitude of 22,300 miles and provide data and voice transport services.
gigabit Ethernet A version of Ethernet that operates at 1,000 Mbps.
global positioning system (GPS) The array of satellites used for radio location around the world. In the telephony world, GPS satellites provide an accurate timing signal for synchronizing office equipment.
Go-Back-N A technique for error correction that causes all frames of data to be transmitted again, starting with the errored frame.
gozinta Goes into.
gozouta Goes out of.
GPRS General-packet radio service; another add-on for GSM networks that is not enjoying a great deal of success in the market yet. Stay tuned.
graded index fiber (GRIN) A type of fiber in which the refractive index changes gradually between the central axis of the fiber and the outer layer, instead of abruptly at the core-cladding interface.
groom and fill Similar to add-drop, groom and fill refers to the ability to add (fill) and drop (groom) payload components at intermediate locations along a network path.
gross domestic product (GDP) The total market value of all the goods and services produced by a nation during a specific period of time.
GSM Global system for mobile communications; the wireless access standard used in many parts of the world that offers two-way paging, short messaging, and two-way radio in addition to cellular telephony.
GUI Graphical user interface; the computer interface characterized by the "click, move, drop" method of file management.

H

half-duplex Two-way transmission, but only one direction at a time.
haptics The science of providing tactile feedback to a user electronically. Often used in high-end virtual-reality systems.
headend The signal origination point in a cable system.
header error correction (HEC) In ATM, the header field used to recover from bit errors in the header data.
header In ATM, the first five bytes of the cell. The header contains information used by the network to route the cell to its ultimate destination. Fields in the cell header include the Generic Flow Control field, Virtual Path Identifier field, Virtual Channel Identifier field, Payload Type Identifier field, Cell Loss Priority field, and Header Error Correction field.
hertz (Hz) A measure of cycles per second in transmission systems.
hop count A measure of the number of machines a message or packet has to pass through between the source and the destination. Often used as a diagnostic tool.
hybrid fiber coax A transmission system architecture in which a fiber feeder penetrates a service area and is then cross-connected to coaxial cable feeders into the customers' premises.
hybrid loop An access facility that uses more than one medium. For example, hybrid-fiber coax (HFC, defined previously) or hybrids of fiber and copper twisted-pair wire.

I

ILEC Incumbent local exchange carrier; an RBOC.
income statement Used to report a corporation's revenues, expenses, and net income (profit) for a particular defined time period. Sometimes called a profit and loss (P&L) statement or statement of operations, the income statement charts a company's performance over a period of time. The results are most often reported as *earnings per share* and *diluted earnings per share.* Earnings per share is defined as the proportion of the firm's net income that can be accounted for on a per-share basis of outstanding common stock. It is calculated by subtracting preferred dividends from net income and dividing the result by the number of common shares that are outstanding. Diluted earnings per share, on the other hand, takes into

account earned or fully vested stock options that haven't yet been exercised by their owner and shares that would be created from the conversion of convertible securities into stock.

indefeasible rights of use (IRU) A long-term capacity lease of a cable. IRUs are identified by channels and available bandwidth and are typically granted for long periods of time.

index of refraction A measure of the ratio between the velocity of light in a vacuum and the velocity of the same light in an optical fiber. The refractive index is always greater than one and is denoted n.

inductance The property of an electric circuit by which an electromotive force is induced in it by a variation of current flowing through the circuit.

inductive coupling The transfer of electromagnetic energy from one circuit to another as a result of the mutual *inductance* between the circuits. Inductive coupling may be intentional, such as in an impedance matcher that matches the impedance of a transmitter or a receiver to an antenna to guarantee maximum power transfer, or it may be unplanned, as in the annoying power line inductive coupling that occasionally takes place in telephone lines, often referred to as crosstalk or hum.

information Data that has been converted to a manipulable form.

infrared (IR) The region of the spectrum within which most optical transmission systems operate, found between 700 nm and 0.1 mm.

injection laser A semiconductor laser (synonym).

inside plant Telephony equipment that is outside of the central office.

intermodulation A fiber nonlinearity that is similar to four-wave mixing, in which the power-dependent refractive index of the transmission medium allows signals to mix and create destructive sidebands.

interoperability 1. Characterized by the ability to logically share information between two communicating devices and be able to read and understand the data of the other. 2. In SONET and SDH, the ability of devices from different manufacturers to send and receive information to and from each other successfully.

intrinsic loss Loss that occurs as the result of physical differences in the two fibers being spliced.

ISDN Integrated services digital network; a digital local loop technology that offers moderately high bit rates to customers.

isochronous A word used in timing systems that means there is constant delay across a network.

ISP Internet service provider; a company that offers Internet access.
ITU International Telecommunications Union; a division of the United Nations that is responsible for managing the telecomm standards development and maintenance processes.
ITU-TSS ITU Telecommunications Standardization Sector; the ITU organization responsible for telecommunications standards development.

J

jacket The protective outer coating of an optical fiber cable. The jacket may be polyethylene, Kevlar, or metallic.
JPEG Joint Photographic Experts Group; a standards body tasked with developing standards for the compression of still images.
jumper An optical cable assembly, usually fairly short, that is terminated on both ends with connectors.

K

knowledge Information that has been acted upon and modified through some form of intuitive human thought process.
knowledge management The process of managing all that a company knows about its customers in an intelligent way so that some benefit is attained for both the customer and the service provider.

L

lambda A single wavelength on a multichannel DWDM system (λ).
LAN emulation (LANE) In ATM, a service that defines the ability to provide bridging services between LANs across an ATM network.
LAN Local area network; a small network that has the following characteristics: privately owned, high speed, low error rate, and physically small.
large core fiber Fiber that characteristically has a core diameter of 200 microns or more.

laser An acronym for "light amplification by the stimulated emission of radiation." Lasers are used in optical transmission systems because they produce coherent light that is almost purely monochromatic.

laser diode (LD) A diode that produces coherent light when a forward-biasing current is applied to it.

LATA Local access and transport area; the geographic area within which an ILEC is allowed to transport traffic. Beyond LATA boundaries the ILEC must hand traffic off to a long-distance carrier.

LEO Slow Earth orbit satellite; satellites that orbit pole-to-pole instead of above the equator and offer near instantaneous response time.

liability Obligations the firm has against its own assets. Accounts payable, for example, are a liability and represent funds owed to someone or to another company that is outside the corporation, but that are balanced by some service or physical asset that has been provided to the company.

light-emitting diode (LED) A diode that emits incoherent light when a forward-bias current is applied to it. LEDs are typically used in shorter-distance, lower-speed systems.

lightguide A term that is used synonymously with optical fiber.

line overhead (LOH) In SONET, the overhead that is used to manage the network regions between multiplexers.

line sharing A business relationship between an ILEC and a CLEC in which the CLEC provides logical DSL service over the ILEC's physical facilities.

linewidth The spectrum of wavelengths that make up an optical signal.

load coil A device that tunes the local loop to the voiceband.

local loop The pair of wires (or digital channel) that runs between the customer's phone (or computer) and the switch in the local central office.

long-term debt Debt that is typically due beyond the one-year maturity period of short-term debt.

loose tube optical cable An optical cable assembly in which the fibers within the cable are loosely contained within tubes inside the sheath of the cable. The fibers are able to move within the tube, thus allowing them to adapt and move without damage as the cable is flexed and stretched.

loss The reduction in signal strength that occurs over distance, usually expressed in decibels.

M

M13 A multiplexer that interfaces between DS1 and DS3 systems.
mainframe A large computer that offers support for very large databases and large numbers of simultaneous sessions.
MAN Metropolitan area network; a network larger than a LAN that provides high-speed services within a metropolitan area.
Manchester coding A data transmission code in which data and clock signals are combined to form a self-synchronizing data stream, in which each represented bit contains a transition at the midpoint of the bit period. The direction of transition determines whether the bit is a zero or one.
market cap(italization) Market cap is the current market value of all outstanding shares that a company has. It is calculated by multiplying the total number of outstanding shares by the current share price.
material dispersion A dispersion effect caused by the fact that different wavelengths of light travel at different speeds through a medium.
MDF Main distribution frame; the large iron structure that provides physical support for cable pairs in a central office between the switch and the incoming/outgoing cables.
message switching An older technique that sends entire messages from point to point instead of breaking the message into packets.
metasignaling virtual channel (MSVC) In ATM, a signaling channel that is always on. It is used for the establishment of temporary signaling channels as well as channels for voice and data transport.
metropolitan optical network (MON) An all-optical network deployed in a metro region.
microbend Changes in the physical structure of an optical fiber caused by bending that can result in light leakage from the fiber.
midspan meet In SONET and SDH, the term used to describe interoperability. See also *interoperability.*
modal dispersion *See* multimode dispersion.
mode A single wave that propagates down a fiber. Multimode fiber allows multiple modes to travel, whereas single-mode fiber allows only a single mode to be transmitted.
modem A term from the words "modulate" and "demodulate." Its job is to make a computer appear to the network like a telephone.

modulation The process of changing or modulating a carrier wave to cause it to carry information.
MPEG Moving Picture Experts Group; a standards body tasked with crafting standards for motion pictures.
MPLS A level-three protocol designed to provide quality of service across IP networks without the need for ATM, by assigning QoS labels to packets as they enter the network.
MTSO Mobile telephone switching office; a central office with special responsibilities for handling cellular services and the interface between cellular users and the wireline network.
multidwelling unit (MDU) A building that houses multiple residence customers such as an apartment building.
multimode dispersion Sometimes referred to as modal dispersion, multimode dispersion is caused by the fact that different modes take different times to move from the ingress point to the egress point of a fiber, thus resulting in modal spreading.
multimode fiber Fiber that has a core diameter of 62.5 microns or greater, wide enough to allow multiple modes of light to be simultaneously transmitted down the fiber.
multiplexer A device that has the ability to combine multiple inputs into a single output as a way to reduce the requirement for additional transmission facilities.
multiprotocol over ATM (MPOA) In ATM, a service that allows IP packets to be routed across an ATM network.
multitenant unit (MTU) A building that houses multiple enterprise customers such as a high-rise office building.
mutual inductance The tendency of a change in the current of one coil to affect the current and voltage in a second coil. When voltage is produced because of a change in current in a coupled coil, the effect is mutual inductance. The voltage always opposes the change in the magnetic field produced by the coupled coil.

N

near-end crosstalk (NEXT) The problem that occurs when an optical signal is reflected back toward the input port from one or more output ports. This problem is sometimes referred to as isolation directivity.
net income Another term meaning bottom-line profit.

network attached storage (NAS) An architecture in which a server accesses storage media via a LAN connection. The storage media are connected to another server.

noise An unpredictable impairment in networks. It cannot be anticipated; it can only be corrected after the fact.

nondispersion shifted fiber (NDSF) Fiber that is designed to operate at the low-dispersion second operational window (1,310 nm).

nonzero dispersion-shifted fiber (NZDSF) A form of single-mode fiber that is designed to operate just outside the 1,550 nm window so that fiber nonlinearities, particularly FWM, are minimized.

numerical aperture (NA) A measure of the ability of a fiber to gather light, NA is also a measure of the maximum angle at which a light source can be from the center axis of a fiber in order to collect light.

O

OAM&P Operations, administration, maintenance, and provisioning; the four key areas in modern network management systems. OAM&P was first coined by the Bell System and continues in widespread use today.

OC-n Optical carrier level n, a measure of bandwidth used in SONET systems. OC-1 is 51.84 Mbps; OC-n is n times 51.84 Mbps.

operating expenses (OPEX) Those expenses that must be accounted for in the year in which they are incurred.

optical amplifier A device that amplifies an optical signal without first converting it to an electrical signal.

optical burst switching (OBS) A technique that uses a one-way reservation technique so that a burst of user data, such as a cluster of IP packets, can be sent without having to establish a dedicated path prior to transmission. A control packet is sent first to reserve the wavelength, followed by the traffic burst. As a result, OBS avoids the protracted end-to-end setup delay and also improves the utilization of optical channels for variable-bit-rate services.

optical carrier level n (OC-n) In SONET, the transmission level at which an optical system is operating.

optical isolator A device used to selectively block specific wavelengths of light.

optical time domain reflectometer (OTDR) A device used to detect failures in an optical span by measuring the amount of light

reflected back from the air-glass interface at the failure point.
OSS Operations support systems; another term for OAM&P.
outside plant Telephone equipment that is outside of the central office.
overhead That part of a transmission stream that the network uses to manage and direct the payload to its destination.

P

packet A variable-size entity normally carried inside a frame or cell.
packet switching The technique for transmitting packets across a wide area network (WAN).
path overhead In SONET and SDH, that part of the overhead that is specific to the payload being transported.
payload In SONET and SDH, the user data that is being transported.
payload type identifier (PTI) In ATM, a cell header field that is used to identify network congestion and cell type. The first bit indicates whether the cell was generated by the user or by the network, while the second indicates the presence or absence of congestion in user-generated cells or in flow-related operations, administration, and maintenance information in cells generated by the network. The third bit is used for service-specific, higher-layer functions in the user-to-network direction, such as to indicate that a cell is the last in a series of cells. From the network to the user, the third bit is used with the second bit to indicate whether the OA&M information refers to segment or end-to-end-related information flow.
PBX Private branch exchange; literally a small telephone switch located on a customer premises. The PBX connects back to the service provider's central office via a collection of high-speed trunks.
PCM Pulse code modulation; the encoding scheme used in North America for digitizing voice.
phase modulation The process of causing an electromagnetic wave to carry information by changing or modulating the phase of the wave.
photodetector A device used to detect an incoming optical signal and convert it to an electrical output.
photodiode A semiconductor that converts light to electricity.

photon The fundamental unit of light, sometimes referred to as a quantum of electromagnetic energy.
photonic The optical equivalent of the term "electronic."
pipelining The process of having multiple unacknowledged outstanding messages in a circuit between two communicating devices.
pixel Contraction of the terms "picture" and "element." The tiny color elements that make up the screen on a computer monitor.
planar waveguide A waveguide fabricated from a flat material such as a sheet of glass, into which are etched fine lines used to conduct optical signals.
plenum The air space in buildings found inside walls, under floors, and above ceilings. The plenum spaces are often used as conduits for optical cables.
plenum cable Cable that passes fire-retardant tests so that it can legally be used in plenum installations.
plesiochronous In timing systems, a term that means "almost synchronized." It refers to the fact that in SONET and SDH systems, payload components frequently derive from different sources and therefore may have slightly different phase characteristics.
pointer In SONET and SDH, a field that is used to indicate the beginning of the transported payload.
polarization The process of modifying the direction of the magnetic field within a light wave.
polarization mode dispersion (PMD) The problem that occurs when light waves with different polarization planes in the same fiber travel at different velocities down the fiber.
preform The cylindrical mass of highly pure fused silica from which optical fiber is drawn during the manufacturing process. In the industry, the preform is sometimes referred to as a "gob."
private line A dedicated point-to-point circuit.
protocol A set of rules that facilitates communications.
proximity-coupling smart card A card that is designed to be readable at a distance of approximately 4 to 10 inches from the reader. These devices are often used for sporting events and other large public gatherings that require access control across a large population of attendees.
pulse spreading The widening or spreading out of an optical signal that occurs over distance in a fiber.
pump laser The laser that provides the energy used to excite the dopant in an optical amplifier.

PVC Permanent virtual circuit; a circuit provisioned in frame relay or ATM that does not change without service-order activity by the service provider.

Q

Q.931 The set of standards that defines signaling packets in ISDN networks.
quantize The process of assigning numerical values to the digitized samples created as part of the voice digitization process.
quick ratio Calculated by dividing the sum of cash, short-term investments, and accounts receivable for a given period by the current liabilities for the same period. It measures the degree of a firm's liquidity.
quiet zone The area on either side of the Universal Product Code (UPC) that has no printing.

R

RAM Random access memory; the volatile memory used in computers for short-term storage.
Rayleigh scattering A scattering effect that occurs in optical fiber as the result of fluctuations in silica density or chemical composition. Metal ions in the fiber often cause Rayleigh scattering.
RBOC Regional Bell operating company; today called an ILEC.
refraction The change in direction that occurs in a light wave as it passes from one medium into another. The most common example is the bending that is often seen to occur when a stick is inserted into water.
refractive index A measure of the speed at which light travels through a medium, usually expressed as a ratio compared to the speed of the same light in a vacuum.
regenerative repeater A device that reconstructs and regenerates a transmitted signal that has been weakened over distance.
regenerator A device that recreates a degraded digital signal before transmitting it on to its final destination.
remote terminal (RT) In loop carrier systems, the multiplexer located in the field. It communicates with the central office terminal (COT).

repeater *See* regenerator.
resilient packet ring (RPR) A ring architecture that comprises multiple nodes that share access to a bidirectional ring. Nodes send data across the ring using a specific MAC protocol created for RPR. The goal of the RPR topology is to interconnect multiple nodes in a ring architecture that is media independent for efficiency purposes.
retained earnings Represents the money a company has earned less any dividends it has paid out. This figure does not necessarily equate to cash; more often than not it reflects that amount of money the corporation has reinvested in itself rather than paid out to shareholders as stock dividends.
return on investment (ROI) Defined as the ratio of a company's profits to the amount of capital that has been invested in it. This calculation measures the financial benefit of a particular business activity relative to the costs of engaging in the activity. The profits used in the calculation of ROI can be calculated before or after taxes and depreciation, and can be defined either as the first year's profit or as the weighted average profit during the lifetime of the entire project. Invested capital, on the other hand, is typically defined as the capital expenditure required for the project's first year of existence. Some companies may include maintenance or recurring costs as part of the invested capital figure, such as software updates. A word of warning about ROI calculations: Because there are no hard and fast rules about the absolute meanings of profits and invested capital, using ROI as a comparison of companies can be risky because of the danger of comparing apples to tractors, as it were. Be sure that comparative ROI calculations use the same bases for comparison.
ROM Read-only memory, or memory that cannot be erased; often used to store critical files or boot instructions.

S

scattering The backsplash or reflection of an optical signal that occurs when it is reflected by small inclusions or particles in the fiber.
SDH Synchronous digital hierarchy; the European equivalent of SONET.
section overhead (SOH) In SONET systems, the overhead that is used to manage the network regions that occur between repeaters.
sector A quadrant on a disk drive to which data can be written; used for locating information on the drive.

Securities and Exchange Commission (SEC) The federal government agency that is responsible for regulation of the securities industry.
selective retransmit An error correction technique in which only the errored frames are retransmitted.
self-phase modulation (SPM) The refractive index of glass is directly related to the power of the transmitted signal. As the power fluctuates, so too does the index of refraction, causing waveform distortion.
shareholder equity Claims that shareholders have against the corporation's assets.
sheath One of the layers of protective coating in an optical fiber cable.
signaling The techniques used to set up, maintain and tear down a call.
signaling virtual channel (SVC) In ATM, a temporary signaling channel used to establish paths for the transport of user traffic.
simplex One-way transmission *only*.
single-mode fiber (SMF) The most popular form of fiber today, characterized by the fact that it allows only a single mode of light to propagate down the fiber.
slotted ALOHA A variation on ALOHA in which stations transmit at predetermined times to ensure maximum throughput and minimal collisions.
soliton A unique waveform that takes advantage of nonlinearities in the fiber medium, the result of which is a signal that suffers essentially no dispersion effects over long distances. Soliton transmission is an area of significant study at the moment, because of the promise it holds for long-haul transmission systems.
SONET Synchronous optical network; a multiplexing standard that begins at DS3 and provides standards-based multiplexing up to gigabit speeds. SONET is widely used in telephone company long-haul transmission systems and was one of the first widely deployed optical transmission systems.
source The emitter of light in an optical transmission system.
Spatial Reuse Protocol (SRP) A media-independent MAC layer protocol that operates over two counter-rotating, optical rings. The dual-ring architecture provides for data survivability in the event of a failed node or a breached ring. SRP uses bandwidth very efficiently by having packets traverse only the part of the ring necessary to get from the source to the destination node.

SS7 Signaling system 7, the current standard for telephony signaling worldwide.
standards The published rules that govern an industry's activities.
statement of cash flows The statement of cash flows illustrates the manner in which the firm generated cash flows (the sources of funds) and the manner in which it employed those cash flows to support ongoing business operations.
steganography A cryptographic technique in which encrypted information is embedded in the pixel patterns of graphical images. The technique is being closely examined as a way to enforce digital-watermarking capabilities and digital-signature capabilities.
step index fiber Fiber that exhibits a continuous refractive index in the core, which then "steps" at the core-cladding interface.
stimulated Brillouin scattering (SBS) A fiber nonlinearity that occurs when a light signal traveling down a fiber interacts with acoustic vibrations in the glass matrix (sometimes called photon-phonon interaction), causing light to be scattered or reflected back toward the source.
stimulated Raman scattering (SRS) A fiber nonlinearity that occurs when power from short-wavelength, high-power channels is bled into longer-wavelength, lower-power channels.
storage area network (SAN) A dedicated storage network that provides access to stored content. In a SAN, multiple servers may have access to the same servers.
store and forward The transmission technique in which data is transmitted to a switch, stored there, examined for errors, examined for address information, and forwarded on to the final destination.
strength member The strand within an optical cable that is used to provide tensile strength to the overall assembly. The member is usually composed of steel, fiberglass, or Aramid yarn.
supply chain The process by which products move intelligently from the manufacturer to the end user, assigned through a variety of functional entities along
the way.
supply-chain management The management methodologies involved in the supply-chain management process.
surface-emitting diode A semiconductor that emits light from its surface, resulting in a low-power, broad-spectrum emission.
synchronous A term that means that both communicating devices derive their synchronization signal from the same source.

synchronous transmission signal level 1 (STS-1) In SONET systems, the lowest transmission level in the hierarchy. STS is the electrical equivalent of OC.

T

T1 The 1.544 Mbps transmission standard in North America.
T3 In the North American Digital Hierarchy, a 44.736 Mbps signal.
tandem A switch that serves as an interface between other switches and typically does not directly host customers.
TDMA Time division multiple access; a digital technique for cellular access in which customers share access to a frequency on a round-robin, time division basis.
telecommunications The science of transmitting sound over distance.
terminal multiplexer In SONET and SDH systems, a device that is used to distribute payload to or receive payload from user devices at the end of an optical span.
tight buffer cable An optical cable in which the fibers are tightly bound by the surrounding material.
time division multiplexing The process of assigning timeslots to specific users.
token ring A LAN technique, originally developed by IBM, that uses token passing to control access to the shared infrastructure.
total internal reflection The phenomenon that occurs when light strikes a surface at such an angle that all of the light is reflected back into the transporting medium. In optical fiber, total internal reflection is achieved by keeping the light source and the fiber core oriented along the same axis so that the light that enters the core is reflected back into the core at the core-cladding interface.
transceiver A device that incorporates both a transmitter and a receiver in the same housing, thus reducing the need for rack space.
transponder 1. A device that incorporates a transmitter, a receiver, and a multiplexer on a single chassis. 2. A device that receives and transmits radio signals at a predetermined frequency range. After receiving a signal, the transponder rebroadcasts it at a different frequency. Transponders are used in satellite communications and in location (RFID), identification, and navigation systems. In the case of RFID, the transponder is the tag that is affixed to the product.

treasury stock Stock that was sold to the public and later repurchased by the company on the open market. It is shown on the balance sheet as a negative number that reflects the cost of the repurchase of the shares rather than the actual market value of the shares. Treasury stock can later be retired or resold to improve earnings-per-share numbers if desired.
twisted pair The wire used to interconnect customers to the telephone network.

U

UPS Uninterruptible power supply; part of the central office power plant that prevents power outages.

V

venture capital (VC) Money used to finance new companies or projects, especially those with high earnings potential. They are often characterized as being high-risk ventures.
vertical cavity surface emitting laser (VCSEL) A small, highly efficient laser that emits light vertically from the surface of the wafer on which it is made.
vicinity-coupling smart card A card designed to operate at a read range of up to 3 or 4 feet.
virtual channel (VC) In ATM, a unidirectional channel between two communicating devices.
virtual channel identifier (VCI) In ATM, the field that identifies a virtual channel.
virtual container In SDH, the technique used to transport subrate payloads.
virtual path (VP) In ATM, a combination of unidirectional virtual channels that make up a bidirectional channel.
virtual path identifier (VPI) In ATM, the field that identifies a virtual path.
virtual private network A network connection that provides private-like services over a public network.
virtual tributary (VT) In SONET, the technique used to transport subrate payloads.

voice/telephony over ATM (VTOA) In ATM, a service used to transport telephony signals across an ATM network.
voiceband The 300-to-3,300 Hz band used for the transmission of voice traffic.

W

WAN Wide area network; a network that provides connectivity over a large geographical area.
waveguide A medium that is designed to conduct light within itself over a significant distance, such as optical fiber.
waveguide dispersion A form of chromatic dispersion that occurs when some of the light traveling in the core escapes into the cladding, traveling there at a different speed than the light in the core is traveling.
wavelength The distance between the same points on two consecutive waves in a chain—for example, from the peak of wave one to the peak of wave two. Wavelength is related to frequency by the equation where lambda (λ) is the wavelength, c is the speed of light, and f is the frequency of the transmitted signal.
wavelength-division multiplexing (WDM) The process of transmitting multiple wavelengths of light down a fiber.
window A region within which optical signals are transmitted at specific wavelengths to take advantage of propagation characteristics that occur there, such as minimum loss or dispersion.
window size A measure of the number of messages that can be outstanding at any time between two communicating entities.

X, Y, Z

zero dispersion wavelength The wavelength at which material and waveguide dispersion cancel each other.

APPENDIX C

OSI Overview

Perhaps the most well known family of protocols is the International Organization for Standardization's Open Systems Interconnection Reference Model, usually called the OSI Model for short. Shown in Figure C-1 and comprising seven layers, it provides a logical way to study and understand data communications and is based on the following simple rules. First, each of the seven layers must perform a clearly defined set of responsibilities, which are unique to that layer, to guarantee the requirement of functional modularity. Second, each layer depends upon the services of the layers above and below to do its own job, as we would expect, given the modular nature of the model. Third, the layers have no idea how the layers around them do what they do; they simply know that they do it. This is called transparency. Finally, there is nothing magic about the number seven. If the industry should decide that we need an eighth layer on the model, or that layer five is redundant, then the model will be changed. The key is functionality. An ongoing battle is being fought within the ranks of OSI Model pundits, for example, over whether a requirement actually exists for *both* layers six and seven, because many believe them to be so similar functionally that one or the other is redundant. Others question whether a need for layer five *really* exists, the functions of which are considered by many to be superfluous and redundant. Whether it ever actually happens is not important. The fact that it *can* is what matters.

Figure C-1
OSI Model

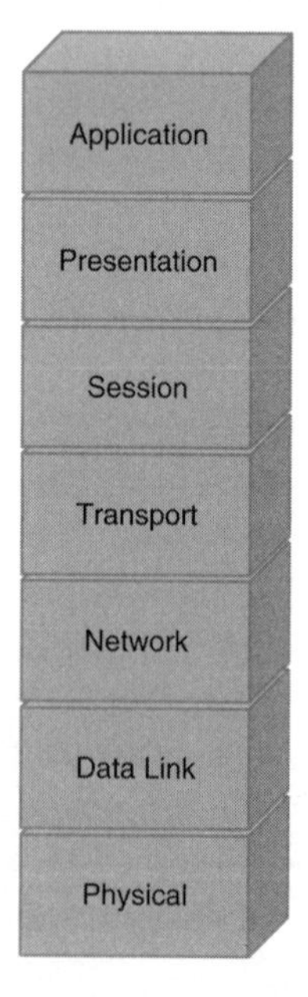

It is important to understand that the OSI Model is nothing more than a conceptual way of thinking about data communications. It isn't hardware; it isn't software. It merely simplifies and groups the processes of data transmission so that they can be easily understood and manipulated. Let's look at the OSI Model in a little more detail (see Figure C-1).

As mentioned earlier, the model is a seven-layer construct within which each layer is tightly dependent upon the layers surrounding it. The Application Layer, at the top of the model, speaks to the actual application process that creates the information to be transported by the network; it is closest to the customer and the customer's processes, and is therefore the most customizable and manipulable of all the layers. It is highly open to interpretation. On the other end of the spectrum, the Physical Layer dwells within the confines of the actual network, is totally standards dependent. There is minimal room here for interpretation; a pulse is either a one or a zero—there's nothing in between. Physical Layer standards therefore tend to be highly commoditized, whereas Application Layer standards tend to be highly specialized. This becomes extremely important as the service provider model shifts from delivering commodity bandwidth to providing customized services—even if they're mass customized—to the customer base. Service providers are clawing their way up the OSI food chain to get as close to the Application Layer end of the model as they can, because that's where the money is.

The functions of the model can be broken into two pieces, as illustrated by the dashed line in Figure C-2 between layers three and four that divides the model into the *chained layers* and the *end-to-end layers.*

The chained layers comprise layers one through three: the Physical Layer, the Data Link Layer, and the Network Layer. They are responsible for providing a service called *connectivity.* The end-to-end layers on the other hand comprise the Transport Layer, the Session Layer, the Presentation Layer, and the Application Layer. They provide a service called *interoperability.* The difference between the two services is important.

Connectivity is the process of establishing a physical connection so that electrons can flow correctly from one end of a circuit to the other. There is little intelligence involved in the process; it occurs, after all, pretty far down in the protocol ooze of the OSI Model. Connectivity is critically important to network people—it represents their lifeblood. Customers, on the other hand, are typically aware of the criticality of connectivity only when it isn't there for some reason. No dial tone? Visible connectivity. Can't connect to the ISP? Visible connectivity. Dropped call on a cell phone? Visible connectivity.

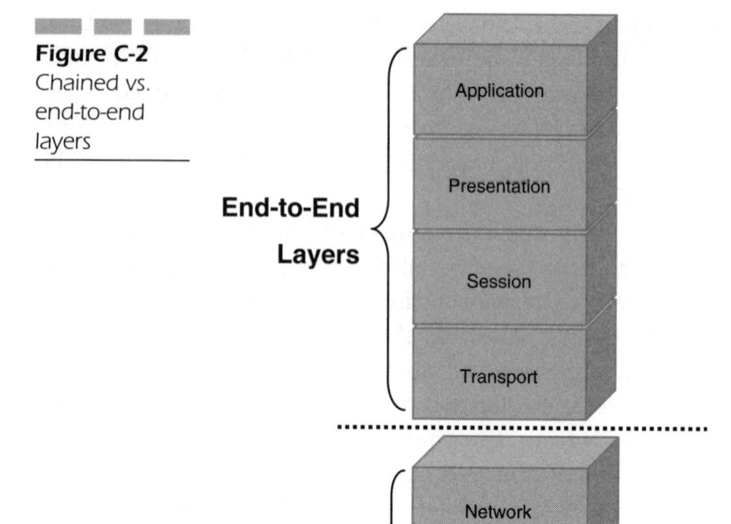

Figure C-2 Chained vs. end-to-end layers

Interoperability, however, is something that customers are much more aware of. Interoperability is the process of guaranteeing *logical connectivity* between two communicating processes over a physical network. It's wonderful that the lower three layers give a user the ability to spit bits back and forth across a wide area network (WAN). But what do the bits mean? Without interoperability, that question cannot be answered. For example, in our e-mail application, the e-mail application running on the PC and the e-mail application running on the mainframe are logically incompatible with each other for any number of reasons that will be discussed shortly. They can certainly swap bits back and forth, but without some form of protocol intervention, the bits are meaningless. Think about it: If the PC shown on the left side of Figure C-3 creates an e-mail message that is compressed, encrypted, ASCII-encoded, and shipped across logical channel 17, do the intermediate switches that create the path over which the message is transmitted care? Of course not. Only the transmitter and receiver of the message that house the applications that will have to interpret it care about such things. The intermediate

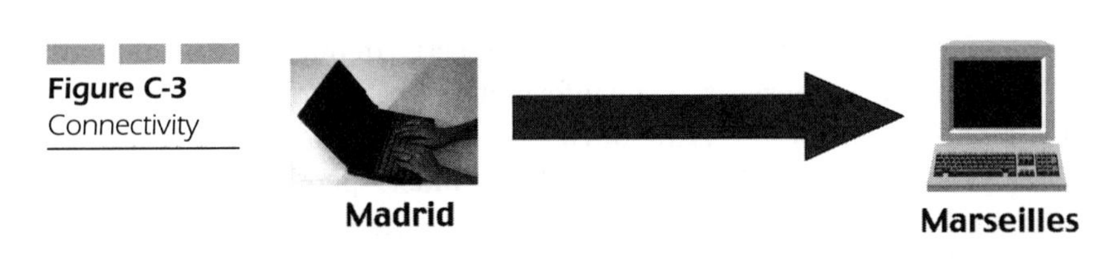

Figure C-3
Connectivity

switches care that they have electrical connectivity, that they can see the bits, that they can determine whether they are the *right* bits, and whether they are the intended recipient or not. Therefore, the end devices, the sources and sinks of the message, must implement all seven layers of the OSI Model, because they must not concern themselves only with connectivity issues, but also with issues of interoperability. The intermediate devices, however, care only about the functions and responsibilities provided by the lower three layers. Interoperability, because it only has significance in the end devices, is provided by the end-to-end layers—layers four through seven. Connectivity on the other hand is provided by the chained layers, layers one through three, because those functions are required in every link of the network chain—hence the name.

Layer by Layer

The OSI Model relies on a process called *enveloping* to perform its tasks. If we return to our earlier e-mail example, we find that each time a layer invokes a particular protocol to perform its tasks, it wraps the user's data in an envelope of overhead information that tells the receiving device about the protocol used. For example, if a layer uses a particular compression technique to reduce the size of a transmitted file, and a specific encryption algorithm to disguise the content of the file, then it is important that the receiving device be made aware of the technique employed so that it knows how to decompress and decrypt the file when it receives it. Needless to say, quite a bit of overhead must be transmitted with each piece of user data. The overhead is needed, however, if the transmission is to work properly. So as the user's data passes down the so-called stack from layer to layer, additional information is added at each step of the way as illustrated by the series of envelopes. In summary then, the message to be transported is handed to layer seven, which performs Application Layer functions and then attaches a header to the beginning of the message that explains the functions performed by that layer so that the receiver can interpret the message correctly. In our illustration, that header function is represented by information written on the envelope at each layer. When the receiving device is finally handed the message at the Physical Layer, each succeeding layer must open its own envelope until the kernel—the message—is exposed for the receiving application. Thus, OSI protocols really do work like a nested Russian doll. After peeling back layer after layer of the network onion, the core message is exposed.

Layer 7: The Application Layer

The network user's application (Eudora, Outlook, Outlook Express, PROFS, and so on) passes data down to the uppermost layer of the OSI Model, called the Application Layer. The Application Layer provides a set of highly specific services to the application that have to do with the *meaning* or *semantic content* of the data. These services include file transfer, remote file access, terminal emulation, network management, mail services, and data interoperability. This interoperability is what allows our PC user and our mainframe-based user to communicate; the Application Layer converts the application-specific information into a common, canonical form that can be understood by both systems. A canonical form is a form that can be understood universally. The word comes from *canon,* which refers to the body of officially established rules or laws that govern the practices of a church. The word also means an accepted set of principles of behavior that all parties in a social or functional grouping agree to abide by. Let's examine a real-world example of a network-oriented canonical form.

When network hardware manufacturers build components—switches, multiplexers, cross-connect systems, modem pools—for sale to their customers, they do so knowing that one of the most important aspects of a successful hardware sale is the inclusion of an element-management system that will allow the customer to manage the device within his or her network. The only problem is that today most networks are made up of equipment purchased from a variety of vendors. Each vendor develops its own element managers on a device-by-device basis, which work exceptionally well for each device. This does not become a problem until it comes time to create a management hierarchy for a large network, shown in Figure C-4, at which time the network management center begins to look a lot like a Macy's television department. (A large network management system is shown in Figure C-5.) Each device or set of devices requires its own display monitor, and when one device in the network fails, causing a waterfall effect, the network manager must reconstruct the entire chain of events to discover what the original causative factor was. This is sometimes called the "Three Mile Island effect." Back in the 1970s when the Three Mile Island nuclear power plant went critical and tried to make Pennsylvania glow in the dark, it became clear to the Monday morning quarterbacks trying to reconstruct the event (and create the "How could this have been prevented" document) that all the

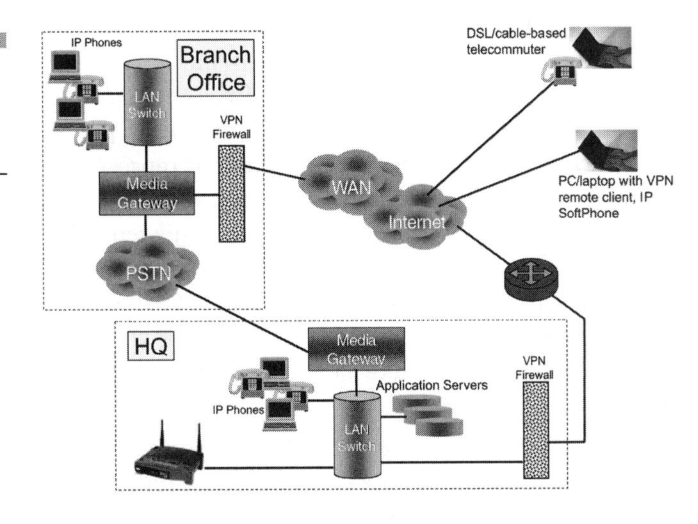

Figure C-4 A large managed network

Figure C-5 A modern network management center

information required to turn the critical failure of the reactor into a non-event was in the control room—buried somewhere in the hundreds of pages of fanfold paper that came spewing out of the high-speed printers scattered all over the control room. No procedure was in place to receive the output from the many managed devices and processes involved in the complex task of managing a nuclear reactor, analyze the output, and hand a simple, easy-to-respond-to decision to the operator.

The same problem is true in complex networks. Most of them have hundreds of managed devices with simple associated element-management systems that generate primitive data about the health and welfare of each device. The information from these element managers is delivered to the network management center, where it is displayed on one of many monitors that the network managers themselves use to track and respond to the status of the network. What they *really* need is a single map of the network that shows all of the managed devices in green if they are healthy. If a device begins to approach a preestablished threshold of performance, the icon on the map that represents that device turns yellow, and if it fails entirely it turns red, yells loudly and automatically reports and escalates the trouble. In one of his many books on American management practices, University of Southern California Professor Emeritus Warren Bennis observes that "the business of the future will be run by a person and a dog. The person will be there to feed the dog; the dog will be there to make sure the person doesn't touch anything." Clearly that model applies here.

So how can this ideal model of network management be achieved? Every vendor will tell you that its element-management system is the best element manager ever created. None of them are willing to change the user interface that they have created so carefully. Using a canonical form, however, there is no reason to. All that has to be done is to exact an agreement from every vendor that stipulates that while the vendor does not have to change its user interface, it must agree to speak the technological equivalent of a "universal language" on the back side of its device. That way the users still get to use the interface they have grown accustomed to, but on the network side, every management system will talk to every other management system using a common and widely accepted form. Again, it's just like a canonical language: If people from five different language groups need to communicate, they have a choice. They can each learn everybody else's language (four additional languages per person) or they can agree on a canonical language (Esperanto), which reduces the requirement to a single language each. An example is shown in Figure C-6.

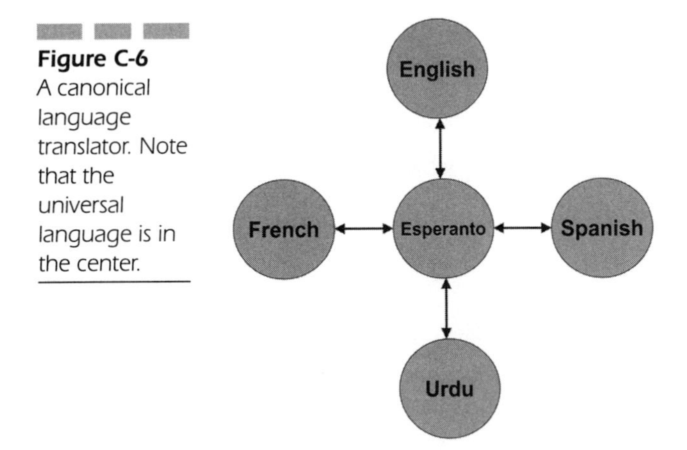

Figure C-6 A canonical language translator. Note that the universal language is in the center.

In network management, there are several canonical forms. The most common are ISO's *Common Management Information Protocol* (CMIP), the IETF's *Simple Network Management Protocol* (SNMP), and the *Object Management Group's Common Object Request Brokered Architecture* (CORBA). As long as every manager device agrees to use one of these on the network side, every system can talk to every other system.

Other canonical forms found at the Application Layer include ISO's X.400/X.500 Message Handling Service (Quiz: Where would you find these?) and the IETF's *Simple Mail Transfer Protocol* (SMTP) for e-mail applications; ISO's *File Transfer, Access, and Management* (FTAM), the IETF's *File Transfer Protocol* (FTP), and the *Hypertext Transfer Protocol* (HTTP) for file transfer; and a host of others. Note that the services provided at this layer are highly specific in nature: They perform a limited subset of tasks.

Layer 6: The Presentation Layer

For our e-mail example, let's assume that the Application Layer converts the PC-specific information to X.400 format and adds a header that will tell the receiving device to look for X.400-formatted content. This is not a difficult concept; think about the nature of the information that must be included in any e-mail encoding scheme. Every system must have a field for the following:

- Sender (From)
- Recipient (To)
- Date
- Subject
- Cc
- Bcc
- Attachment
- Copy
- Message body
- Signature (optional)
- Priority
- Various other miscellaneous fields

The point is that the number of defined fields is relatively small, and as long as each mail system knows what the fields are and where they exist in the coding scheme of the canonical standard, it will be able to map its own content to and from X.400 or SMTP. Problem solved.

Once the message has been encoded properly as an e-mail message, the Application Layer passes the now slightly larger message down to the Presentation Layer. It does this across a layer-to-layer interface using a simple set of commands called *service primitives*.

The Presentation Layer provides a more general set of services than the Application Layer provided, which have to do with the structural *form* or *syntax* of the data. These services include code conversion, such as seven-bit ASCII to eight-bit EBCDIC translation; compression, using such services as PKZIP, British Telecom Lempel-Ziv, the various releases of the *Moving Picture Experts Group* (MPEG), the *Joint Photographic Experts Group* (JPEG), and a variety of others; and encryption, including *pretty good privacy* (PGP) and *public key infrastructure* (PKI). Note that these services can be used on any form of data: Spreadsheets, word-processed documents, and rock music can be compressed and encrypted.

Compression is typically used to reduce the number of bits required to represent a file through a complex manipulative mathematical process that identifies redundant information in the image, removes it, and sends the resulting smaller file off to be transmitted or archived.

Encryption, on the other hand, is used when the information contained in a file is deemed sensitive enough to require that it be hidden from all but those eyes with permission to view it. Encryption is one aspect of a very old science called cryptography. Cryptography is the science of writing in code; its first known use dates to 1900 B.C. when an

Egyptian scribe used nonstandard hieroglyphs to capture the private thoughts of a customer. Some historians feel that cryptography first appeared as the natural result of the invention of the written word; its use in diplomatic messages, business strategy, and battle plans certainly supports the theory.

In data communications and telecommunications, encryption is required any time the information to be conveyed is sensitive and the possibility exists that the transmission medium is insecure. This can occur over any network, although the Internet is most commonly cited as being the most insecure of all networks.

All secure networks require a set of specific characteristics if they are to be truly secure. The most important of them are as follows:

- *Privacy and confidentiality:* The ability to guarantee that no one can read the message except the intended recipient
- *Authentication:* The guarantee that the identity of the recipient can be verified with full confidence
- *Message integrity:* Assurance that the receiver can confirm that the message has not been changed in any way during its transit across the network
- *Nonrepudiation:* A mechanism to prove that the sender really sent this message and it was not sent by someone pretending to be the sender

Cryptographic techniques, including encryption, have two responsibilities: They ensure that the transmitted information is free from theft or any form of alteration, and they provide authentication for both senders and receivers. Today, three forms of encryption are most commonly employed: secret key (or symmetric) cryptography, public-key (or asymmetric) cryptography, and hash functions. How they work is beyond the scope of this book, but numerous resources are available on the topic.

Layer 5: The Session Layer

We have now left the Presentation Layer. Our e-mail message is encrypted, compressed, and may have gone through an ASCII-to-EBCDIC code conversion before descending into the complexity of the Session Layer. As before, the Presentation Layer added a header containing information about the services it employed.

For being such an innocuous layer, the Session Layer certainly engenders a lot of attention. Some believe that the Session Layer could be eliminated by incorporating its functions into the layer above or the layer below, thus simplifying the OSI Model. The bottom line is that it *does* perform a set of critical functions that cannot be ignored.

First of all, the Session Layer ensures that a logical relationship is created between the transmitting and receiving applications. It guarantees, for example, that our PC user in Madrid receives his or her mail and *only* his or her mail from the mainframe, which is undoubtedly hosting large numbers of other e-mail users. This requires the creation and assignment of a logical session identifier.

Many years ago, I recall an instance when I logged into my e-mail account and found to my horror that I was actually logged into my vice president's account. Needless to say I backpedaled out of there as fast as I could. Today I know that this occurred because of an execution glitch in the Session Layer.

Layer five also shares responsibility for security with the Presentation Layer. You may have noticed that when you log in to your e-mail application, the first thing the system does is ask for a login ID, which you dutifully enter. The ID appears in the appropriate field on the screen. When the system asks for your password, however, the password does not appear on the screen. The field remains blank or is filled with stars, shown graphically in Figure C-7. This is because the Session Layer knows that the information should not be displayed. When it receives the correct login ID, it sends a command to the terminal (your PC) asking you to enter your password. It then immediately sends a second message to the terminal telling it to turn off local echo so that your keystrokes are not echoed back on to the screen. As soon as the password has been transmitted, the Session Layer issues a command to turn local echo back on again, allowing you to once again see what you type.

Another responsibility of the Session Layer that is particularly important in mainframe environments is a process called *checkpoint restart*. This is a process that is analogous to the autosave function that is avail-

Figure C-7 Session Layer turns off echo to protect user.

able on many PC-based applications today. I call it the Hansel and Gretel function: As the mainframe performs its many tasks during the online day, the Session Layer keeps track of everything that has been done, scattering a trail of digital bread crumbs along the way as processing is performed. Should the mainframe fail for some reason (the dreaded ABEND), the Session Layer will provided a series of recovery checkpoints. As soon as the machine has been rebooted, the Session Layer performs the digital equivalent of walking back along its trail of bread crumbs. It finds the most recent checkpoint and uses that point as its most recent recovery data, thus eliminating the possibility of losing huge amounts of recently processed information.

So, the Session Layer may not be the most glamorous of the seven layers, but its functions are clearly important. As far as standards go, the list is fairly sparse; see the ITU-T's X.225 standard for the most comprehensive document on the subject.

After adding a header, layer five hands the steadily growing *protocol data unit*, or PDU, down to the Transport Layer. This is the point where we first enter the network. Until now, all functions have been software based and in many cases a function of the operating system.

Layer 4: The Transport Layer

The Transport Layer's job is simple: to guarantee end-to-end, error-free delivery of the entire transmitted message—not bits, not frames or cells, not packets, but the entire message. It does this by taking into account the nature and robustness of the underlying physical network over which the message is being transmitted, including the following characteristics:

- Class of service required
- Data transfer requirements
- User interface characteristics
- Connection management requirements
- Specific security concerns
- Network management and reporting status data

There are two basic network types: dedicated and switched. We will examine each in turn before discussing Transport Layer protocols.

Dedicated networks are exactly what the name implies: always-on network resources, often dedicated to a specific customer and providing very high-quality transport service. That's the good news. The bad news is that dedicated facilities tend to be expensive, particularly because the customer pays for them whether the customer is using the facility or not. Unless the customer is using it 100 percent of the time, the network is costing them money. The other downside of a dedicated facility is susceptibility to failure: Should a terrorist backhoe driver decide to take the cable out, no alternative route exists for the traffic to take. It requires some sort of intervention on the part of the service provider that is largely manual. Furthermore, dedicated circuits tend to be inflexible, again because they are dedicated.

Switched resources, on the other hand, work in a different fashion and have their own set of advantages and disadvantages to consider. First and foremost, they require an understanding of the word "virtual."

When a customer purchases a dedicated facility, the customer literally owns the resources between the two communicating entities as shown in Figure C-8. Either the circuit itself is physically dedicated to the customer (common in the 1980s), or a timeslot on a shared physical resource such as a T-carrier is dedicated to them. Data is placed on the timeslot or the circuit, and it travels to the other end of the established facility—very simple and straightforward. No possibility exists for a misdelivered message, because the message has only a single possible destination. Imagine turning on the spigot in your front yard to water the plants and having water pour out of your neighbor's hose—it would be about that ridiculous.

In a switched environment, things work quite differently. In switched networks the only thing that is actually dedicated is a timeslot, because everything in the network that is physical is shared among many different users. Imagine what a wonderful boon to the service providers this technology is: It gives them the ability to properly support the transport requirements of large numbers of customers while selling the same physical resources to them, over and over and over again. Imagine!

To understand how this technology works, please refer to Figure C-9. In this example, device D on the left needs to transmit data to device K

Figure C-8
Point-to-point circuit

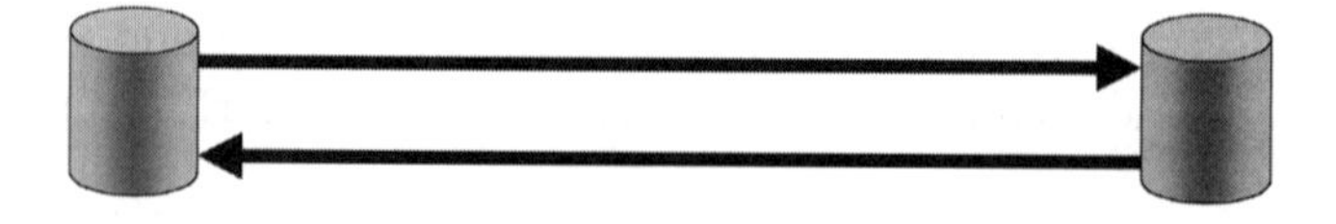

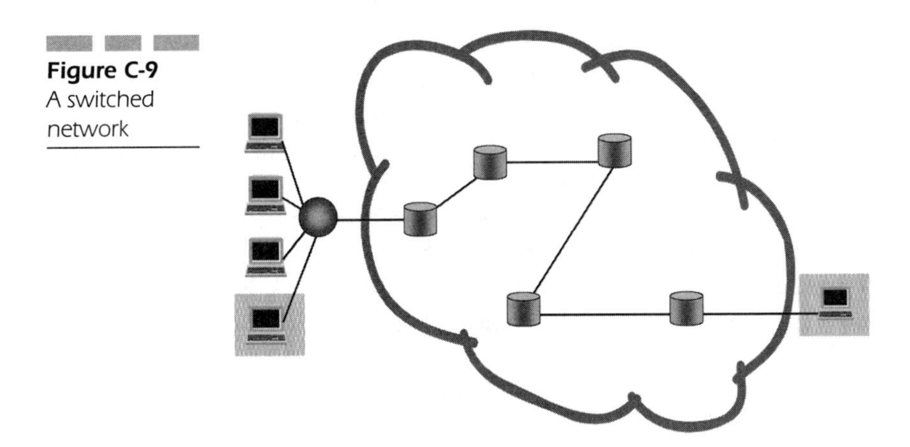

Figure C-9 A switched network

on the right. Notice that access to the network resources (the switches in the cloud) is shared with three other machines. In order for this to work, each device must have a unique identifier so that the switches in the network can distinguish among all the traffic streams that are flowing through them. This identifier, often called a virtual circuit identifier, is assigned by the Transport Layer as one of its many responsibilities. As examples, X.25, frame relay, and ATM are all switched-network technologies that rely on this technique. In X.25 and in asynchronous transfer mode (ATM), the identifier is called a *virtual circuit identifier*; in frame relay, it is called a *data link connection identifier* (DLCI, pronounced "Delsey"). Each of these will be described in greater detail later in the book.

When device D generates its message to device K for transport across the network, the Transport Layer packages the data for transmission. Among other things, it assigns a logical channel that the ingress switch uses to uniquely identify the incoming data. It does this by creating a unique combination of the unique logical address with the shared physical port to create an entirely unique virtual circuit identifier.

When the message arrives at the first switch, the switch enters the logical channel information in a routing table that it uses to manage incoming and outgoing data. There can be other information in the routing table as well, such as quality of service (QoS) indicators. These details will be covered later when we discuss the Network Layer (layer three).

The technology that a customer uses in a switched network is clearly not dedicated, but it gives the appearance that it is. This is called *virtual circuit service,* because it gives the appearance of being there when in fact it isn't. *Virtual private networks* (VPNs), for example, give a customer the appearance that they are buying private network service. In a sense they are: They do have a dedicated logical facility. The difference is that they share the physical facilities with many other users, which allows the service provider to offer the transport service for a lower cost. Furthermore, secure protocols protect each customer's traffic from interception. VPNs are illustrated in Figure C-10.

As you may have intuited by now, the degree of involvement that the Transport Layer has varies with the type of network. For example, if the network consists of a single, dedicated, point-to-point circuit, then very little could happen to the data during the transmission, because the data would consist of an uninterrupted, single-hop stream—that is, there are no switches along the way that could cause pieces of the message to go awry. The Transport Layer, therefore, would have little to do to guarantee the delivery of the message.

However, what if the architecture of the network is not as robust as a private-line circuit? What if this is a packet network, in which case the message is broken into segments by the Transport Layer and the segments are independently routed through the fabric of the network? Furthermore, what if there is no guarantee that all of the packets will take the same route through the network wilderness? In that case, the route actually consists of a *series* of routes between the switches, like a string of sausage links. In this situation no guarantee can be made that the components of the message will arrive in sequence. In fact, there is no guarantee that they will arrive at all! The Transport Layer therefore has a major responsibility to ensure that all of the message components arrive and that they carry enough additional information in the form of yet another header—this time on each packet—to allow them to be properly resequenced at the destination. The header, for example, contains sequence numbers that the receiving Transport Layer can use to reassemble the original message from the stream of random packets.

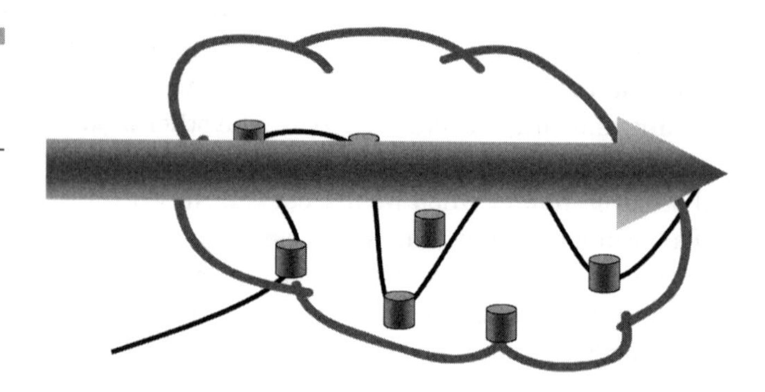

Figure C-10 Virtual private network (VPN)

Consider the following scenario: A transmitter fires a message into the network, where it passes through each of the upper layers until it reaches the originating Transport Layer, which segments the message into a series of five packets, labeled one of five, two of five, three of five, and so on. The packets enter the network and proceed to make their way across the wilderness of the network fabric. Packets one, two, three, and five arrive without incident, although they do arrive out of order. Packet four, unfortunately, gets caught in a routing loop in New Mexico. The receiving Transport Layer, tasked with delivering a complete, correct message to the layers above it, puts everything on hold while it awaits the arrival of the errant packet. The layer, however, will wait only so long. It has no idea where the packet is; it does, however, know where it is *not*. After some predetermined period of time the receiving Transport Layer assumes that the packet isn't going to make it and initiates recovery procedures that result in the retransmission of the missing packet.

Meanwhile, the lost packet has finally stopped and asked for directions, extricated itself from the traffic jams of Albuquerque, and it has made its way to the destination. It arrives, covered with dust, an "I've Seen Crystal Caverns" bumper sticker on its trailer, expecting to be incorporated into the original message. By this time, however, the packet has been replaced with the resent packet. Clearly, some kind of process must be in place to handle duplicate packet situations, which happen rather frequently. The Transport Layer then becomes the center point of message integrity.

Transport Layer standards are diverse and numerous. ISO, the ITU-T, and the IETF publish recommendations for layer four. The ITU-T publishes X.224 and X.234, which detail the functions of both connection-oriented and connectionless networks. ISO publishes ISO 8073, which defines a transport protocol with five layers of functionality ranging from TP0 through TP4, as shown here:

- Class 0 (TP0): Simple class
- Class 1 (TP1): Basic error recovery class
- Class 2 (TP2): Multiplexing class
- Class 3 (TP3): Error recovery and multiplexing class
- Class 4 (TP4): Error detection and recovery class

TP0 has the least capability; it is roughly equivalent to the IETF's *User Datagram Protocol* (UDP), which will be discussed a bit later. TP4 is the most common ISO Transport Layer protocol and is equivalent in capability to the IETF's *Transmission Control Protocol* (TCP). It provides an ironclad transport function and operates under the assumption that the network is wholly unreliable and must therefore take extraordinary steps to safeguard the user's data.

Before we descend into the wilds of the Network Layer, let's introduce the concepts of switching and routing. Modern networks are often represented as a cloud filled with boxes representing switches or routers. Depending upon such factors as congestion, cost, number of hops between routers, and other considerations, the network selects the optimal end-to-end path for the stream of packets created by the Transport Layer. Depending upon the nature of the Network Layer protocol that is in use, the network will take one of two actions. It will either establish a single path over which all the packets will travel, in sequence, or the network will simply be handed the packets and told to deliver them as it sees fit. The first technique, which establishes a seemingly dedicated path, is called *connection-oriented service*. The other technique, which does *not* dedicate a path, is called *connectionless service*. We will discuss each of these in turn. Before we do, however, let's discuss the evolution of switched networks.

Switched Networks

Modern switched networks typically fall into one of two major categories: *circuit switched*, in which the network preestablishes a path for the transport of traffic from a source to a destination, as is done in the traditional telephone network, and *store-and-forward networks*, in which the traffic is handed from one switch to the next as it makes its way across the network fabric. When traffic arrives at a switch in a store-and-forward network, it is stored, examined for errors and destination information, and forwarded to the next hop along the path—hence the name, store and forward. Packet switching is one form of store-and-forward technology.

Store-and-Forward Switching

The first store-and-forward networks were invented and used by the early Greeks and Romans. Indeed, Mycenae learned of the fall of Troy because of a line of signal towers between the two cities that used fire in each tower to transmit information from tower to tower. An opening on the side of each tower could be alternately opened and blocked, and using a rudimentary signaling code, short messages could be sent between them in a short period of time. A message could be conveyed across a large country such as France in a matter of hours, as Napoleon discovered and used to his great advantage.

The earliest *modern* store-and-forward networks were the telegraph networks. When a customer handed over a message in the form of a yellow flimsy paper that was to be transmitted, the operator would transmit the message in code over the open-wire telegraph lines to the next office, where the message printed out on a streaming paper tape. On the tape would appear a sequence of alternating pencil marks and gaps, or spaces, combinations of which represented characters—a mark represented a one, whereas a space represented a zero. A point of historical interest is that the terminology "mark" and "space" are common parlance in modern networks: In T-carrier, the encoding scheme is called *alternate mark inversion* (AMI), because every other one alternates in polarity from the ones that surround it. Similarly, *alternate space inversion* is used in signaling schemes such as on the ISDN D-Channel.

At any rate, the entire message would be delivered in this fashion, from office to office to office, ultimately arriving at its final destination, a technique called *message switching*. Over time, of course, the process became fully mechanized and the telegraph operators disappeared.

One major problem existed with this technique. What happened if the message, upon arrival, was found to be corrupt, or if it simply did not arrive for some odd reason? In that case, the entire message would have to be resent at the request of the receiver. This added overall delay in the system and was awfully inefficient because in most cases only a few characters were corrupted. Nevertheless, the entire message was retransmitted. Once the system was fully mechanized, it meant that the switches had to have hard drives on which to store the incoming messages, which added yet more delay because hard drives are mechanical devices and by their very nature relatively slow. Improvements didn't come along until the advent of *packet switching*.

Packet Switching

With the arrival of low-cost, high-speed solid-state microelectronics in the 1970s, it became possible to take significant steps forward in switching technology. One of the first innovative steps was *packet switching*. In packet switching, the message that was transmitted in its entirety over the earlier message-switched store-and-forward networks is now broken into smaller, more manageable pieces that are numbered by the Transport Layer before being passed onto the network for routing and delivery. This innovation offers several advantages. First, it eliminates the need for the mechanical, switch-based hard drive, because the small packets can now be handled blindingly fast by solid-state memory. Second, should a packet arrive with errors, it and it alone can be discarded and replaced. In message-switched environments, an unrecoverable bit error resulted in the inevitable retransmission of the entire message—not a particularly elegant solution. Packet switching, then, offers a number of distinct advantages.

As before, of course, there are also disadvantages. No longer is there (necessarily) a physically or logically dedicated path from the source to the destination, which means that the ability to guarantee quality of service (QoS) on an end-to-end basis is severely restricted. There are ways around this as you will see in the section that follows, but they are often costly and *always* complex. This is one of the reasons that IP telephony is having a difficult time achieving widespread deployment. It works fine in controlled, relatively small corporate environments where traffic patterns can be scrutinized and throttled as required to maintain QoS. In the public IP environment (read *the Internet*), however, no way exists to ensure that degree of control. Will it work? Of course. Is it dependable? Absolutely not. A customer might be willing to call a friend with it, but they would be less inclined to use the service for a business call—not because it's bad, but because it's not dependable or predictable. Until it is, the old circuit-switched telephone network will continue to enjoy its century (or two) in the sun. The day will certainly come, but for now it isn't ready for prime time.

Packet switching can be implemented in two very different ways. We'll discuss them now.

Connection-Oriented Networks

When Meriwether Lewis and William Clark left St. Louis with the Corps of Discovery in 1803 to travel up the Missouri and Columbia Rivers to the Pacific Ocean, they had no idea how to get where they were going. They traveled with and relied on a massive collection of maps, instruments, transcripts of interviews with trappers and Native American guides, an awful lot of courage, and the knowledge of Sacagawea, the wife of independent French-Canadian trader Toussaint Charbonneau, who accompanied them on their journey. As they made their way across the wilderness of the northwest, they marked trees every few hundred feet by cutting away a large and highly visible swath of bark, a process known as "blazing." By blazing their trail, others could easily follow them without the need for maps, trapper lore, or guides. They did not need to bring compasses, sextants, chronometers, or local guides; they simply followed the well-marked trail.

If you understand this concept, then you also understand the concept of connection-oriented switching, sometimes called *virtual circuit switching,* one of the two principal forms of switching technologies. When a device sends packets into a connection-oriented network, the first packet, often called a *call setup packet* or *discovery packet*, carries embedded in it the final destination address that it is searching for. Upon arrival at the first switch in the network, the switch examines the packet, looks at the destination address, and selects an outgoing port that will get the packet closer to its destination. It has the ability to do this because presumably, somewhere in the recent past, it has recorded the port of arrival of a packet from the destination machine and concludes that if a packet arrived on that port from the destination host, then a good way to get closer to the destination is to go out the same port on which the arriving packet came in. The switch then records in its routing tables an entry that dictates that all packets originating from the same source (the source being a virtual circuit address/physical port address combination that identifies the logical source of the packets) should be transmitted out the same switch port. This process is then followed by every switch in the path, from the source to the destination. Each switch makes table entries, similar to the blazes left by the Corps of Discovery.[1]

With this technique, the only packet that requires a complete address is the initial one that blazes the trail through the network wilderness. All subsequent packets carry nothing more than a short identifier—a virtual circuit address—that instructs each switch they pass through how to handle them. Thus, all the packets with the same origin will follow the same path through the network. Consequently they will arrive in order and will all be delayed the same amount of time as they traverse the network. The service provided by connection-oriented networks is called *virtual circuit service*, because it simulates the service provided by a dedicated network. The technique is called connection-oriented because the switches perceive that there is a relationship, or connection, between all of the packets that derive from the same source.

---

[1] The alternative is to have a network administrator manually preconfigure the routes from source to destination. This guarantees a great deal of control but also obviates the need for an intelligent network.

As with most technologies, a downside can be found to connection-oriented transmission: In the event of a network failure or heavy congestion somewhere along the predetermined path, the circuit is interrupted and will require some form of intervention to correct the problem because the network is not self-healing from a protocol point of view. Network management schemes and stopgap measures are certainly in place to reduce the possibility that a network failure might cause a service interruption, but in a connection-oriented network these measures are external. Nevertheless, because of its ability to emulate the service provided by a dedicated network, connection-oriented services are widely deployed and very successful. Examples include frame relay, X.25 packet-based service, and ATM. All will be discussed in detail later in the book.

Connectionless Networks

The alternative to connection-oriented switching is *connectionless switching*, sometimes called *datagram service.* In connectionless networks there is no predetermined path from the source to the destination. There is no call setup packet; all data packets are treated independently, and the switches perceive no relationship between them as they arrive—hence the name, "connectionless." Every packet carries a complete destination address, because it cannot rely on the existence of a preestablished path created by a call setup packet.

When a packet arrives at the ingress switch of a connectionless network, the switch examines the packet's destination address. Based on what it knows about the topology of the network, congestion, cost of individual routes, distance (sometimes called *hop count*), and other factors that affect routing decisions, the switch will select an outbound route that optimizes whatever parameters the switch has been instructed to concern itself with. Each switch along the path does the same thing.

For example, let's assume that the first packet of a message, upon arrival at the ingress switch, would normally be directed out physical port number 7, because based upon current known network conditions that port provides the shortest path (lowest hop count) to the destination. However, upon closer examination, the switch realizes that although port 7 provides the shortest hop count, the route beyond the port is severely congested. As a result, the packet is routed out port 13, which results in a longer path but avoids the congestion. And because no preordained route through the network exists, the packet will simply have to get directions when it arrives at the next switch.

Now, the second packet of the message arrives. Because this is a connectionless environment, however, the switch does not realize that the packet is related to the packet that preceded it. The switch examines the destination address on the second packet, and then proceeds to route the packet as it did with the preceding one. This time, however, upon examination of the network, the switch finds that port 7, the shortest path from the source to the destination, is no longer congested. It therefore transmits the packet on port 7, thus ensuring that packet two will in all likelihood arrive before packet one! Clearly, this poses a problem for message integrity and illustrates the criticality of the Transport Layer, which, you will recall, provides end-to-end message integrity by reassembling the message from a collection of out-of-order packets that arrive with varying degrees of delay because of the vagaries of connectionless networks.

Connectionless service is often called "unreliable" because it fails to guarantee delay minimums, sequential delivery, or, for that matter, any kind of delivery. This causes many people to question why network designers would rely on a technology that guarantees so little. The answer lies within the layered protocol model. Although connectionless networks do not guarantee sequential delivery or limits on delay, they *will* ultimately deliver the packets. Because they are not required to transmit along a fixed path, the switches in a connectionless network have the freedom to route around trouble spots by dynamically selecting alternate pathways, thus ensuring delivery, albeit somewhat unpredictable. If this process results in out-of-order delivery, no problem: That's what the Transport Layer is for. Data communications are a team effort and require the capabilities of many different layers to ensure the integrity and delivery of a message from the transmitter to the receiver. Thus, even an "unreliable" protocol has distinct advantages.

An example of a well-known connectionless protocol is the Internet Protocol, or IP. It relies on the TCP, to guarantee end-to-end correctness of the delivered message. There are times, however, when the foolproof capabilities of TCP and TCP-like protocols are considered overkill. For example, network management systems tend to generate large volumes of small messages on a regularly scheduled basis. These messages carry information about the health and welfare of the network and about topological changes that routing protocols need to know about if they are to maintain accurate routing tables in the switches. The problem with these messages is that they (1) are numerous and (2) often carry information that hasn't changed since the *last* time the message was generated, 30 seconds ago.

TCP and TP4 protocols are extremely overhead-heavy compared to their lighter-weight cousins UDP and TP1. Otherwise, they would not be able to absolutely, positively guarantee the delivery of the message. In some cases, however, there may not be a need to absolutely, positively guarantee delivery. After all, if I lose one of those status messages, no problem: It will be generated again in 30 seconds anyway. The result of this is that some networks choose not to employ the robust and capable protocols available to them, simply because the marginal advantage they provide doesn't merit the transport and processing overhead they create in the network. Thus, connectionless networks are extremely widely deployed. After all, those 500 million (or so) Internet users must be reasonably happy with the technology.

Let's now examine the Network Layer. Please note that we are now entering the realm of the chained layers, which you will recall are used by all devices in the path—end-user devices as well as the switches or routers themselves.

Layer 3: The Network Layer

The Network Layer, which is the uppermost of the three chained layers, has two key responsibilities: routing and congestion control. We will also briefly discuss switching at this layer, even though many books consider it to be a layer two process. So for the purists in the audience, please bear with me—there's a method to my madness.

When the telephone network first started its remarkable growth path at the sunrise of the twentieth century, there was no concept of switching. If a customer wanted to be able to speak with another customer, he or she had to have a phone in the house with a dedicated path to the other person's home. The other person required another phone and phone line—and you quickly begin to see where this is leading. Figure C-11 illustrates the problem: The telephone network's success would bring on the next ice age, blocking the sun with all the aerial wire the telephone network would be required to deploy. Consider this simple mathematical model: In order to fully interconnect, or mesh, five customers as shown in Figure C-12 so that any one of them can call any other, the telephone company would have to install 10 circuits, according to the equation $n(n-1)/2$, where n is the number of devices that wish to communicate. Extrapolate that out to the population of even a small city—say, 2,000 people. That boils down to 3,997,999 circuits that would have to be installed—all to allow 2,000

Figure C-11 Aerial telephone wire (Courtesy Lucent Technologies)

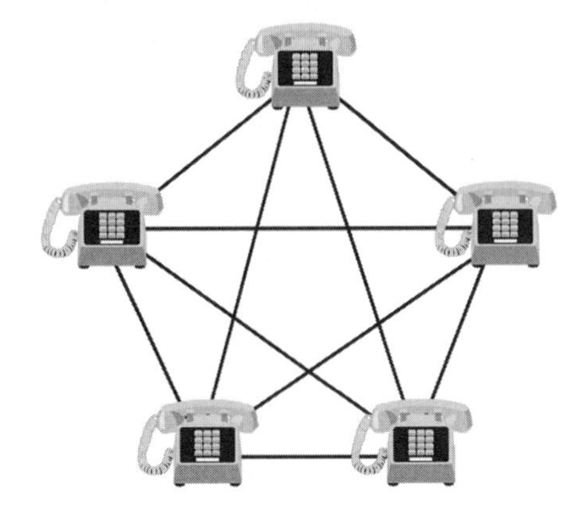

Figure C-12 Meshed network; five users, ten circuits

people to call each other. Obviously, some alternative solution was greatly needed. That solution was the switch.

The first "switches" did not arrive until 1878 with the near-disastrous hiring of young boys to work the cord boards in the first central offices. John Brooks, author of *Telephone: The First 100 Years,* offers the following:

> The year of 1878 was the year of male operators, who seem to have been an instant and memorable disaster. The lads, most of them in their late teens, were simply too impatient and high spirited for the job, which, in view of the imperfections of the equipment and the inexperience of the

> subscribers, was one demanding above all patience and calm. According to the late reminiscences of some of them, they were given to lightening the tedium of their work by roughhousing, shouting constantly at each other, and swearing frequently at the customers. An early visitor to the Chicago exchange said of it, "The racket is almost deafening. Boys are rushing madly hither and thither, while others are putting in or taking out pegs from a central framework as if they were lunatics engaged in a game of fox and cheese. It was a perfect bedlam."

Later in 1878 the boys were grabbed by their ears and removed from their operator positions, replaced quickly—and, according to a multitude of accounts, to the enormous satisfaction of the customers—by women, shown in Figure C-13.

These operators were in fact the first switches. Now, instead of needing a dedicated circuit running from every customer to every other customer, each subscriber needed a single circuit that ran into the central exchange, where it appeared on the jack field in front of an operator (each operator managed approximately 100 positions, the optimum number according to Bell System studies). When a customer wanted to make a call, he or she would crank the handle on the phone, generating current which would cause a flag to drop, a light to light, or a bell to ring in front of the operator. Seeing the signal the operator would plug a headset into the customer's jack and announce that they were ready to receive the number to be dialed. The operator would then simply cross-connect the caller to the called party and then wait for the receiver to be picked up on the other end. The operator would periodically monitor the call and pull the patch cord down when the call was complete.

This model meant that instead of needing 3,997,999 circuits to provide universal connectivity for a town of 2,000 people, 2,000 were needed—a rather *significant* reduction in capital outlay for the telephone company, wouldn't you say? Instead of looking like Figure C-12, the network now looked like Figure C-14.

Over time, manual switching slowly disappeared, replaced by mechanical switches initially followed by more modern all-electronic switches. The first true mechanical switches didn't arrive until 1892 when Almon Strowger's step-by-step switch was first installed by his company, Automatic Electric.

Strowger's story is worth telling, because it illustrates the serendipity that characterized so much of this industry's development. It seems that Almon Strowger was not an inventor, nor was he a telephone person. He was, in fact, an undertaker in a small town in Missouri. One day, he came to the realization that his business was (okay, I won't say dying) declin-

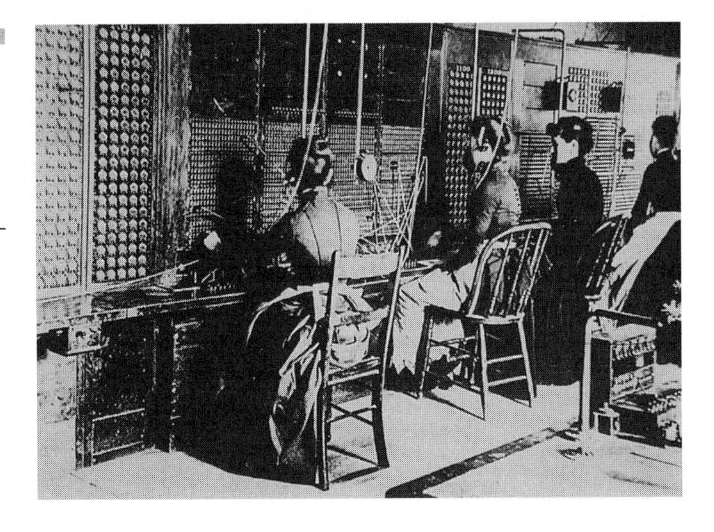

Figure C-13 First women operators (Courtesy Lucent Technologies)

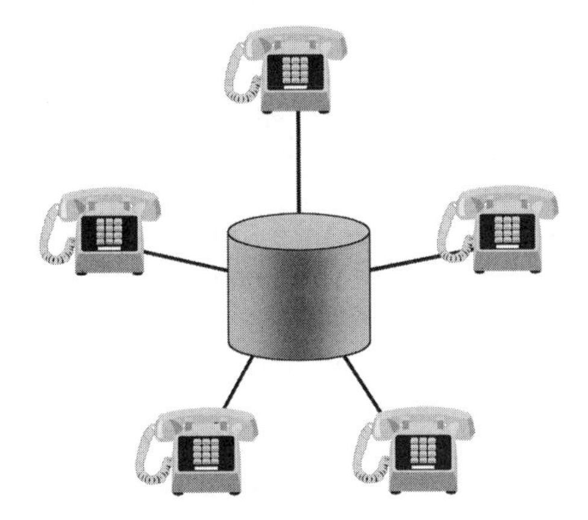

Figure C-14 Switched network; five users, five circuits

ing, and upon closer investigation determined that the town's operator was married to his competitor! As a result, any calls that came in for the undertaker naturally went to her husband—and *not* to Strowger.

To equalize the playing field, Strowger called upon his considerable talents as a tinkerer and designed a mechanical switch and dial telephone, which is still in use today in a number of developing countries.

The bottom line to all this is that switches create temporary end-to-end paths between two or more devices that wish to communicate with each other. They do it in a variety of ways; circuit-switched networks create a "virtually dedicated path" between the two end points and offer constant end-to-end delay, making them acceptable for delay-sensitive applications or connections with long hold times. Store-and-forward networks, particularly packet networks, work well for short, bursty messages with minimal delay sensitivity.

To manage all this, however, is more complicated than it would seem at first blush. First of all, the switches must have the ability to select not only a path, but the *best* path, based on QoS parameters. This constitutes intelligent routing. Second, they should have some way of monitoring the network so that they always know its current operational conditions. Finally, should they encounter unavoidable congestion, the switches should have one or more ways to deal with it.

Routing Protocols

So, how are routing decisions made in a typical network? Whether they are connectionless or connection oriented, the routers and switches in the network must take into account a variety of factors to determine the best path for the traffic they manage. These factors fall into a broad category of rule sets called *routing protocols*. For reference purposes, please refer to the tree shown in Figure C-15.

Once the Transport Layer has taken whatever steps are necessary to prepare the packets for their transmission across the network, the packets are passed to the Network Layer.

The Network Layer has two primary responsibilities in the name of network integrity: *routing* and *congestion control*. Routing is the process of intelligently selecting the most appropriate route through the network for the packets; congestion control is the process that ensures that the packets are minimally delayed (or at least equally delayed) as they make their way from the source to the destination. We will begin with a discussion of routing.

Routing Responsibilities

Routing protocols are divided into two main categories—*static routing protocols* and *dynamic routing protocols*. Static routing protocols are those that require a network administrator to establish and maintain them. If a routing-table change is required, the network administrator must manually make the change. This ensures absolute security but is labor intensive and therefore less frequently used except in highly secure environments (e.g., in military and health care situations) or network

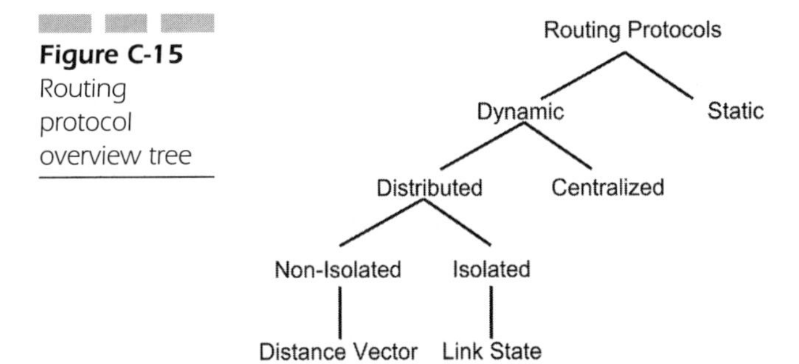

Figure C-15 Routing protocol overview tree

architectures that are designed around static routing because the routes are relatively stable anyway (e.g., IBM's Systems Network Architecture, SNA). More common are dynamic routing protocols, where network devices make their own decisions about optimum route selection. They do this in the following general way: They pay attention to the network around them and collect information from their neighbors about the best routes to particular destinations based on such parameters as the least number of hops, least delay, lowest cost, or highest bandwidth. Then they archive those bits of information in tables, and finally they selectively flush the tables periodically to ensure that the information contained in them is always as current as possible. Because dynamic routing protocols assume intelligence in the switch and can therefore reduce the amount of human intervention required, they are commonly used and are in fact the most widely deployed routing protocols.

Dynamic routing protocols are further divided into two subcategories, *centralized* and *distributed*. Centralized routing protocols concentrate the route decision-making processes in a single node, thus ensuring that all nodes in the network receive the same and most current information possible. When a switch or router needs routing information that is not contained in its own table, it sends a request to the root node asking for direction. Significant downsides exist to this technique: By concentrating the decision-making capability in a single node, the likelihood of a catastrophic failure is dramatically increased. If that node fails, the entire network's ability to seek optimal routing decisions fails. Second, because all nodes in the network must go to that central device for routing instructions, a significant choke point can result.

Several options can reduce the vulnerability of a single point of failure. The first, of course, is to distribute the routing function. This conflicts with the concept of centralized routing, but only somewhat. Consider the Internet, for example. It uses a sort of hybrid of centralized and distributed routing protocols in its *domain name server* (DNS) function. A limited number of devices are tasked with the responsibility of maintaining knowledge of the topology of the network and the location of domains, providing something of a AAA trip-planning service for data packets.

Another option is to designate a backup machine tasked with the responsibility to take over in the event of a failure of the primary routing machine. This technique, used in the early Tymnet packet networks, relied on the ability of the primary machine to send a "sleeping-pill packet" to the backup machine, with these instructions: "Pay attention to what I do, memorize everything I learn, but just sit in the corner and be a potted plant. Take no action, make no routing decisions—just learn. Let the sleeping pill keep you in a semicomatose state. If, for some reason, I fail, I will stop sending the sleeping pills, at which time you will wake up and take over until I recover." An ingenious technique, but overly complex and far too prone to failure for modern network administrators. Distributed routing protocols are far more common today.

In distributed routing protocol environments, each device collects information about the topology of the network and makes independent routing decisions based upon the information it accumulates. For example, if a router sees a packet from source X arrive on port 12, it knows that somewhere out port 12 it will find destination X. It doesn't know how far out there necessarily, just that the destination is somewhere out there over the digital horizon. Thus, if a packet arrives on another port looking to be transmitted to X, the router knows that by sending the packet out port 12 it will at least get closer to its destination. It therefore makes an entry in its routing tables to that effect, so that the next time a packet arrives with the same destination, the switch can consult its table and route the packet quickly.

These routing protocols are analogous to the process of stopping and asking for directions on a road trip (or not), reputedly one of the great male-female differentiators—right after who controls the TV remote. Anthropologists must have a field day with this kind of stuff. According to apocryphal lore, women have no problem whatsoever stopping and asking for directions, whereas men are loathe to do so—one of those silly threats to the manhood things. Anyway, back to telecomm. If you were planning a road trip across the country, you could do so using one of two philosophies. You could go to AAA or a travel agent and have them plan out the entire route for you, or you could do the Jack Kerouac thing and simply get in the car and drive. Going to AAA seems to be the simplest option because, once the route is planned, all you have to do is follow the directions—Lewis's blazed trail, as it were. The downside is that if you make it as far as Scratch Ankle, Alabama (yes, it's a real place) and the road over which you are supposed to travel is closed, you are stuck—you have to stop and ask for directions anyway.

The alternative is to simply get in the car, drive to the nearest gas station, and tell them that you are trying to get to Dime Box, Texas. The attendant will no doubt tell you the following: "I don't know where Dime Box is, but I know that Texas is down in the Southwest. So if you take this highway here to Kansas City, it'll get you closer. But you'll have to ask for better directions when you get there." The next gas station attendant may tell you, "Well, the quickest way to central Texas is along this highway, but it's rush hour and you'll be stuck for days if you go that way. I'd take the surface street. It's a little less comfortable, but there's no congestion." By stopping at a series of gas stations as you traverse the country and asking for help you will eventually reach your destination, but the route may not be the most efficient. That's okay, though, because you will never have to worry about getting stuck with bad directions. Clearly the first example of these is connection-oriented travel; the second is connectionless. Connection-oriented travel is a far more comfortable, secure way to go; connectionless is riskier, less sure, more flexible, and much more fun. Obviously, distributed routing protocols are centrally important to the traveler as well as to the gas station attendant who must give him or her reliable directions, and equally important to routers and switches in connectionless data networks.

Distributed routing protocols fall into two categories: *distance vector* and *link state*. Distance vector protocols rely on a technique called *table swapping* to exchange information about network topology with each other. This information includes destination/cost pairs that allow each device to select the least cost route from one place to another. On a scheduled basis, routers transmit their entire routing tables on all ports to all adjacent devices. Each device then adds any newly arrived information to its own tables, thus ensuring currency. The only problem with this technique is that it results in a tremendous amount of management traffic (the tables) being sent between network devices, and if the network is relatively static—that is, changes in topology don't happen all that often—then much of the information is unnecessary and can cause serious congestion. In fact, it is not uncommon to encounter networks that have more management traffic than actual user traffic traversing their circuits. What's wrong with this picture? Distance vector works well in small networks where the number of multihop traverses is relatively low, thus minimizing the impact of its bandwidth-intensive nature.

Distance vector protocols have that name because of the way they work. Recovering physicists will remember that a vector is a measure of something that has both direction and magnitude associated with it. The name is appropriate in this case, because the routing protocol optimizes on a direction (port number) and a magnitude (hop count).

Since networks are growing larger, traffic routinely encounters route solutions with large hop counts. This reduces the effectiveness of distance vector solutions. A better solution is the link state protocol. Instead of transmitting entire routing tables on a scheduled basis, link state protocols use a technique called *flooding* to transmit to adjacent devices only the changes that occur *as they occur*. This results in less congestion and more efficient use of network resources and reduces the impact of multiple hops in large-scale networks.

Both distance vector and link state protocols are in widespread use today. The most common distance vector protocols are the *Routing Information Protocol* (RIP), Cisco's *Interior Gateway Routing Protocol* (IGRP), and Cisco's *Border Gateway Protocol* (BGP). Link state protocols include *Open Shortest Path First* (OSPF), commonly used on the Internet, as well as the *Netware Link Services Protocol* (NLSP), which are used to route IPX traffic.

Clearly, both connection-oriented and connectionless transport techniques, as well as their related routing protocols, have a place in the modern telecommunications arena. As QoS becomes such a critical component of the service offered by network providers, the importance of both routing and congestion control becomes apparent. We now turn out attention to the second area of responsibility at the Network Layer, *congestion control*.

Congestion Control Responsibilities

At its most fundamental level, congestion control is a mechanism for reducing the volume of traffic on a particular route through some form of load balancing. No matter how large, diverse, or capable a network is, some degree of congestion is inevitable. It can result from sudden unexpectedly high-utilization levels in one area of the network, from failures of network components, or from poor engineering. In the telephone network, for example, the busiest calling day of the year in the United States is Mother's Day. To reduce the probability that a caller will not be able to complete a call to Mom, network traffic engineers take extraordinary steps to load balance the network. For example, when subscribers on the East Coast are making long-distance calls at 9:00 in the morning, West Coast subscribers haven't even turned on their latte machines yet.

Network resources in the West are underutilized during that period, so engineers route East Coast traffic westward and then hairpin it back to its destination, to spread the load across the entire network. As the day gets later, they reduce the volume of westward-bound traffic to ensure that California has adequate network resources for its own calls.

Two terms are important in this discussion. One is congestion; the other is delay. The terms are often used interchangeably, but they are not the same thing, nor are they always related to one another.

Years ago I lived in the San Francisco area where traffic congestion is a way of life. I often had to drive across the many bridges that crisscross San Francisco Bay, the Suisun Straits, or the Sacramento River. Many of those bridges require drivers to stop and pay a toll, resulting in localized delay. The time it takes to stop and pay the toll is mere seconds—yet traffic often backs up for miles as a result of this local phenomenon, causing a widespread effect.

This is the relationship between the two: Local delay often results in widespread congestion. And congestion is usually caused by inadequate buffer or memory space. Increase the number of buffers—the lanes of traffic on the bridge, if you will—and congestion drops off. Open another line or two at Home Depot ("No waiting on line seven!")—and congestion drops off.

The various players in the fast-food industry manage congestion in different ways—and with dramatically different results. Without naming them, some use a single queue with a single server to take orders, a technique that works well until the lunch rush begins. Then things back up dramatically. Others use multiple queues with multiple servers, a technique that is better, except that one queue can experience serious delays should someone place an order for a standard item or try to pay with a credit card. That line then experiences serious delay. The most effective restaurants stole an idea from the airlines, and use a single queue with multiple servers. This keeps things moving because the instant a server is available, the next person in line is served.

Remember when Jeff Goldblum, the chaos theoretician in *Jurassic Park,* talked about the Butterfly Effect? How a butterfly flapping its wings in the Amazon Basin can kick off a chain of events that affect weather patterns in New York City? That aspect of Chaos Theory contributes greatly to the manner in which networks behave—and the degree to which their behavior is immensely difficult to predict under load.

So how is congestion handled in data networks? The simplest technique, used by both frame relay and ATM, is called packet discard. In the event that traffic gets too heavy based on preestablished management parameters, the switches simply discard excess packets. They can do this because of two facts: First, networks are highly capable and the switches rarely have to resort to these measures, and second, the devices on the ends of the network are intelligent and will detect the loss of information and take whatever steps are required to have the discarded data resent. As drastic as this technique seems, it is not all that catastrophic. Modern networks are heavily dependent on optical fiber and highly capable digital switches; as a result, packet discard, although serious, does not pose a major problem for the network. And even when it is required, recovery techniques are fast and accurate.

Because congestion occurs primarily as the result of inadequate buffer capacity, one solution is to preallocate buffers to high-priority traffic. Another is to create multiple queues with varying priority levels. If a voice or video packet arrives that requires high-priority, low-delay treatment, it will be deposited into the highest-priority queue for instantaneous transmission. This technique holds great promise for the evolving all-services Internet, because it's greatest failing is its inability to deliver multiple, dependable, sustainable grades of service quality.

Other techniques are available that are somewhat more complex than packet discard, but do not result in loss of data. For example, some devices, when informed of congestion within the network, will delay transmission to give the network switches time to process their overload before receiving more. Others will divert traffic to alternate, less-congested routes, or they will trickle packets into the network, a process known as "choking." Frame relay, for example, has the ability to send what is called a *choke packet* into the network, implicitly telling the receiving device that it should throttle back to reduce congestion in the network. Whether it does or not is another story, but the point is that the network has the intelligence to invoke such a command when required. This technique is now used on freeways in large cities: Traffic lights are installed on major onramps that meter the traffic onto the roadway. This results in a dramatic reduction in congestion. Other networks have the intelligence to diversely route traffic, thus reducing the congestion that occurs in certain areas.

Clearly, the Network Layer provides a set of centrally important capabilities to the network itself. Through a combination of network protocols, routing protocols, and congestion control protocols, routers and switches provide granular control over the integrity of the network.

Back to the E-Mail Message

Let us return now to our e-mail example. The message has been divided into packets by the Transport Layer and delivered in pieces to the Network Layer, which now takes whatever steps are necessary to ensure that the packets are properly addressed for efficient delivery to the destination. Each packet now has a header attached to it that contains routing information and a destination address. The next step is to get the packet correctly to the next link in the network chain—in this case, the next switch or router along the way. This is the responsibility of the Data Link Layer.

Layer 2: The Data Link Layer

The Data Link Layer is responsible for ensuring bit-level integrity of the data being transmitted. In short, its job is to make the layers above believe that the world is an error-free and perfect place. When a packet is handed down to the Data Link Layer from the Network Layer, it wraps the packet in a *frame*. In fact, the Data Link Layer is sometimes called the *frame layer.* The frame built by the Data Link Layer comprises several fields (see Figure C-16) that give the network devices the ability to ensure bit-level integrity and proper delivery of the packet, now encased in a frame, *from switch to switch*. Please note that this is different from the Network Layer, which concerns itself with routing packets *to the final destination*. Even the addressing is unique: Packets contain the address of the ultimate destination, used by the network to route the packet properly; frames contain the address of the next link in the network chain (the next switch), used by the network to move the packet along, switch by switch.

Figure C-16 Data Link Layer frame

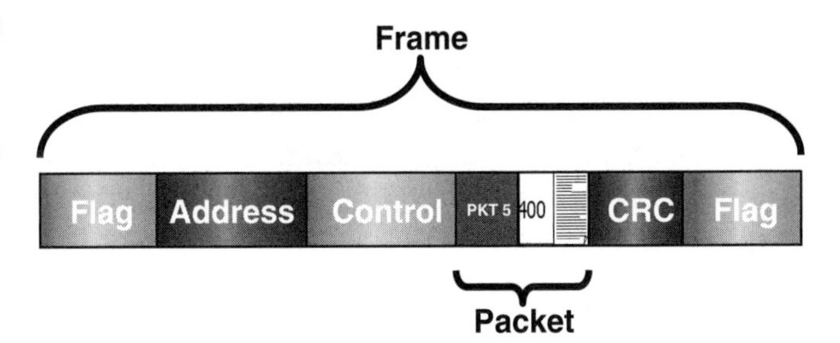

As the diagram illustrates, the beginning and end fields of the frame are called *flags*. These fields, made up of a unique series of bits (0111110), can occur only at the beginning and end of the frame. They are never allowed to occur within the bitstream inside the frame through a process that we will describe momentarily. These flags are used to signal to a receiving device that a new frame is arriving or that it has reached the end of the current frame.[2] This is why a flag's unique bit pattern can never be allowed to occur naturally within the data itself because it could indicate to the receiver (falsely) that this is the end of the current frame. If the flag pattern *does* occur within the bitstream, it is disrupted by the transmitting device through a process called *bit stuffing* or *zero-bit insertion*, in which an extra zero is inserted in the middle of the flag pattern, based on the following rule set. When a frame of data is created at an originating device, the very last device to touch the frame—indeed, the device that actually adds the flags—is called a *universal synchronous/asynchronous receiver-transmitter* (USART). The USART, sometimes called an integrated data link controller (IDLC), is a chipset that has a degree of embedded intelligence. This intelligence is used to detect (among other things) the presence of a false flag pattern in the bitstream around which it builds the frame. Because a flag comprises a zero followed by six ones and a final zero, the IDLC knows that it can never allow that particular pattern to exist between any two real flags. So, as it processes the incoming bitstream, it looks for that pattern and makes the following decision: *If I see a zero followed by five ones, I will automatically and without question insert a zero into the bitstream at that point.*

This of course destroys the integrity of the message, but it doesn't matter. At the receive device, the IDLC monitors the incoming bits. As the frame arrives it sees a *real* flag at the beginning of the frame, an indication that a frame is beginning. As it monitors the bits flowing by it *will* find the zero followed by five bits, at which point it knows, beyond a shadow of a doubt, that the very next bit is a zero—which it will promptly remove, thus restoring the integrity of the original message.

The receiving device has the ability to detect the extra zero and remove it before the data moves up the protocol stack for interpretation. This bit-stuffing process guarantees that a false flag will never be interpreted as a final flag and acted upon in error.

The next field in the frame is the *address field*. This field identifies the address of the next switch in the chain to which the frame is directed,

---

[2]Indeed, the final flag of one frame is the beginning flag of the *next* frame.

and it changes at every node. The only address that remains constant is the destination address, safely embedded in the packet itself.

The third field found in many frames is called the *control field*. It contains supervisory information that the network uses to control the integrity of the data link. For example, if a remote device is not responding to a query from a transmitter, the control field can send a "mandatory response required" message that will allow it to determine the nature of the problem at the far end. It is also used in hierarchical multipoint networks to manage communications functions. For example, a multiplexer may have multiple terminal devices attached to it, all of which routinely transmit and receive data. In some systems, only a single device is allowed to "talk" at a time. The control field can be used to force these devices to take turns. This field is optional; some protocols do not use it.

The final field we will cover is the *cyclic redundancy check*, or CRC, field. The CRC is a mathematical procedure used to test the integrity of the bits within each frame. It does this by treating the zeroes and ones of data as a binary number (which, of course, it is) instead of as a series of characters. It divides the "number" by a carefully crafted polynomial value that is designed to *always* yield a remainder following the division process. The value of this remainder is then placed in the CRC field and transmitted as part of the frame to the next switch. The receiving switch performs the same calculation, and then compares the two remainders. As long as they are the same, the switch knows that the bits arrived unaltered. If they are different, the received frame is discarded and the transmitting switch is ordered to resend the frame, a process that is repeated until the frame is received correctly. This process can result in a transmission delay, because the Data Link Layer will not allow a bad frame to be carried through the network. Thus, the Data Link Layer converts errors into delay.

Error-Recovery Options

There are a number of techniques in common use that allow receiving devices to recover from bit errors. The simplest of these is frame discard, the technique used by frame-relay and ATM networks. In frame discard environments, an errored frame is simply discarded—period. No other form of recovery takes place within the network. Instead, the end devices (the originator and receiver) have the end-to-end responsibility to detect that a frame is missing and take whatever steps are necessary to generate a second copy. Reasons for this strategy will be discussed in the section on fast packet services.

A second common technique is called *forward error correction (FEC)*. FEC is used when (1) no backward channel is available over which to request the resend of an errored packet, or (2) the transit delay is so great that a resend would take longer than the application would allow, such as in a satellite hop over which an application is transmitting delay-sensitive traffic. Instead, FEC systems transmit the application data with additional information that allows a receive device not only to determine that an error has occurred, but to fix it. No resend is required.

The third and perhaps most common form of error detection and correction is called *detect and retransmit.* Detect-and-retransmit systems use the CRC field to detect errors when they occur. The errored frames are then discarded, and the previous switch is ordered to resend the errored frame. This implies a number of things: The frames must be numbered, some form of positive and negative acknowledgment system must be in place, the transmitter must keep a copy of the frame until its receipt has been acknowledged, and some facility must be in place to allow the receiver to communicate upstream to the transmitter.

Two recovery techniques are commonly utilized in synchronous systems. To understand them, we must first introduce a couple of transmission protocols used to meter the transmission of frames between switches.

In early communications systems (1970s), the network was known to be relatively hostile to data transmission. After all, if noise occurred during a voice conversation, no problem—it was a simple matter to ignore it provided it wasn't too bad. In data, however, a small amount of noise could be catastrophic, easily capable of destroying a long series of frames in a few milliseconds. As a result, early data systems such as IBM's *Binary Synchronous Communications (BISYNC)* used a protocol called *Stop-and-Wait* that would permit only a single frame at a time to be outstanding without acknowledgment from the receiver. This was obviously terribly inefficient, but in those early days it was as good as it got. Thus, if a major problem occurred, the maximum number of frames that would ever have to be resent was one.

As time passed and network quality improved, designers got brave and began to allow multiple unacknowledged frames to be outstanding, a process called *pipelining.* In pipelined systems, a maximum *window size* is agreed upon that defines the maximum number of frames that can ever be allowed to be outstanding at any point in time. If the number is reached, the window closes, closing down the pipeline and prohibiting other frames from being transmitted. As soon as one or more frames clear the receiver, that many new frames are allowed into the pipeline by the sliding window. Obviously, this protocol is reliant on the ability to number the frames so that the system knows when the maximum window size has been reached.

Okay, back to error recovery. We mentioned earlier that there are two common techniques. The first of these is called *selective retransmit.* In selective retransmit environments, if an error occurs, the transmitter is directed to resend *only the errored frame.* This is a complex technique, but is quite efficient.

The second technique is called *Go Back N.* In Go Back N environments, if an error occurs, the transmitter is directed to go back to the errored frame *and retransmit everything from that frame forward.* This technique is less efficient but is far simpler from an implementation point of view.

Let's look at an example. Let's assume that a transmitter generates five frames, which are numbered one of five, two of five, three of five, and so on. Now let's assume that frame three arrives and is found to be errored. In a selective retransmit environment, the transmitter is directed to resend frame three, and *only* frame three. In a Go Back N environment, however, the transmitter will be directed to resend everything from frame three going forward, which means that it will send frames three, four, and five. The receiver will simply discard frames four and five.

So let's review the task of the Data Link Layer. It frames the packet so that it can be checked for bit errors, provides various line-control functions so that the network itself can be managed properly, provides addressing information so that the frame can be delivered appropriately, and performs error detection and (sometimes) correction.

Layer 1: The Physical Layer

Once the CRC is calculated and the frame is fully constructed, the Data Link Layer passes the frame down to the Physical Layer, the lowest layer in the networking food chain. The Physical Layer's job is to transmit the bits, which include the proper representation of zeroes and ones, transmission speeds, and physical connector rules. For example, if the network is electrical, then what is the proper range of transmitted voltages required to identify whether the received entity is a zero or a one? Is a one in an optical network represented as the presence of light or the absence of light? Is a one represented in a copper-based system as a positive or as a negative voltage, or both? Also, where is information transmitted and received? For example, if pin 2 is identified as the transmit lead in a cable, over what lead is data received? All of these physical parameters are designed to ensure that the individual bits are able to maintain their integrity and be recognized by the receiving equipment.

Many transmission standards are found at the Physical Layer, including T1, E1, SONET, SDH, DWDM, and the many flavors of DSL. *T1* and *E1* are longtime standards that provide 1.544 and 2.048 Mbps of bandwidth respectively; they have been in existence since the early 1960s and occupy a central position in the typical network. *SONET*, the Synchronous Optical Network, and *SDH*, the Synchronous Digital Hierarchy, provide standards-based optical transmission at rates above those provided by the traditional carrier hierarchy. *Dense wavelength-division multiplexing*, or DWDM, is a frequency division multiplexing (FDM) technique that allows multiple wavelengths of light to be transmitted across a single fiber, providing massive bandwidth multiplication across the strand. And *DSL* extends the useful life of the standard copper-wire pair by expanding the bandwidth it is capable of delivering as well as the distance over which that bandwidth can be delivered.

OSI Summary

We have now discussed the functions carried out at each layer of the OSI Model. Layers six and seven ensure application integrity, layer five ensures security, and layer four guarantees the integrity of the transmitted message. Layer three ensures network integrity; layer two, data integrity; and layer one, the integrity of the bits themselves. Thus, transmission is guaranteed on an end-to-end basis through a series of protocols that are interdependent upon each other and which work closely to ensure integrity at every possible level of the transmission hierarchy.

So let's now go back to our e-mail example and walk through the entire process.

The Eudora e-mail application running on the PC creates a message at the behest of the human user[3] and passes the message to the Application Layer. The Application Layer converts the message into a format that can be universally understood as an e-mail message, in this case X.400. It then adds a header that identifies the X.400 format of the message.

---

[3]I specifically note "human user" here because some protocols do not recognize the existence of the human in the network loop. In IBM SNA environments, for example, users are devices or processes that use network resources. There are no humans.

The X.400 message with its new header is then passed down to the Presentation Layer, which encodes it as ASCII, encrypts it using PGP, and compresses it using a British Telecom Lempel-Ziv compression algorithm. After adding a header that details all this, it passes the message to layer five.

The Session Layer assigns a logical session number to the message, glues on a packet header identifying the session ID, and passes the steadily growing message down to the Transport Layer. Based on network limitations and rule sets, the Transport Layer breaks the message into 11 packets and numbers them appropriately. Each packet is given a header with address and QoS information.

The packets now enter the chained layers, where they will first encounter the network. The Network Layer examines each packet in turn, and, based on the nature of the underlying network (connection-oriented? connectionless?) and congestion status, queues the packets for transmission. After creating the header on each packet, the packets are handed individually down to the Data Link Layer.

The Data Link Layer proceeds to build a frame around each packet. It calculates a CRC, inserts a Data Link Layer address, inserts appropriate control information, and finally adds flags on each end of the frame. Note that all other layers add a header *only;* the Data Link Layer is the only layer that also adds a trailer.

Once the Data Link frame has been constructed it is passed down to the Physical Layer, which encodes the incoming bitstream according to the transmission requirements of the underlying network. For example, if the data is to be transmitted across a T- or E-carrier network, the data will be encoded *using AMI* and will be transmitted across the facility to the next switch at either 1.544 Mbps or 2.048 Mbps, depending on whether the network is T1 or E1.

When the bitstream arrives at the next switch (not the destination), the bits flow into the Physical Layer, which determines that it can read the bits. The Physical Layer hands the bits up to the Data Link Layer, which proceeds to find the flags so that it can frame the incoming stream of data and check it for errors. If we assume that it finds none, it strips off the Data Link frame surrounding the packet and passes the packet up to the Network Layer. The Network Layer examines the destination address in the packet, at which point it realizes that it is not the intended recipient. So, it passes it back to the Data Link Layer, which builds a new frame around it, calculating a new CRC and adding a new Data Link Layer address as it does so. It then passes the frame back to the Physical Layer for transmission. The Physical Layer spits the bits out of the facility to the next switch, which for our purposes we will assume is the intended destination. The Physical Layer receives the bits and passes them to the Data Link Layer, which checks them for errors. If it finds an errored frame, it requests a resend, but ultimately receives the frame correctly. It then strips off the header and trailer, leaving the original packet. The packet is then passed up to the Network Layer, which after examining the packet address determines that it is in fact the intended recipient of the packet. As a result it passes the packet up to the Transport Layer after stripping off the Network Layer header.

The Transport Layer examines the packet and notices that it has received packet three of 11 packets. Because its job is to assemble and pass entire messages up to the Session Layer, the Transport Layer simply places the packet into a buffer while it waits for the other 10 packets to arrive. It will wait as long as it has to; it knows that it cannot deliver a partial message because the higher layers are not smart enough to figure out the missing pieces.

Once it has received all 11 packets, the Transport Layer reassembles the original message and passes it up to the Session Layer, which examines the Session Layer header created by the transmitter and notes that this is to be handed to whatever process cares about logical channel number seven. It then strips off the header and passes the message up to the Presentation Layer.

The Presentation Layer reads the Presentation Layer header created at the transmit end of the circuit and notes that this is an ASCII message that has been encrypted using PGP and compressed using BTLZ. It decompresses the message using the same protocol, decrypts the message using the appropriate public key, and, because it is resident in a mainframe, converts the ASCII message to EBCDIC. Stripping off the Presentation Layer header, it hands the message up to the Application Layer. The Application Layer notes that the message is X.400 encoded and is therefore an e-mail message. As a result it passes the message to the e-mail application that is resident in the mainframe system.

The process just described happens every time you hit the SEND button.

Click.

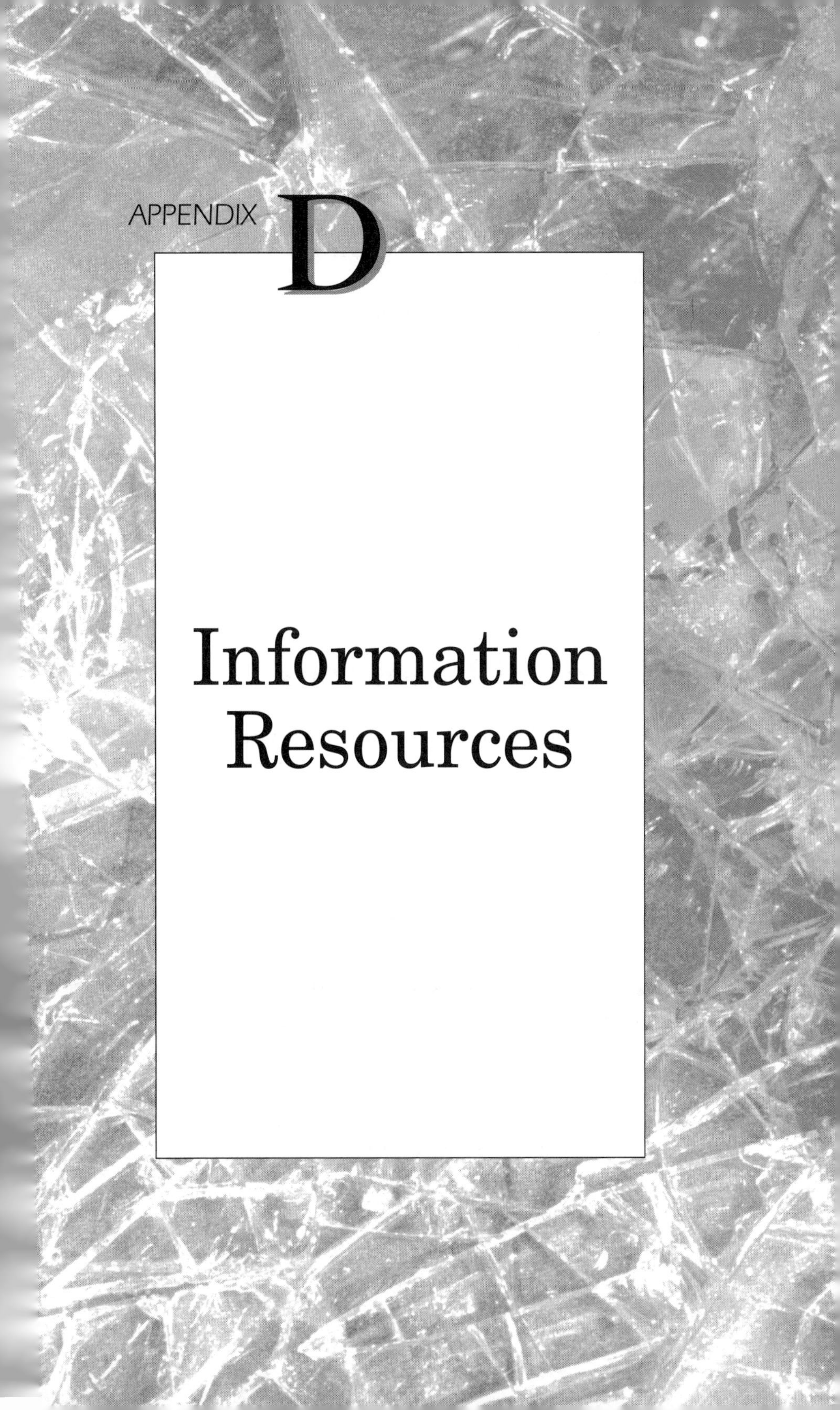

APPENDIX D

Information Resources

Telecom Resources

Telephony Magazine Online www.telephonyonline.com

Telecom Daily Lead (Daily insights from USTA) www.dailylead.com/usta/

Today in Telecom www.igigroup.com/news.html

Network World www.subscribenw.com/nl2

Internet Telephony E-News www.tmcnet.com/enews

Optical Resources

Light Reading (Optical Industry) www.lightreading.com

Financial Resources

The Motley Fool (Investing insights) www.fool.com/community/freemail/

Forbes Online www.forbes.com/membership/signup.jhtml

General Industry

McKinsey Quarterly www.mckinseyquarterly.com

CyberAtlas (Online Statistics) http://cyberatlas.internet.com

The Economist Online www.economist.com/members/registration.cfm

NewsScan Daily Send mail to NewsScan-html@NewsScan.com, with the word "subscribe" as the subject.

Wainhouse Research Bulletin (Videoconferencing industry) www.wainhouse.com/bulletin

Information Week www.informationweeksubscriptions.com
E-Marketer Newsletter www.emarketer.com/news/newsletter.php
Darwin Observer www2.darwinmag.com/connect/newsletters.cfm
CRM and Knowledge Management www.crm-forum.com

Wireless Resources

CTIA News www.wow-com.com/news/

Computer Industry Resources

ComputerWorld Magazine www.cwrld.com/nl/sub.asp

Asia-Pac Resources

Telecom Asia www.emarketer.com/news/newsletter.php
Asia Telecomm Newsletter www.export.gov/infotech

CALA Resources

Latin America Telecom Newsletter www.telecom.ita.doc.gov

EMEA Resources

EuroNet Newsletter www.euronetmag.com/euronet/

Paid Subscriptions

I also subscribe to the *Wall Street Journal Online* ($49/year), the *Harvard Business Review* ($118/year), *Business Communications Review* ($60/year), *The Atlantic Monthly* ($29/year), *Foreign Affairs Journal* ($50/year), and *The Economist* ($30/year).

Additionally, I read (somewhat sporadically) *World Trade Magazine*, *Wired Magazine*, and anything else that catches my attention on the newsstand.

INDEX

A

B

C

D

E

H

I

N

O

P

Q

R

S

T

U

V

W

X

Y

Z

ABOUT THE AUTHOR

Steven Shepard is the president of the Shepard Communications Group in Williston, Vermont. A professional author and educator with 22 years of varied experience in the telecommunications industry, he has written books and magazine articles on a wide variety of topics. He is the author of *Telecommunications Convergence: How to Profit from the Convergence of Technologies, Services, and Companies* (McGraw-Hill: New York, 2000); *A Spanish-English Telecommunications Dictionary* (Shepard Communications Group: Williston, Vermont, 2001); *Managing Cross-Cultural Transition: A Handbook for Corporations, Employees and Their Families* (Aletheia Publications: New York, 1997); *An Optical Networking Crash Course* (McGraw-Hill: New York, February 2001); *SONET and SDH Demystified* (McGraw-Hill: New York, 2001), *Telecomm Crash Course* (McGraw-Hill: New York, October 2001); *Telecommunications Convergence, Second Edition* (McGraw-Hill: New York, February 2002); *Videoconferencing Demystified* (McGraw-Hill: New York, April 2002); *Metro Networking Demystified* (McGraw-Hill: New York, October 2002); *The Shepard Report: Charting a Path in Uncertain Times* (SCG: March 2004); and *RFID* (McGraw-Hill: New York, July 2004). *VoIP Crash Course* will be released in mid-2005. Steve is also the Series Editor of the McGraw-Hill *Portable Consultant* book series.

Mr. Shepard received his undergraduate degree in Spanish and Romance Philology from the University of California at Berkeley and his masters degree in International Business from St. Mary's College. He spent 11 years with Pacific Bell in San Francisco in a variety of capacities including network analysis, computer operations, systems standards development, and advanced technical training, followed by 9 years with Hill Associates, a world-renowned telecommunications education company, before forming the Shepard Communications Group. He is a Fellow of the Da Vinci Institute for Technology Management of South Africa, a member of the Board of Directors of the Regional Educational Television Network, and a member of the Board of Trustees of Champlain College in Burlington, Vermont. He is also the resident director of the University of Southern California's Executive Leadership and Advanced Management Programs in Telecommunications and adjunct faculty member at the University of Southern California, The Garvin School of International Management (Thunderbird University), the University of Vermont, Champlain College, and St. Michael's College. He is married and has two children.

Mr. Shepard specializes in international issues in telecommunications with an emphasis on strategic technical sales, convergence and optical networking, the social implications of technological change, the development of multilingual educational materials, and the effective use of multiple delivery media. He has written and directed more than 40 videos and films, and written technical presentations on a broad range of topics for more than 70 companies and organizations worldwide. He is fluent in Spanish and routinely publishes and delivers presentations in that language. Global clients include major telecommunications manufacturers, service providers, software development firms, multinational corporations, universities, advertising firms, and regulatory bodies.